FOREWORD

There is much current debate about the form of contractual arrangement most suited to the needs of the industry and its clients. Significantly, there is only limited exposure of the demands being placed on those individuals responsible for running the site and the problems they face in co-ordinating the various inputs to the project. No less importantly they have to maintain an appropriate understanding of advancing technology, accommodate continual changes in contractual requirements and legislation and be familiar with developments in management issues.

The Institute's Directorate of Professional Services supports those responsible for site management through its annual *Building Manager of the Year Awards* and through the *Technical Information Service*. The latter has contributed some 23 past papers — suitably up-dated — included in this fourth volume.

In addition there are five specially commissioned contributions. Michael Hatchett takes a philosophical look at the demands on the site manager in the next century and points out the industry's lack of investment in this key area of management. Brian Atkin and his colleagues at Reading University, give the benefit of their experience and forward thinking in regard to the viability of automation and robotics on site. Colin Gray builds on his reseach study on the changing role of specialists and trade contractors in exploring the relationship between sub-contractors and site management. David Brooks from his position as chairman of the relevant BSI committee outlines the salient features of BS8000 which is concerned with the vital issue of workmanship. Finally, but not least, Dr Golob of the Health and Safety Commission, identifies how site management and building management in general can make a positive contribution to reducing the appalling toll of deaths and injuries.

The papers incorporated in this volume hopefully all make a contribution to the individuals CPD. Many will be appropriate for degree students and those taking courses leading to membership of a professional institution concerned with construction. However, the overall aim is to support the man, or woman, concerned with the management of the site.

To complement the papers there is a classified bibliography, providing a comprehensive reference source to all matters concerned with the practice of site management for the period 1984 to 1989.

The support of the individual authors in updating their papers is gratefully acknowledged, as is that of those individuals and organisations who amended papers where the original author was unable to contribute. Without their contributions this book would not have been possible.

Peter Harlow
January 1992

CONTENTS

FOREWORD   iii

ABOUT THE AUTHORS   vii

SITE MANAGEMENT
Meeting the challenge of the twenty-first century   Michael Hatchett   1
Site management and sub-contracting   Colin Gray   5
The clerk of works and the site management team   R.J. Martin   14
Labour-only sub-contracting   David Langford   20
Variation orders, site works orders and site records   D.F. Turner   32

SITE OPERATIONS
Modular building systems and the role of construction automation systems and robotics   Brian L. Atkin et al   42
Transporting and placing of concrete   John Illingworth   49
Good practice in laying floor screeds   E.G. Mold & K.F. Endean   64
Demolition and the main contractor   K.C. Kelsall & The Institute of Demolition Engineers   76
Setting out   C.E. Baker & B.M. Sadgrove   85
Formwork and site management   John G. Richardson   92
Vandalism on building sites   Andrew Charlett   102

MANAGEMENT TECHNIQUES
Communication   Richard Biggs   116
Modern site control methods suitable for the larger project   R. Oxley & J. Poskitt   123
Use of perations research on site   A.T. Baxendale   132
Methods of measuring production times for construction work   A.D.F. Price & Frank Harris   143
Measuring site productivity by work sampling   A.T. Baxendale   157
Productivity, the key to control   Malcolm Horner   166

INDUSTRIAL RELATIONS
Site management and construction health and safety   L. Golob   176
Absenteeism in the building industry   Trevor Burch   180
Responsibility of contractors, subcontractors and others under the Health and Safety at Work etc Act 1974   B.R. Norton   190

QUALITY ISSUES
Workmanship: a test of supervision skills — an introduction to workmanship on building sites   David Brooks   203

Achieving quality brickwork on site Peter Allars 209

Quality control and tolerances for internal finishes in buildings Ian Rankin 222

MATERIALS CONTROL

Storage on site W.W. Abbott 240

An approach to reducing materials waste on site John R Skoyles 251

MATERIALS

Steelwork today Peter Allen 259

INFORMATION ON SITE MANAGEMENT 271

ABOUT THE AUTHORS

ABBOTT W W, PhD, MPhil, FCIOB, FRICS

After several years experience as a quantity surveyor and estimator with John Laing Construction, *Dr Abbott* became Principal Lecturer and Senior Course Tutor in the Department of Civil Engineering and building at Coventry Polytechnic. His major research interests are aimed at achieving product quality on site and preventing waste of materials.

ALLARS P, FCIOB

Peter Allars was educated at High Wycombe Royal Grammar School and from there went on to East Ham School of Building in 1943 where be eventually took the Higher National Certificate and LIOB examinations. He started work as a quantity surveyor and then moved into site management, progressing from there into contract management and finally general management. He retired as a Director of Shepherd Construction in 1988 and now lives in Lincolnshire. He was the Chairman of the BDA Quality Brickwork Award Scheme for its first two formative years.

ALLEN P H, CEng, MICE, MWeldI

Peter Allen trained as a civil engineer but with a structural bias and in particular steelwork. He has worked for Simon Carves, Sir William Halcrow & Partners, BI Construction Co. Ltd and the London Borough of Bromley, during which time he was engaged on soft ground tunneling, the design of masts and towers, unit construction bridges, industrial and public buildings using light and heavy steelwork, reinforced concrete, load bearing brickwork and structural timber. He has also carried out structural investigations of schools and public buildings involving full scale load testing. Mr Allen joined BCSA in 1979 and represents them on BSI Committees, particularly those concerned with the construction of steel buildings. With the advent of the single European Market in 1992 he is heavily involved in the preparation of European and international standards on these aspects. He also has responsibility for safety aspects within the industry, arranging courses and publications for both those within the constructional steelwork industry and those who specify its use, and dealing with queries from the public at large.

ATKIN B L, BSc, MPhil, PhD, FRICS, MCIOB

Brian Atkin is senior lecturer, Department of Construction Management and Engineering at the University of Reading. He leads the University's construction robotics programmes. His background in industry reflects work experience gained in construction management and in quantity surveying practice. He holds a first degree in building economics and has pursued two higher degrees in the fields of engineering infrastructure and time-cost planning. Dr Atkin's other research interests include CAD in a structural engineering design context, and the emergence of intelligent buildings.

ATKINSON, P, BSc(Eng), ACGI, CEng, FIEE

Peter Atkinson is senior lecturer, Department of Engineering at the University of Reading. He graduated from Imperial College, London University as an electrical engineer in 1955. After four years with British Aerospace he held a number of lecturing posts before joining Reading University as a founder member of the Cybernetics Department in 1964. In 1969 he became a senior lecturer, later transferring to the Engineering Department where he is now Director of Studies for the enhanced BEng(Hons) degree in electronic engineering. His teaching and research interests are primarily in control engineering. He is currently serving as a member of the council of the Institution of Electrical Engineers.

BRIDGEWATER, C E, BSc(Eng), MSc, ACGI, AMIEEE

Colin Bridgewater is Senior Research Fellow, Department of Cybernetics, University of Reading, He is a civil engineering graduate with industrial experience gained in a consulting engineer's office. He holds a Masters degree in knowledge-based systems and has an active interest in the application of AI techniques within the robotics field. Mr Bridgewater is pursuing a doctoral programme in the field of construction robotics on a part-time basis.

IBANEZ-GUZMAN, J, BSc, MSEE

Javier Ibanez-Guzman is a research officer, Department of Construction Management, at the University of Reading. He is a cybernetics graduate and has wide industrial experience in electrical systems development. He has studied in Pennsylvania, under a Fulbright Scholarship, where he gained a Masters degree in electronics engineering. Mr Ibanez-Guzman is pursuing a doctoral programme in the field of construction robotics on a part-time basis.

BAYES, J R, MA(Cantab)

John Bayes is senior scientific officer, Department of Engineering, at the University of Reading. He graduated in the Mechanical Sciences Tripos with a specialisation in mechanics and control theory. He has worked in mechanical design and development and also was works manager of a company with a dozen numerically controlled machine tools. Mr Bayes research and teaching interests include applications of computers to manufacturing (CA-DAM, CNC tools and robotics), mechanisms and mechanical engineering design.

BAXENDALE A T, BSc(Hons), MPhil, MCIOB

Tony Baxendale worked for a number of years for national contractors as a site engineer and planning engineer before taking up lecturing. He is Principal Lecturer in construction management at Bristol Polytechnic with a particular interest in information systems for production control. As current Honorary Secretary of the Association of Researchers in Construction Management he is concerned with encouraging research and disseminating findings to the industry at a national and international level. In addition to lecturing at undergraduate and post-graduate level, he has been involved with many CPD courses for site management.

BIGGS R J, MSc, FCIOB, MBIM, MAPM

Richard Biggs began his career as an apprentice carpenter and joiner but soon moved into site management progressing in ten years spent with the Bovis Group to contracts manager. Six years with a provincial builder followed before he took a year out to gain his Master's Degree at Aston. Returning to contracting he added general management and marketing experience in six years before taking up Project Management with Martin Barnes in 1981. For the last three years he has worked as an independent building consultant and part time as project manager for London and Wessex Developments. He is also associated as a consultant with the Wright Project Management Company.

BROOKS D R, FCIOB

David Brooks was until recently Chairman of the Site Management Practice Committee and a member of the Professional Practice Board, National Council and one of the judges of the Building Manager of the Year Award. He is also Chairman of BSI Technical Committee B/146 which prepared the new Code of Practice BS8000 Basic standards of workmanship on building sites. He started his career as an apprentice management trainee, with a formative spell as a labourer and still retains his keen interest in the effectiveness of site management as the key to the success of companies and the reputation of the industry.

BURCH T, PhD, MSc, FCIOB MBIM
Trevor Burch spent his early years in the industry as an apprentice bricklayer and craftsman with George Wimpey. Several years in site management followed before becoming a lecturer in Building Technology and Management with concurrent research interests in human behaviour. Dr Burch is currently Head of Construction at Chesterfield College of Technology and Arts.

CHARLETT A J, BA, MPhil, MCIOB
Andrew Charlett spent his early career as a management trainee with Holland Hannen & Cubitts and this was followed by posts as planning surveyor and planning engineer with Shepherd Construction and David Charles (Northampton) respectively. He joined Nene College as a lecturer in building studies in 1976, moving to Trent Polytechnic (now Nottingham Polytechnic) in 1981 as a senior lecturer in building technology management. Mr Charlett was awarded a QEII Silver Jubilee Scholarship in 1982 to research security on building sites.

ENDEAN K F, BSC, MCIOB
Kenneth Endean was employed until recently as Company Senior Engineer by G Percy Trentham Ltd and also works as a freelance technical writer. His particular interests include temporary works and the investigation of unconventional building problems.

HIGGS, T A, MSc, CEng, MIStructE
Trevor Higgs is Deputy Director of Technical Services at the Building Employers Confederation. His initial training was as a structural engineer with a firm of consulting engineers who undertook a complete range of building work. A period in research and development with a large steel fabricator followed. After a short period with a loca authority he joined BEC as a building consultant. His particular responsibilities cover house building, roof construction and matters relating to Building Regulations.

GOLOB L, BSc, PhD
After obtaining a BSc (Chemistry) at Leeds University and PhD(Chemical Physics) at Southampton University, *Dr Golob* joined the Health and Safety Executive as HM Inspector of Factories in 1975. In the fifteen years he has been with HSE, he has inspected a variety of industries in the London area including the Heath Services, the Docks, engineering and, latterly, the construction industry. Currently he is working on the HSE's Construction National Interest Group which provides guidance for both inspectors and the industry on the standards to be achieved.

GRAY C, MCIOB
Colin Gray is the course director of the Masters programme in Construction Management at the University of Reading. His research interests are centred on modelling the design, sub-contracting and construction processes as a basis for improving the management of the whole integrated system of construction of modern building projects.

HATCHETT M, MPhil, MCIOB

Michael Hatchett is now a part-time lecturer in Building Studies at the London School of Economics and Political Science. He is also a consultant to LSE Housing and to the Priority Estates Project Ltd, and is specialising in improvements in the management of building works on large housing estates. Over the last 30 years he has been involved in site management and supervisory training and development in the construction industry. Until recently he has been actively involved in the CIOB Site Management Education and Training Scheme, the Site Management Open Tech Scheme and the Technical Information Service. He is now helping the RIBA to develop its own open learning programme.

HORNER R M W, PhD, MICE

Professor *Malcolm Horner* worked for eleven years for Taylor Woodrow Construction Ltd where he was largely concerned with the design and construction of nuclear power systems. In 1977 he joined the University of Dundee where he established the Construction Management Research Unit. He was appointed Head of Department of Civil Engineering in 1985, and to a Personal Chair in Engineering Management in 1986. He has published more than 30 papers in the field of construction management, specialising in labour productivity and project modelling and control. He acts as a consultant to many clients in both the public and private sectors.

ILLINGWORTH J R, BSc(Eng), FCIOB

Now retired after 43 years in the industry. *John Illingworth's* whole career, both in the Services during the war and before and after, was spent in the practical side of construction. He created the Construction Planning Department in George Wimpey plc dealing with building and structural work, which eventually developed on an international basis. The department also dealt with temporary works design, R&D in construction methods and techniques and the provision of work study services. John has published widely and is the author of three books — *Movement and distribution of concrete*, *Site handling equipment*, and *Temporary Works — their role in construction*. ince retirement he has been active in further writing and lecturing on construction technology to both well known contractors and two universities degrees on construction management. He chaired the TRADA steering committee on timber in temporary works and was a member of a number of working parties of the HSE which produced guidance notes on safety in falsework for in situ beams and slabs and site safety and concrete construction.

LANGFORD D, MSc, MPhil, MCIOB, MBIM

David Langford has studied at Bristol Polytechnic, Aston University and Cranfield School of Management and he currently holds the Barr Chair of Construction at the University of Strathclyde. Prior to entering education he was employed for severa years by contractors in a variety of management positions. He is an active researcher and is consultant to several public and private sector construction organisations. His special interests include the human aspects of construction management.

MARTIN R J, PPICW

Now in retirement, *Mr Martin* was clerk of works for 40 years, first entering the profession via the Royal Engineers School of Military Engineering. Most of his career was spent in local government, which he left as a senior clerk of works with Essex County Architect's Department. He ended his career as chief clerk of works with the East Anglian Regional Health Authority. He was President of the Institute of Clerks of Works in 1980 and was long associated with the Institute's Practice Committee.

NORTON B R,

Ben Norton is a solicitor of the Supreme Court and is currently the Deputy Director General of the Federation of Civil Engineering Contractors. He has been with the Federation for over twelve years during which time he has had responsibility for safety, health and welfare. He is a member of the Construction Industry Advisory Committee which advises the Health and Safety Commission on construction matters. He is also a member of the CBI Health and Safety Advisory Committee and sits on numerous specialist committees both within CBI and HSE dealing with safety matters. He was the UK employers' representative at the ILO Conference in 1987 and 1988 when the new Convention on Safety and Health in Construction was being drafted. He was one of the experts who advised the European Commission on the 'Temporary or Mobile Construction Sites Directive' which lays down minimum standards for health and safety on all construction sites throughout the European Community. Prior to joining the Federation he spent 20 years in private practice dealing mainly with personal injury claims and commercial litigation. He has spent most of his professional life dealing with the legal aspects of safety, health and welfare. He spent a brief period as legal adviser to the Health and Safety Executive.

OXLEY R, MSc, MCIOB, MBIM

Ray Oxley is a director of a building company, carrying out new build and refurbishment work. He is responsible for estimating, setting targets and overall planning. Previously Mr Oxley was a Principal Lecturer in Construction Management at Sheffield City Polytechnic where he developed integrated estimating/targeting/control system now available commercially and used extensively within his present company.

POSKITT J, MCIOB, MBIM

Mr Poskitt is a part time senior lecturer at Sheffield City Polytechnic and Director of the same building company as Ray Oxley. He is responsible for control of projects and short term planning. Data are derived from estimates for short term planning and control. Mr Oxley and Mr Poskitt are joint authors of the book, *Management techniques applied to the building industry* which is now in its fourth edition.

PRICE A D F, BSc, PhD, MCIOB

Andrew Price is a Lecturer in Construction Management at Loughborough University. He is also Course Director for both the full and part-time courses in Construction Management run by the Department of Civil Engineering. His main interests relate to construction productivity and he has written several articles on the subject. He has recently obtained the 'New blood' lectureship in construction at Loughborough University of Technology. He previously worked as a Research Assistant on a SERC funded project entitled 'An evaluation of production output for construction labour and plant'. Prior to joining Loughborough he was employed by a building contractor and a local firm of structural engineers.

HARRIS F C, BEng, MSc, PhD, CEng, MICE, MCIOB

Frank Harris is Professor of Construction Studies at Wolverhampton Polytechnic and Head of the School of Construction, Engineering and Technology. He is author of five textbooks and has written extensively reflecting his interests in construction management.

RANKIN I, BSc, MPhil, MCIOB

Ian Rankin has worked with the National House Building Council (NHBC) for many years. Specialising in private sector housing, he has been involved as Technical Secretary in the drafting of technical specifications literature and guidance notes on all aspects of housing design and construction. More recently, he has been responsible for the inspection staff in the South East Region and dealing with the day to day problems of both builders and purchasers. The first part of his paper refers to detailed research work undertaken by him during 1973-74 but which is still relevant and now proven to be useful over the years.

RICHARDSON J G, FICT, FIIM

John G Richardson trained in the Royal Corps of Engineers, worked for 20 years in industry as a formwork designer, formwork supervisor and production and works manager in the precast concrete industry. Subsequently, he was for nearly 20 years a lecturer in construction topics for the Cement and Concrete Association. During the last eight years of this period he worked with British Standards Institution as a technical assessor for quality assurance in the precast industry, work which he currently continues, in tandem with consultancy and training activities. For 20 years he has been an examiner for the City & Guilds of London Institute. He is author of a number of books on formwork and precasting and has lectured on these topics worldwide.

SADGROVE B M, MA, CEng

Mr Sadgrove's early experience was as a site engineer. He moved into research in 1962, first at the Cement and Concrete Association and then at CIRIA. In 1982 he was appointed Director of Research at CIRIA, a post he held for five years. Mr Sadgrove is now Quality Manager with Taylor Woodrow Construction Holdings Ltd.

SKOYLES J R, BSc(Econ) PhD

John Skoyles worked with his father, Ted Skoyles, in the early 80s studying materials waste. More recently he has gained an honours BSc(Econ) from the London School of Economics and a PhD from London University. He is now presently a researcher in the Department of Psychology, University College London.

TURNER D F, BA(Hons), FRICS, MCIOB, ACIArb

Dennis Turner is an independent contract consultant based in the north of England and sometime Head of Building Economics at Robert Gordon's Institute of Technology, Aberdeen. He has wide experience as a practising quantity surveyor in private and central government practice. He is author of six books, including *Building contracts: a practical guide, Building contract disputes* and *Design and build contract practice,* and of various papers dealing with contractual and procedural matters in construction. He advises and lectures regularly on these topics throughout the country.

MEETING THE CHALLENGE OF THE TWENTYFIRST CENTURY

by Michael Hatchett, MPhil, MCIOB

INTRODUCTION

In the mid 1960s I introduced a programme of study at the Bartlett School of Architecture and Planning, University College London, which gave undergraduate students of architecture, building and civil engineering an opportunity to monitor the performance of site management teams on a project of their choice for one academic year. This is the last year in which I shall be involved in this course and it is appropriate to look at the principal findings and what lessons these might have for those concerned about the future effectiveness of site management in the UK.

Perhaps a qualifying word or two should be stated at the beginning. Each year students on this course have monitored some 20 sites, and therefore over the period site management performance on some 500 sites in and around London has been observed with varying degrees of interest and skill. Therefore, this has been an insignificant sample of UK construction sites during this period. I have not recorded all the main contractors or the key participants in every project, but as one would expect from our industry there have been some extremely well run projects and some have lacked all but the most rudimentary signs of management. It may be that the more forward looking would find nothing new in this appraisal, but the evidence suggests that most companies need to give much greater thought to the development of efficient site management teams on most of their projects.

MANY PROJECTS HAVE SITE MANAGEMENT TEAMS THAT ARE TOO SMALL

The most common finding has been the relatively small size of most site management teams. On many projects the whole site management responsibility has been vested in a team of three or less. It seems an historical failing of the UK construction industry to try and save costs by having minimal management teams on most sites. The evidence suggests that most site management teams are over stretched and find difficulty in keeping up with the pace of work. The most common evidence for this proposition is the lack of attention that is given to site and contract administration. Many site management teams do not keep their records up to date; progress reports, safety records and drawing registers are often neglected so that those off site do not have an accurate picture fed back to them of the current site situation. Sometimes this is a defence mechanism but more often it is simply that the staff do not have time to do everything that is associated with their role, and administration is the first aspect of their work to suffer. It is often a very false economy to believe that routine site administration is best performed by busy site managers, yet many companies provide no administrative support whatsoever to their site management teams.

THE TRADITIONAL ROUTE INTO SITE MANAGEMENT WEAKENING

The well established traditional port of entry into site management has been through the crafts. Thus apprenticeship, followed by journeymanship and first line supervision have provided the most common route into site management. With the current demise of the general contractor and the increase in sub-contracting there are fewer following this route. Hence the interest in technician and graduate entry into site management. However, this is not as easy as it sounds. Many technicians and graduates have little site experience and often see site management as a necessary but undesirable phase in their path towards higher management. The CIOB Site Management Education and Training Scheme has provided much needed support for the concept of qualified and professional site managers, but far too few people are involved in the scheme. The industry has known for many years that it needs better qualified and experienced site management personnel, yet it finds systematic training and career development within site management difficult to achieve, with only a very few firms having a coherent policy in this aspect of their work.

If the industry wants more graduates to enter site management, then what is the role they will be required to play and how should potential employers seek to attract graduates into site management as a career? Little attention seems to be directed to this issue and as a consequence many graduates have little indication of the challenges provided by site management or the rewards that come from effectively managing the building process on site. Less than 20% of the site monitored saw graduates as potential site managers.

A MORE STRATEGIC APPROACH TO TEAM BUILDING NEEDED

For many projects there seems to exist a rather mechanistic approach to the development of site management teams and to the development of the principal roles and responsibilities within the team. Traditionally, site managers are still depicted as being progress chasers at loggerheads with buyers, plant managers and surveyors. Many site management teams seem either to exist in name only or to be established on adversarial principles. Much more attention needs to be given to the part site managers should play in the selection and briefing of sub-contractors and suppliers, in the choice of plant, in the design of temporary works, in programming and in the recovery of costs. All too often site managers are simply expected to chase sub-contractors and suppliers, and to provide evidence of poor performance, sometimes a long time after work has been completed, in order to justify claims initiated by surveyors. Site managers seem to be expected to operate only in response mode. It has been apparent for many years that UK site managers are generally not encouraged to take either a strategic or an initiating role. On many of the sites monitored a classic picture emerged of very real differences of perception between site managers and surveyors regarding the key management issues that faced the site management team.

A lot more attention needs to be given by employers to the role of site management personnel in the development of coherent policies regarding procurement, plant management, temporary works and cost recovery when constructing and briefing management teams for individual projects. Thus, more attention needs to be given to site management team building and to establishing mutually supporting roles within these teams. This is made much more difficult by the increase in sub-contracting and in the use of agency staff because site management teams are now more organisationally fragmented than they have been.

THE CONTRACTUAL MINEFIELD

One of the principal areas of uncertainty which has developed over the last decade has been the increasing variety and complexity of contractual arrangements. It is now, regretfully, very common to find that site managers do not seem to know even the most basic facts about the contracts that operate on the sites they are managing. Very often, over the last decade, we have been unable to find which forms of contract have been in use, and site managers have confirmed to us that they have not possessed copies of the terms and conditions of the sub-contracts for most of the work on their sites. This results in many cases of self inflicted wounds and instances where site managers make up their own rules as they go along, confident that the sub-contractors with whom they are working know even less about the contract conditions than they do. From time to time this results in chaos and animosity. Much needs to be done to improve the site administation of sub-contracts.

THE SETTING-UP PERIOD

One of the surprising findings of our site monitoring work has been the mobility of site management personnel. Very few site management team members remain on site for the whole of the construction period. On one project lasting two and a half years some 70 site engineers were used. One of the consequences of this high level of turnover coupled with the lack of adequate site administration and established contract procedures has been an increase in the number of physical errors apparent on site and a lengthening of the time it takes for site management personnel to recognise the existence of problems. This finding has led us to the view that it is absolutely essential for a new site management team to have a setting up period in which it can become familiar with the principal features of the project and establish the basic systems and procedures for efficient site management. It is also apparent that key site roles require a hand-over period during which the outgoing team member can brief the incoming person. Without these opportunities to establish their roles there seems little chance that new site management teams or new members of existing teams will ever be able to cope effectively with the demands of the job.

SIMULATION OF CONSTRUCTION SEQUENCES

Most construction projects contain some unique features and most site management team members face new challenges on each project. Frequently two standard responses have been apparent to these new challenges. First and foremost has been the strategy of delay coupled with over reaction. The issue is ignored until the last moment when all available resources are used to overcome the problem. The second strategy has been to marginalise the issue, either by sub-contracting or by delegation, if this is possible.

Programming has been both an asset and a liability in this respect. Programming can be an asset by identifying issues that need to be considered and tasks that need to be completed. It is, however, a liability when a single 'solution' is presented as the preferred programme. Data processing is now so user friendly that we should be producing many simulations of construction sequences, looking at many options and involving a wide range of skills in the exploration of these options. The UK construction industry is one of the lowest industrial spenders on data processing, and if one takes out that which is spent on payroll processing the remainder is insignificant. We have to make major strides here in order to emulate the Japanese.

Re-active Rather than pro-active
Not given detail of costs or contracts

PROJECT BASED LEARNING SYSTEMS

Finally I want to highlight the abysmal lack of project based training which takes place in the construction industry. Many high street shops, with more stable staffing and less variable transactions, spend much more time in store based staff training than the construction industry spends on site based training. We are still cast in the mould of the craft guilds, whereby a trained and experienced craftsman, operative or manager can undertake the full range of tasks appropriate to the occupation. This premise no longer holds true. Most site management teams discover the appropriate action by wasteful trial and error. Site based training for the key members of the site management team should be essential. There should be weekly training sessions and these should be directly related to the project needs. I am convinced that the costs of such training would be much less than the benefits, and moreover, these benefits would show within the project period.

CONCLUSION

Meeting the challenge of the 21st century, suggests from the evidence that I have collected, the following:

- more resources directed towards site management not less;
- more attention given to the routes into site management and to the career development of site management personnel;
- more attention to the development of site management team work;
- more attention to contract procedures and associated communications systems;
- more thought given to the initial site setting up process and to the creation of more stable site management teams, less mobility of site management personnel between projects;
- greater use of data processing to simulate the construction process and to explore more options;
- the inclusion within the life of each project of a budget for project based training which would be project related and which would run in parallel with the construction process.

It is my considered opinion that unless we address these issues, and there are many more then to consider, we shall get trapped in the chaos of complexity and uncertainty to which we seem to have only two responses; either sack the site manager or claim against client, sub-contractor or supplier. None of these responses seem to be an appropriate strategy for a major industrial sector or for its lead professionals.

SITE MANAGEMENT AND SUB-CONTRACTING

by Colin Gray, MCIOB

INTRODUCTION

At first glance the role of the modern day contractor's site manager is relatively simple. Appoint the right sub-contractors, give them the responsibility under a carefully worded contract, monitor the contract and watch the work done with minimal risk. Unfortunately, practise is not that simple. Building has become more complex both technically and administratively. Designers are using a greater number of specialists to help them design the detail of the building and, therefore, there is a vast range of components to be brought on to the site and fixed using a wider range of specialist skills. There is also an increasing sophistication by clients, contractors and sub-contractors in money management, thereby creating a greater awareness of the financial implications of the way people deal with each other and, consequently, the risks they are prepared to take. Both of these pressures have led to a concentration on the detail of the construction contract and to where to place the risk of failure. These risks have largely been devolved to sub-contractors who in turn can only attempt to reduce the risks by the careful management of their own task. This seemingly narrow minded view by the sub-contractors of their role is the stark reality of construction and all exhortation by clients, the design team and contractors to the contrary will produce little real response.

In reality the modern site manager will have to work largely within this conventional situation to develop a management strategy which maximises the strengths of the sub-contractors and minimises the risks that they face in order that they do not retrench into a narrow interpretation of their obligations. Whilst the ideal would be to break out of this essentially restrictive approach it would take a fundamental rethinking of approach and attitude by all to achieve it.

THE SITE MANAGEMENT OBJECTIVE — QUALITY PRODUCTION

There can only be one primary objective and that is to produce a building of the highest possible quality within the constraints of cost and time. For the purpose of this paper quality is defined as doing it right first time. This objective can only be achieved if the whole system of information and component supply is designed to allow the tradesman to fix the material and complete the work properly. It is claimed that there are many reasons why this cannot be achieved. Blame is directed at the lack of skilled workers, idle workers, poor sub-contractors, low morale, poor motivation, poor pay, onerous contract conditions and so on without looking deeply at the underlying causes. In practice individuals are trying very hard, but the system in which they work is seemingly designed to defeat them. Modern site management must, therefore, understand the system of construction and manipulate it to achieve the conditions that the worker needs to do his task properly and only once.

MANAGEMENT MUST RECOGNISE THE TOTAL SYSTEM OF BUILDING

The proportion of a building that is manufactured on site is becoming less as building is evolving into a series of assembly operations. The modern components that are being assembled have gone through a complex design, manufacture and assembly operation and are delivered in perfect condition. Each part of this system must function perfectly and be oriented to the needs of the site operations of fitting the component with the other components, which in turn will be the result of another system of production. Site assembly is thus the bringing together of the results of the whole range of systems of design and manufacture, each of which involved countless people doing countless tasks. If any of these tasks are wrong then the impact on the site is profound, as it will be very difficult to rectify the component quickly and so continue with the site work. The probability of failure because of the increasing complexity of the system, is, therefore, increasing all the time.

SITE WORK THROUGH OTHER ORGANISATIONS

Site management no longer has direct control of the workforce as the workforce is now virtually entirely sub-contracted in one form or another. It is not really necessary to differentiate between nominated and domestic sub-contractors in this context. Whatever the site management wants has to be achieved through the co-operation of the sub-contractors' management. In principle, having accepted the order for the work, the sub-contractor has the responsibility to complete the work as he sees fit. Practically, he has to fit in with other sub-contractors, but in essence he has sole responsibility for his own work. If site management requests that he does something which he does not want to do, they have no direct power and have to attempt to coerce him into performance. Site management can only identify the potential problems sufficiently early and make everyone aware of them and propose, cajole, influence and monitor any action which they deem necessary to avoid the problem. Managing in these circumstances requires experienced people who can first understand the whole production system and can identify the bottle-necks and problems before they emerge and secondly create a working environment which is conducive to co-operation and 'give and take' between the sub-contractor and all others on the project.

A NEW DEFINITION OF THE SITE MANAGEMENT TASK

In order to achieve the quality site production that is needed, site management must concentrate on understanding the whole system of construction and ensuring that it is focussed on the production aims of the site operations.

Modifying the system for production

A complete understanding of each sub-contractor's whole design and procurement system is necessary as well as an understanding of how each part will be executed. Only then can the system be examined to see what is the likelihood of it being delivered to site. Where there are deficiencies or where problems are identified, then action to make sure it is going to work can be taken.

Making the system work - checks and controls

Having looked at the system there is no point in allowing it to continue without checking whether each stage is properly supplying the next stage. A simple method of monitoring which rapidly identifies where the system is failing needs to be created. The monitoring must be focussed on the real issues of production, ie, the timely delivery of correct information and/or components to each step in the sequence of the system.

Understanding the interfaces within the system

A system by its nature is composed of many parts. A system in building will interface with many other systems, leading to a great many interfaces. The most logical interface is that where the different technologies come together for the fixing and mutual support. These are not difficult to identify. The site management task is to ensure that both sides of an interface understand the needs and demands of each other and are sympathetic to them. This also involves the design team and requires that they correctly specify across the interfaces so that the systems can function smoothly. An example is the specification of compatible tolerances which are sympathetic to the installation needs of both systems.

Helping people to do their jobs - removing barriers
As site management does not have direct control of the workforce a different approach to obtaining production from the workforce must be taken. Nearly all studies on productivity emphasise that the worker can work continuously and productively if he is kept continuously fed with information and materials. However, as soon as this flow is interrupted then productivity drops, as does motivation. Consequently, there is little point in just demanding a higher level of productivity because the worker is largely powerless to achieve it. The technology of the building process also seems to put barriers in the way by breaking up work flow, eg, arbitrary bay sizes in concrete pours. It is site management's task to identify and remove them.

UNDERSTANDING SUB-CONTRACTORS' MOTIVATION
The majority of sub-contracts are secured on the basis of price sought through intensive competition. There can be no other expectation than that the sub-contractor will attempt at all times to retain his profit and, where the opportunity arises, to maximise it. The sub-contractor will attempt to do this by removing uncertainty and risk. The easiest way to achieve this is to minimise the effect of the individual site on his total business operations by creating a situation which allows him to minimise his workforce, to keep it at a constant level and to keep it working continuously. Basically these ideas mean that work, which to do quickly would require a fluctuating workforce, takes a relatively long time and may affect another trade.

Over the last five years project construction times have been progressively reducing as clients have sought rapid delivery to suit their needs. Consequently, site management has been faced with buying sub-contracts with less and less time before the start on site and consequently with less time in which to organise their operations. Sub-contractors have been pressed into starting on site as early as the site will allow and once on site to adopt a flexible attitude to when they can do their work. They are expected to have a workforce which can change in number daily to meet the varying operational needs. Sub-contractors as part of their risk avoidance have used sub-sub-contracting and labour-only to manage and absorb the variations in their own labour force.

Thus, the sub-contractor's ideals are compromised, but he is naturally reluctant to allow this as it may impact upon his profits. As many sub-contractors are very small organisations this causes extreme operating difficulties leading to a knock on effect and wide variability in performance on site. The most severe consequences are at the operative level where incentive and motivation are reduced and, consequently, the quality of the work is poor.

If the causes are examined in detail it can be seen that through intelligent analysis and sound practice the majority of the problems can be resolved. The basic problems are:

(a) interference
   There are several causes of interference. The first is technological interference where the next stage of the operation cannot be started because another trade's work has to be done first. The next is location where the work has been completed in one place and a

physical move is necessary to another floor or part of the building. Another is where materials are not available and so work stops whilst the operative either has to go and find the item or waits until a delivery occurs.

(b) interface control
   This is at the point of connection between two components where the fixing base for one item is not in place when the next trade requires it.

(c) poor completion
   Achieving total completion appears to be very difficult and so there is an increasing list of items which are 95% complete.

(d) damage
   Where the worker does not have pride in his own work then he will not have it for other work. In consequence there is a large amount of avoidable damage.

These problems initiate the downward spiral in productivity, pride in the work and morale. It is the most pressing challenge facing the site manager.

THE IMPORTANCE OF TASK DESIGN

It is a fact that the British construction worker is neither indolent nor stupid. Comparative studies with construction industries in other countries have shown him to be less productive at his task but, that his task is complex and not well managed. Studies have also shown that the average construction worker is only productive for 40% of his time; the rest is spent moving from one task to another or waiting for materials and instruction. In other words there would be no shortage of skilled manpower now or in the future if the tasks were well organised and managed. Whilst this has been known since the middle of the 1960s it is only with the advent of the 'managed' form of contract that it has been taken seriously. Value engineering and buildability focus on three stages in order to raise productivity: the design of the task to be performed; the working conditions of the operative; the organisation of the working environment.

The problem is largely overcome by management action in the early design stages. It is this need which is driving the greater use of complete work packages so that the sub-contractor can manipulate the design of his work to achieve the high levels of productivity that are needed. It can be done. There are numerous examples where a radical rethinking of construction activity, eg, lift assembly, has streamlined the site activity.

Task design

The average time spent per visit by a tradesman to a traditionally built house is one and a half hours. On a recent major office project the dry lining fixers visited the same core over twenty times to complete the cladding around the other trades. These are typical examples of task design that is the outcome of conventional design team thinking. The cladding operations were rationalised down to six operational visits once the sub-contractor was given the opportunity to propose alternative designs. Task rationalisation can only occur during design if it is to have any significant effect. The sub-contractor must be in a position to contribute this type of thinking, both by having the right expertise available and being employed at the right time. His own designers must be fully aware of the practical problems faced by the site teams. In addition to satisfying the design team's requirements they must achieve the following to enable the site teams to work steadily and continuously:

- rationalise the work into large units in order to increase the work time from a few hours to a one day minimum of continuous activity for each operative;

- remove tasks which are unnecessary by redesign. This is what every manufacturer is now involved in to increase his productivity. The Japanese are masters at this;

- remove the need for a dependency on other contractor's components to provide the fixing base, thus removing the need for breaks and revisits;

- assemble, off-site, complex areas which are difficult to organise on-site, eg, toilet pods, lift cars, lift motor rooms, plant rooms, bathrooms and air handling areas.

Task design should allow, on each single visit, for the work to be finished completely. There can be no pride taken in the outcome of a piece of work if it is never finished. It is usually the result of poor organisation that pieces are left off or the item is so minor that it is left until someone happens to be going past that site.

THE WORKING CONDITIONS
Site work is, by its nature wet, dirty and physically demanding. Yet work of a high degree of accuracy is expected under these conditions. To a degree it will be difficult to change many of these intrinsic problems, but if the workforce is to be retained then the site conditions will have to be comparable with other industrial conditions. Once again there are examples where this problem has been taken seriously. At Broadgate for example the whole of the area around the buildings was concreted to give a good road surface and was cleaned daily. Even during winter it was possible to walk comfortably around the site in town shoes.

The organisation of the working environment
The major problem is to give the worker a sense of purpose. Erratic movement from one place to another disorients him and is a prime demotivator. If the time between each movement could be increased there is the opportunity to give the worker a horizon and clearer objective to aim for. This apparent slowing down of activity should give a greater level of productivity. The disciplined management of site work such as that in the work area concept could help significantly to resolve the potential problems of poor worker motivation and work quality, but it can only function if the task design is sympathetic.

A work area (Figure 1) is a defined space which is occupied for a specific time period. It is physically defined by barriers or tape so that other gangs do not trespass into the area. Only certain people, for the purposes of safety and management can enter the area. The area should allow a team to work continuously for a period of either one week or half a week. In the UK this is either six or three days. It is recognised that there are difficulties in this arrangement, particularly in balancing the workload of one gang with that of another in the same physical space. In practice there may have to be a significant change in the composition of the gangs to achieve the two aims of constant time and space. On high rise construction this is relatively easy to achieve as the floor defines the space and the rate of construction and is either a floor a week or two floors a week. However, for low rise, deep plan construction this simplicity must be introduced in another way. Experience of this approach has shown that the earlier trades have worked to the one week period, but that the later finishing trades have more naturally worked to a two day period.

Site Management and sub-contracting

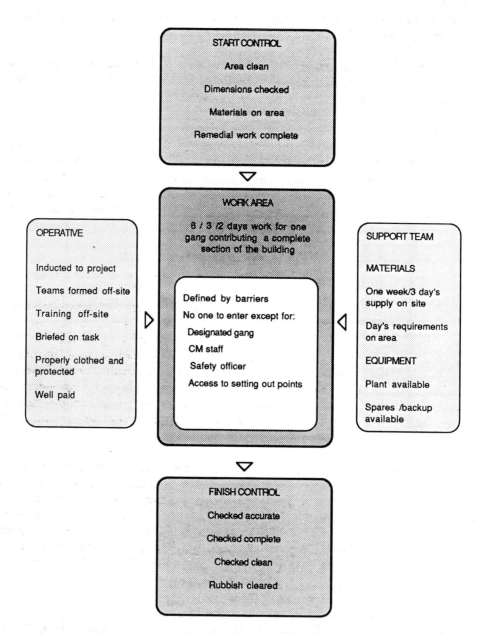

Figure 1 Work area control strategy for effective site operations

Once the work area is defined a discipline for starting and finishing can be more clearly established.

Start control
The sub-contractor has to ensure that before his team enters the work area it is clean and that all preceding work has been checked and approved.

Any problems either in the dimensions or quality of work are to be corrected by the preceding trade. All materials for the day's work are to be ready on the work area. This approach ensures that there are no barriers to a quick start to the day's work.

Finish control
It is equally important that work in the area is totally complete before the gang quits the work area at the end of the possession period. All the checks for completeness, accuracy and cleanliness must be made in sufficient time to allow for any remedial action to be taken. Finally, all rubbish is to be cleared away and the floor swept. The quality of the hand-over is a matter of agreement between the outgoing trade, the incoming trade and the site manager.

THE TOOLS FOR THE JOB
Having established good working conditions the major problem of avoiding interruptions through lack of materials and equipment must be addressed. To keep the gang working effectively a dedicated support team is required. Its role is to ensure that the materials arrive in the right order and at the right time. This is eased if buffer stocks of three days or enough for one work area are held on site. The support team's role is vitally important and requires the ability to plan in detail and to look sufficiently far ahead to guarantee that all the materials will be available. This is far greater than the traditional fetch and carry role of labourers and carries real responsibility. The same requirements are for the plant and equipment to be functioning at peak performance throughout the work period. Arrangements, therefore, have to be made for either stand-by machines for critical operations or ready access to repair crews.

Once again this is the ideal situation and must be tempered with the practicalities. It will depend upon the size of the site and the size of each work package and the relative time-scale, whether it is more practical for there to be a materials handling gang provided by each sub-contractor or a general purpose gang for all the site. The problem with the general purpose gang is that it is never right. There are always conflicting demands upon its activities and blame for the deficiencies of others can be laid off against the contactor. It is a difficult problem which the site manager must resolve in the context of achieving the work area control that is advocated here.

Communication with the workforce
The work area control imposes a discipline that is lacking on many sites and when faced with it most workers find it to be quite daunting and quite unrealistic when compared with their past experience. This is the major challenge - to break the barrier to acceptance that this way of work will be to their benefit as well as to the smooth running of the site. The site manager must make sure that employees at all levels are convinced and motivated to make it work and equally they must convince the workforce that they are going to make it work. This can only be done by the site management talking continuously to the workers about what they are doing to make it work, where the problems are and the action that they have taken to solve

them. The ideal way is to get the teams together before they start the day's work and keep them informed about the importance of their task, keeping it controlled, working safely, what the next task is and the action management have taken to resolve yesterday's problems. This requires a frankness and honesty which is often unusual, but it is vitally important as it shows that the site management is totally interested in the production activity, is constantly working to make it better and to support the best endeavours of each worker.

OFF-SITE PLANNING AND MANAGEMENT

Once the tasks and the work areas have been designed and work initiated, the role of the site manager moves to the control of the external environment so that the carefully structured, highly productive work environment is not upset. This entails understanding the off-site supply system and managing it to meet the demands of the site. It sounds easier than it is because most supply systems are working to their own interpretation of the site's needs and combining it with their own production needs. The site manager must make sure that the needs of the site are understood and being worked to. If necessary he must visit the factories and design studios to get the production message across to everyone concerned with the project. Control systems which highlight, at an early stage, potential problems are needed as well as identifying a course of action to bring things back on track. The onus is on site management to think, often months ahead, so that the materials supply is uninterrupted, that the components come in the right order and there are no vital bits missing.

CONCLUSIONS

The site manager's role is changing rapidly from that of a technician to that of a manipulator of organisations to produce a building. Understanding the detail of the technology is as important as is setting the standards for quality and this must be supplemented by wide organisational skills. Productivity must be raised as must quality. In fact, everything must be better. The site manager must set the right targets and then set about making sure that all the sub-contractors understand them and work within them. Everyone must focus on site production. Every activity must focus on how it is to maximise site production. Problems are activities which detract from site production.

Where more of the key activities are being designed by the sub-contractor there is a greater chance of the production issues of 'buildability' and work area design being incorporated in the detail design. It is not until task design is taken seriously that any of the other essentials to increasing productivity will work. The work area framework by concentrating on keeping teams apart, starting work properly, completing work and resisting pressure to compromise the work area allows absolute control of the work. This clarifies the management task, but clearly shows up the weaknesses in the traditional contractor's site management capability. Work area control places big demands on planning - both on and off site, disciplined working, not accepting low standards of work, avoiding rework and achieving completion. These demands not only raise the calibre of people needed to manage under this discipline, but also require them to change their attitudes away from the acceptance of traditional levels of performance to understanding the potential in the new way of working.

A lot is talked about of attitudes and the need to change them to achieve many of the goals outlined above. Attitudes will not change in isolation. Only if the work is within a framework which allows the system of construction to be understood and to function properly is change possible. Any barriers such as contracts or payment procedures which place doubt on the fact that everyone on site is there to produce a high quality building without wasted effort must be removed. High productivity can only be achieved through a totally sympathetic system. It is the site manager's role to create such a system.

Further reading
MANN N R (1989) Keys to excellence - the Deming philosophy. Mercury Books.
For the total system view of production processes and an understanding of why quality and productivity are so important and interlinked and how difficult it is to achieve without a total systems view.

GRAY C (1986) Buildability - the construction contribution. Occasional paper 29. CIOB. For an introduction to the issues of task design.

GRAY C and FLANAGAN R (1989) The changing role of specialist and trade contractors. CIOB. For an overview of the way sub-contracting is structured and the issues which have to be managed both by the sub-contractor and contractors.

THE CLERK OF WORKS AND THE SITE MANAGEMENT TEAM

Revised by R J Martin, PPICW

INTRODUCTION

Until quite recently the supervision of contracts on behalf of the client followed a traditional pattern which was typified by an architect acting under the provisions of the JCT Standard Form of Contract. This is now longer the case and alternative forms of contract are being increasingly used. It is now by no means certain that the person exercising the supervisory role on behalf of the client will be an architect by profession. However, that supervisory role still carries with it the responsibility to the client for ensuring that the work is carried out generally in accordance with the contract documents and it remains the case that where frequent or constant inspection/supervision is required a clerk of works should be appointed.

The Supervising Officer (SO) will expect the clerk of works to carry out the duties of inspection on his own initiative and will rely upon him to ensure that the work is carried out in accordance with the contract documents. To do this the clerk of works must have the full support of the Supervising Officer and he for his part must take care to keep the SO fully informed and to deserve that support.

When work begins on site the clerk of works makes contact with the contractor's site management and begins a relationship that will affect the project for good or ill. Unfortunately, the relationship is often poor because of misunderstandings about the relative role each has to play in the construction process. This paper is designed to help remove these misunderstandings by detailing the functions and responsibilities of the clerk of works.

APPOINTMENT OF A CLERK OF WORKS

The authority for the clerk of works activities on site derives from a clause in the contract under which the work is being carried out, for example, Clause 12 of the 1980 edition of the JCT standard form of contract. However the terms of employment will vary; some clerks of works will be permanently employed by a local authority or other public body and will become increasingly knowledgeable about the type of work their employer needs. Others may be permanently employed by private professional practices or they may, in both the public and private sectors, be appointed for one job only. Permanent employment by a local authority or other public body, which represented some 90% of all clerks of works employment in the post war public building boom, is now in retreat with the disbandment and privatisation of many such public sector departments. Increasingly, the clerk of works will be self employed, in private practice on his own account.

Where employment is for one job only it is usual to take care to appoint a clerk of works with experience of the type of work involved.

On very large or complex structures there may be a number of clerks of works, each of whom will have responsibility for a different part or element of the structure. In such circumstances communications and directions will be issued through the senior clerk of works nominated by the Supervising Officer.

The appointment should preferably begin at least a month before the contract for the building works is let. This enables the clerk of works to work with the Supervising Officer and ensures continuity from contract planning and design through to full time supervision of the works on site.

CONTRACTUAL RESPONSIBILITIES

While the basic function of the clerk of works is the inspection of the works in progress, the forms of contract for building works vary greatly in the definition of his duties and the degree of authority vested in him. At one end of the spectrum, the GC/Works/1 form of contract confers a great deal of authority on the clerk of works to act directly in the matter of works and materials not complying with contract requirements. On the other hand, the JCT 81 With Contractor's Design form does not admit of his existence by name and the authority for his presence on site, contained in Clause 11, while granting access does not state for what purpose. This must be clarified separately by the employer's agent.

The form of contract which is probably used more than any other is the 1980 JCT Standard Form. Clause 12 sets out his duties as follows:

'The employer shall be entitled to appoint a clerk of works whose duty shall be to act solely as an inspector on behalf of the Employer under the directions of the Architect/Supervising Officer and the Contractor shall afford every reasonable facility for the performance of that duty. If any direction is given to the Contractor by the clerk of works the same shall be of no effect unless given in regard to a matter of which the Architect/Supervising Officer is expressly empowered by the Conditions to issue instructions and unless confirmed in writing by the Architect/Supervising Officer within 2 working days of such direction being given. If any such direction is so given and confirmed then as from the date of issue of that confirmation it shall be deemed to be an Architect's/ Supervising Offficer's instruction'.

Clearly, the clause is in two parts. One provides for the appointment of the clerk of works by the employer and the other lays down the procedure if a clerk of works issues directions to the contractor. The function of the clerk of works is to inspect work in progress, to see that it is in accordance with the contract documents or to the Supervising Officer's variation orders or written instructions issued within the terms of the contract.

The clerk of works should record in detail the work done and any deviations from the terms of the contract and report these objectively to the Supervising Officer.

It has been said that the function of a clerk of works is:

  to inspect in detail;
  to interpret accurately;
  to record completely;
  to report precisely.

Directions in writing will be issued by the clerk of works but need not be complied with unless they affect matters specifically within the power of the Supervising Officer and they are confirmed in writing by the Supervising Officer within two working days of their being given.

DUTIES OF THE CLERKS OF WORKS

The duties associated with inspection are not defined in the various contract clauses and

The clerk of works and the site management team

it is apparent that they will vary from contract to contract, site to site, Supervising Officer to Supervising Officer, employer to employer and contractor to contractor.

In practical terms, the clerk of works can inspect any part of the building under construction at any time, as well as any piece of material to be used in the works, whether it is on site, or at any yard, factory or workshop, to ensure its compliance with the appropriate specification. Samples can be requested and sent for testing to an approved testing laboratory to substantiate manufacturers' or suppliers' claims.

Defective work and materials not complying with the terms of the contract should be identified by the clerk of works and reported to the Supervising Officer so that directions can be given to the contractor for the work to be remedied or the material replaced, or removed from site.

Specific duties may include the following:

(a) inspecting all work in progress;

(b) checking the site grid and setting-out as the work proceeds;

(c) inspecting all materials upon delivery;

(d) informing the Supervising Officer of outstanding information required by the contractor;

(e) informing the Supervising Officer of any discrepancy in the drawings and other contract documentation;

(f) assisting the Supervising Officer with practical solutions to on-site detailing when required;

(g) testing and submitting for test, materials used in the works;

(h) informing the contractor and Supervising Officer about work which does not conform to contract documentation;

(i) examining the contractor's progress schedule, checking and recording progress and delays.

(j) endorsing daywork vouchers in respect of time and materials;

(k) liaising with consultant engineers about trades and sub-contractors under their direction;

(l) checking steelwork, and reinforcement cover, spacings and fixings;

(m) assisting the quantity surveyor with measurement as required - particularly foundations, drainage and other work to be covered up;

(n) attending all site meetings;

(o) providing regular weekly reports on site progress to the Supervising Officer;

(p) preparing or assisting the Supervising Officer in listing all defects or omissions prior to take-over of the building.

He must not incur extra costs, undertake financial commitments, vary procedures laid down in the contract documents by the Supervising Officer or interpret details which may not be clear, without authority.

RECORDS

As part of his duties the clerk of works will need to compile and maintain records of the following in addition to his daily diary:
(a) working conditions and weather, including air temperature throughout the day;
(b) site labour employed;
(c) plant standing, or in use, on site;
(d) directives passed to the contractor;
(e) visitors to the site;
(f) drawings and information received from the Supervising Officer, consultants and contractor;
(g) delays;
(h) principal deliveries to site and general particulars of shortages;
(i) verbal instructions received from Supervising Officer and consultants;
(j) material tests;
(k) work done outside the terms of contract and for which variation orders have, or should be issued.

QUALIFICATIONS AND QUALITIES OF CLERK OF WORKS

Where a clerk of works is employed by a public body he may sometimes have to undertake duties that lie outside those which may be regarded as normal. In such circumstances the clerk of works may be asked to assume extra responsibilities. Employed for many different types of work, from civil engineering to building maintenance and as the traditional estate clerk of works, the qualities required for a successful clerk of works are consequently numerous. A sound and wide knowledge of building construction is essential, as is the ability to communicate and to maintain good personal relationships.

Many clerks of works have first-hand experience of managing a site and this stands them in good stead. It enables them to appreciate the difficulties experienced by site management, and at the same time can give them greater confidence; this promotes closer co-operation with the site manager. Formal training can be provided by City and Guilds and TEC leading to the Final Examinations of the Institute of Clerks of Works, but this must be linked to the ability to keep pace with rapidly changing and advancing techniques and a willingness to learn about new materials and methods.

Personal qualities are of great importance. The effective clerk of works must:
(a) be able to command authority and to act fairly and justly when using it;
(b) be courteous to men and management;
(c) be tactful when dealing with site problems;
(d) have integrity;
(e) avoid the use of bad language;
(f) check all facts before making decisions. He must always be factually correct when reporting to the Supervising Officer.

COMMUNICATIONS BETWEEN SITE MANAGEMENT AND CLERKS OF WORKS

It is obvious that the responsibilities and powers of the clerk of works are such that he can affect the progress of a contract for good or ill. Comprehension of this fact should make the site manager seek the clerk of works' co-operation from the beginning. He should establish the standard of workmanship laid down in the contract documents and should consult the clerk of works before work on any particular section begins and seek approval of the materials and methods proposed.

Any disregard of instructions by the contractor must be dealt with firmly and speedily by the clerk of works.

While it may be difficult for the clerk of works to object to work in progress on certain occasions, he will not be fulfilling his obligations to the client where he allows defective work or unacceptable practices to continue without comment immediately such a situation becomes apparent.

The clerk of work should avoid giving personal views on the way the job is run; he should confine his remarks to the standard and time aspects of the work.

Where a large number of sub-contractors are working on a site the clerk of works should satisfy himself that they have been engaged within the terms of the main contract. Any fall-off in the quality of work by sub-contractors should first be raised with the main contractor and if remedial action is not speedily undertaken the matter reported to the Supervising Officer.

To maintain good relationships with site management the clerk of works should check the drawings for errors before issue to the contractor if possible and if not as soon after issue as is humanly possible, and certainly before the drawing is used on site.

When a new Supervising Officer is appointed, the clerk of works can be particularly helpful in briefing him about the project and any particular problem being experienced.

Continuous consultation between the clerk of works and site management is essential in building up mutual respect and trust. Much of the clerk of works' authority relies upon his credibility and whilst it is not his responsibility to assume the duties of either the Supervising Officer or the site manager, by demonstrating his ability, integrity, and vigilance, he will soon establish the reasons for his presence on site.

CONCLUSIONS

Close co-operation and the establishment of good communications are essential to a good working relationship between the clerk of works and site management. The extent to which they can be achieved can determine to no small degree, the success or failure of a building project.

BIBLIOGRAPHY

1. INSTITUTE OF CLERK OF WORKS (1989) The clerk of works. pp14
2. GREATER LONDON COUNCIL (1983) Handbook, 3rd edition. Architectural Press pp130.
3. JOHNSTON J E (1975) The clerk of works in the construction industry. Crosby Lockwood Staples. pp349

4. RIBA & INSTITUTE OF CLERK OF WORKS (1984) Clerks of works' manual. RIBA
   Publications (currently being revised). pp64

ACKNOWLEDGEMENT

This paper remains largely as prepared by J E Johnston some seventeen years ago, but
sections have been rewritten to reflect changes which have taken place during that time.

LABOUR-ONLY SUB-CONTRACTING — A REVIEW OF PRACTICE UP TO THE 1980s

By D Langford, MCIOB, MSc, MPhil

SUMMARY

This paper reviews the practice of labour-only sub-contracting (LOSC) in the construction industry. It does not seek to describe the process of hiring individuals or gangs but rather takes a broad view of the impact of such labour practices upon the industry. In particular it notes the many organisational variations of LOSC, how it developed, how much labour is involved in the practice and the arguments for and against its use. Finally, its impact upon the industry is discussed.

INTRODUCTION *important*

No subject so dominated discussion in the construction industry during the 1970s as 'the lump'. No form of employment has dominated the 1980s as has labour-only sub-contracting. In this sense the debate would appear to be settled, the 1970s 'lump' with its perjorative title has metamorphosed into the more sedate labour-only sub-contractor of the 1980s — the practice may be the same but the hostility or enthusiasm for it is more muted. The issues are undoubtedly the same, yet only sporadic discussion addresses the use and abuses of labour-only, the pros and the cons of its use and perhaps more importantly what are the trends in labour and how is the industry to shape up to the changes in the composition of the industry's workforce. From the evidence available it would appear that LOSC is growing either by stealth or by forceful design, more people are involved and more contractors are using labour-only; more tax exemption certificates are in circulation. Yet the practice is not universal — it is strongest in certain regions — why is this? What are the influences that larger numbers have upon training, on the quality of work, on how management seeks to control sites etc? This paper reviews some of the literature available and draws some tenuous conclusions about the prospects of LOSC.

THE DEVELOPMENT OF LABOUR-ONLY SUB-CONTRACTING

There is a temptation to think that labour-only sub-contracting is a phenomena associated with the latter part of the 20th century. However, the available evidence suggests that the practice was widely used during the Industrial Revolution. It would have arisen from the industrial structure whereby clients employed separate trade masters for certain elements of the work. In time the trade masters would employ labour on a piece work basis.

Labour-only was particularly prominent in civil engineering where canals and railways were built on the 'butty system'. (A phrase drawn from piecework in the mining industry where a ganger would pay by the ton a man or gang for the coal it had mined in a day.) Here it was traditional for the client or architect/engineer to buy the materials and to let the labour element to the master craftsman in each trade. The master in turn let the work on a labour-only basis. The practice excited much concern for the emergent trade unions and by the late nineteenth century the issue had surfaced in the debates of the labour and trade union movement.

Marx[1] noted in *Das Kapital* that the piecework system was 'that most suited to the capitalist mode of production' and by 1891 the Trades Union Congress had adopted a resolution which urged 'upon all sectional trades employed in the erection of buildings to use their utmost endeavours to eradicate sub-contracting and scamping in the building trade'. This resolution was the platform from which the trades unions could discuss the issue in its submissions to the Royal Commission on Labour 1892-94[2] where the issue of labour-only sub-contracting was the subject of frequent complaint.

The craft unions were particularly opposed to the use of sub-contracting, which they felt led to the use of men who were not bound by their own rules and labour practices.

The practice of labour-only, whilst present in the early part of the twentieth century, did not really become an issue and there is evidence to suggest that during the inter-war years it was an accepted method of employment for the large numbers of speculative housing developments which were taking place in the London suburbs.

It was not until the 1950s that the trade unions and employers sought to control effectively the use of labour-only. This control was exercised by an industry wide agreement, whereby labour-only sub-contractors were to prove their bone fides by demonstrating to main contractors their financial substance. Local joint committees of the National Joint Council for the Building Industry, along with the Regional and National Conciliation Panels, were invited to arbitrate in disputes about sub-contractors' bone fides. It may be argued that this event set in train the process of 'legitimising' labour-only, for by 1964 the National Working Rule Agreement (NWRA) was amended to lay down conditions for its use. The amendment sought to impose conditions upon labour only sub-contractors, notably that they would abide by the NWRA and that persons under their direction should be covered by an employer's liability insurance policy.

Further, main contractors were to allow trade union officers access to sub-contractors' operatives. This clause was challenged by a leading legal case — *Lothian vs Emerald Construction* — which centered on the differences between a contract of service and a contract for services in respect of self-employed labour.

Despite this institutional acceptance it must be recorded that many sections of the trade union movement, in particular the rank and file, remained hostile to LOSC. However, this hostility is now much more muted because in 1988 the trade unions dropped a long held credo that self-employed persons could not be in trade union membership.

So the perceived need for joint control was stimulated by the growth in the numbers moving to the lump but what events stimulated this shift in employment? The imposition of Selective Employment Tax upon the construction industry was an important influence, and this, coupled with increases in the employers on-costs for labour, encouraged employers to seek labour-only rather than directly employed labour to satisfy acute demands for resources to meet the building boom of the 60s.

In order to assist contractors appointing LOSC the National Federation of Building Trades Employers (now Building Employers Confederation) prepared a model form of contract which many firms used or drew upon as the basis of their own contract forms.

Whilst the practice was formerly frowned upon by the regulating machinery of the industry it may be noted that LOSC was still endemic to the employment policies of the industry. As this paper will go on to show the NWRA influence has had little effect upon those wishing to work as self-employed sub-contractors or on the willingness, if not preference, of employers to use sub-contractors.

Labour-only sub-contracting

This change has meant that the increasing numbers of self-employed could be recruited into the unions. For the leadership of the unions this was a pragmatic attempt to build membership and restore influence in the industry. For employers it retained the structure of collective bargaining which had delivered a stable climate of industrial relations in the industry.

Clause 28 of the National Working Rule Agreement was amended to accommodate union representation of the self-employed.

Indeed, the practice of LOSC has ushered in the changes in the pattern of industrial relations, the Working Rule Agreement being changed in 1988 to accommodate the presence of LOSC within the industry.

Prior to this constitutional acceptance the extent of labour-only sub-contracting became of such concern to the Labour Government of 1974-79 that it established the Construction Industry Manpower Board (CIMB) in 1976 with the brief to investigate casual employment in the industry and a long term objective of formulating proposals for the decasualisation of the industry. This initiative followed many attempts to control sub-contracting by legislation. In particular the Government of 1966-70 introduced the Construction Industry (Contracts) Bill 1970 which sought to regularise employment practices in the industry but this fell because of the General Election of 1970.

The Finance Act of 1971 shifted the focus of the discussion towards tax avoidance by sub-contractors and regularised the tax position by the introduction of '714' cards. (The Act required individuals who were self-employed to satisfy the Inland Revenue that they would pay tax on a self-employed basis. If self-employed persons could not present a card to the main contractor then the main contractor would deduct 30% of the gross earnings and forward this amount to the Inland Revenue.) Further attempts at legislatively outlawing 'labour-only' came with the Labour-only Sub-contracting Bill of 1973. Again, the election of 1974 prevented enactment.

By 1982 the tone of the legislative argument had changed. The Employment Act of 1982 outlawed the concept of using contract clauses to ensure that sub-contractors abided by NJCBI conditions. This effectively gave a filip to the practice; neither clients nor main contractors could insist upon sub-contractors not using self-employed labour. Finally, the most recent intervention is the trade union acceptance of the practice in 1988.

FORMS OF LABOUR-ONLY SUB-CONTRACTING

The 'lump' (or 'the grip' in Scotland) disguises many different methods and patterns of employment; there is a spectrum of arrangements for sub-contracting.

- Bone fide sub-contractors

A bone fide firm is usually a limited liability company where the workers are employed on the basis of contract for service. This pattern is often used for specialist trades in the industry and for building services (eg, plumbing, electricians, etc.), it is universal. Here the use of such specialists is not controversial.

However, there are difficulties in interpreting what is meant by a specialist. For example, tiling and roofing is considered as a specialist trade in the South, whereas in the North of England it is regarded as a legitimate area of direct employment.

The bone fide sub-contractor is at liberty to sub-contract pieces of work to labour-only sub-contractors.

● Labour master
A more informal agreement is where a 'labour master' takes on a sub-contract from a main contractor then employs labour to service the sub-contract. This employment may be legitimate in that the labour master observes the statutory obligations of an employer. However, the evidence is that many workers employed on this basis are paid by a day rate as 'cash in hand' without the necessary documentation or deductions.

● Stable gang
With a stable gang the tradesmen negotiate a piece rate with the main contractor. Typically, members of the gang are responsible for their own national insurance stamp and are consequently self-employed.

● The loner
The self-employed person working alone will be taken on by a main contractor after negotiating his own agreement. This would be based on an hourly rate, day rate or piecework for certain items of work. He may have a '714' card which entitles the worker to gross payment, or the main contractor may need to deduct 30% of the gross pay and forward it to the Inland Revenue.

● Agency labour
In this form of employment the main contractor draws his required labour from a labour agency. The main contractor pays for the operative by the hour and forwards the payment to the agency. The agency, in turn, pays the operative for the hours he has committed to the site.

The practice of 'lump' employment may occur in all of these forms of sub-contracting. The distinguishing feature of what the DOE (1973) has called 'bogus self-employment' is that the employee has a 'contract for services' not a 'contract of service'; in this sense the 'lumper' is an independent contractor and more often has a social and organisational, rather than a contractual relationship, with an 'employer'. The phrase 'the lump' has, however, other connotations; tax avoidance, non-payment of national insurance and other statutory obligations. Whilst these matters are of social concern they are tangential to the argument concerning the extensive use of 'lump' labour in the industry and its impact upon the characteristics of employment in the industry.

USE OF LABOUR-ONLY SUB-CONTRACTORS
There have been many attempts to determine the extent of LOSC. Estimates of the numbers involved have been drawn from samples of employing organisations, the Inland Revenue, trade unions and the former Construction Industry Manpower Board. What has been evident since 1975 is that the extent of labour-only sub- contracting is increasing as a proportion of the workforce employed in the industry. Official records show a decline in the absolute numbers of directly employed operatives in the industry but equally the number of self-employed has dramatically increased. Figure 1 demonstrates the shift.

The 1980 figure of 520,000 self-employed recorded by UCL research[4] obviously includes the bone fide self-employed as well as the 'lump' — nonetheless the increase is the key factor.

Labour-only sub-contracting

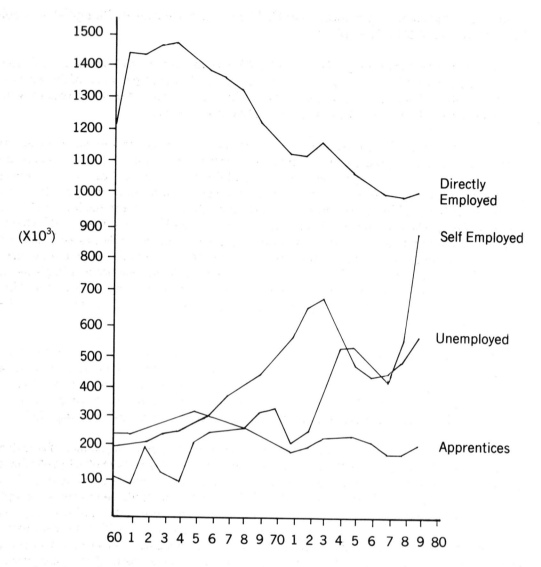

Figure 1 Operative employment, unemployment and number of apprentices
   *Reproduced by permission of Dr Linda Clarke, UCL

The Phelps Brown Report in 1968 estimated LOSC between 165,000 – 200,000, but predicted that it would increase.

By 1975 the Construction Industry Manpower Board[5] estimated that 215,000 workers were employed on the 'lump', yet by October 1976 the CIMB[6] had revised its figures to 335,000. Using the number of '714' certificates issued by the Inland Revenue as the benchmark of those involved, then between April 1978 – April 1982 the number of '714s' issued grew by 21%. The conclusion that can be drawn is that operatives were changing from direct employment to LOSC.

Leopold[7] makes an estimate of the extent of lump labour based on '714' issues, arguing that the Inland Revenue certificates represented 66% of the building 'lump' in 1978 and 60% in

1979. Assuming that these proportions still hold then the current level of 'lump' labour could be in the region of 570,000 — 600,000.

This gives an increase of lump labour of 125,000 over the period 1978-82 against a drop in the directly employed of 199,000 over the same period. As the Department of the Environment estimates private sector direct employment in April 1982 at 581,000 this would suggest the LOSC accounts for some 50% of the operative labour force.

This increase masks the strong regional and trade basis of labour-only sub-contracting. Leopold and Leonard[8] pointed out that there is a negative correlation between direct labour and construction activity in general and house building in particular. Areas with the lowest proportion of direct labour are the most active in house building; conversely Scotland and the North of England have the highest proportion of direct labour and the lowest share of house building. These figures confirmed the findings of the Phelps Brown report[9] where it was found that LOSC was prevalent in London, the Midlands, South Wales and the South West, whereas in the North its use was lower than average. Scotland was said not to use it all. LOSC was prominent in bricklaying and carpentry but its highest use was in plastering — all traditional trades easily measurable and familiar in house-building construction. Thus, the technical characteristics of the project appcar to have an influence upon the extent of lump labour.

The extent of LOSC in private section housing merits explanation. Housing with its predominant small batch production in dispersed sites with traditional construction methods may depress the need for long term direct labour. O'Rourke[10] confirms the use LOSC in housing in Eire with its easily packaged and measured items being ideal vehicles for LOSC. He identifies masonry and plastering work as the most prominent trades using LOSC.

However, more global influence may be at work; certainly, the changing pattern of workload is influential. The workload in the UK fell from £26,000m in 1972 to £20,500m in 1982 (based upon 1980 prices), yet the lump grew despite this trend. This would suggest that the lump is now replacing direct labour, rather than being used in its traditional way, ie, supplementing direct labour for short-term labour uses. This changing pattern of use may have more to do with economic conditions in a region rather than peculiarities of the industry's workload. It may be postulated that the growth in LOSC took place when labour was scarce and that in order to overcome shortages main contractors were able to offer better terms to a marginal group by sub-contracting. This was cheaper than offering across the board increases and would draw individuals to sub-contracting.

Change in employment status would be strongest in areas where employment opportunities were greatest and conversely weakest in areas of higher unemployment. Thus, the regional variability in the use of loss may be a function of the local labour market. Another factor would reinforce this regional difference — namely trade union influence. Where trade union influence is significant LOSC is not the predominant pattern of employment.

O'Rourke[11] sustains this view by quoting international examples. Self-employment is known in Belgium, Sweden, Denmark and the USA but in these countries strong employer and union control moderates the practice. Conversely, a divided and weak trade union movement coupled with labour shortage and a tradition of casual employment encourages sub-contracting.

Labour-only sub-contracting

In the UK the industry is moving to one based upon specialist sub-contractors; between 1978-81 employment with main contractors fell by 45% whilst that by specialist sub-contractors only fell by 9%. It is clear that workers are re-organising themselves into specialist groupings based around trades and this is consistent with the greater number of projects using management contracting, construction management or project management as a method of procurement. So one of two phenomena is occurring. The changes in the industry are driving the labour force into lump employment, or, the preference for being self-employed is forcing organisational and contractual changes upon the industry.

Whatever the cause of these changes their extent is having a profound effect upon the construction industry.

THE CASE FOR AND AGAINST LABOUR-ONLY SUB-CONTRACTORS

The arguments about the use of LOSC may often reflect differences of philosophical outlook. Differences concerning the status and role of labour in construction are likely to surface in the argument over the use of LOSC.

The trade unions have long argued their objections to LOSC and self-employment. Central to these objections is the TUC's[12] view 'that by its very nature it corrupts, leads to indiscipline, destroys morale and fragments the construction process to the point where management techniques become impossible to implement. This leads to inefficiency and high cost which in the long run are an unnecessary strain on the economy of the country as a whole'.

Naturally, the trade unions saw a threat implicit in the way that workers can and do negotiate their own pay deals, rather than rely upon the nationally agreed pay rates. Undoubtedly, the wages drift between the effective pay rate on offer for LOSC and the basic wages agreed by the trade unions posed difficulties for trade union recruitment. The situation became so desperate in the latter 1980s that the union leadership could see little future for the union given the proportion of the industry's workforce in membership. A rapprochement with sub-contractors was inevitable.

The need for a stable pattern of employment is also strongly evidenced in the trade union movement. In its paper on sub-contracting the TUC[13] argued 'alongside casual employment LOSC exists in a widespread scale and if the stable employment situation regarded as normal in every other industry is to be achieved in construction, then LOSC must be eliminated. The continued existence of LOSC via bogus self-employment seems to perpetrate casual methods of working'.

To employers the casual nature of LOSC may offer the necessary flexibility in the use of labour. A NFBTE[14] paper in 1973 claimed that bone fide sub-contractors assisted construction output and reduced labour turnover. The increased output was due to the strong social bonds which may exist between gang members in LOSC. The argument claims that the downtime associated with the running in and running out periods associated with direct labour could be avoided.

However, wider arguments about the nature of labour could also be detected. The NFBTE saw LOSC as a component of manpower planning — LOSC being a mobile labour force. This mobility was not restricted to physical location but the building employers noted that the rate of pay may also be heavily related to market forces and be more responsive to the labour market than fixed and agreed wage rates. More ideological arguments may surface from time to time — LOSC being noted as a liberating force from trade union hegemony, liberation of the individual from the tedium of routine employment etc.

More forcible than any of the above arguments is the economic advanges of using LOSC. To the individual the movement away from Schedule E to Schedule D for taxation purposes is evident. Normal household expenses can be offset against tax. For the employer avoidance of employment on-costs makes LOSC cheaper to employ. A NEDO[15] survey calculated that sub-contracting could save £260 pa per worker by using LOSC. Langford[16] re-evaluated this and suggested that it could be as much as £340 pa. Current calculations suggest savings of £2,125. These savings are, of course, predictated upon payment of the same wages to the LOSC and the directly employed and that output for the two groups is identical. This condition is unlikely to hold up in practice — LOSC are usually offered higher take home pay against an anticipated increase in productivity.

LABOUR-ONLY SUB-CONTRACTING AND ISSUES FOR THE MAIN CONTRACTOR

The debate over LOSC has aired many issues regarding the management of LOSC; in particular the influence of LOSC in the quality of work produced, the provision of a future trained workforce and the effect of LOSC on site safety. At a more immediate level the pattern of relationships which exist between the main contractor and LOSC will be shaped by the terms and conditions of employment. It is unnecessary to restate the various form of sub-contracting but the courts have defined the legal relationship between the main contractor and LOSC as a 'contract of service' not as the conventional sub-contracting arrangement in the 'contract for services'. The leading case of *Fergusen vs John Dawson & Partners* 1976 held that the key factors distinguishing the relationship were that tools and equipment were provided by the main contractor and that the work of LOSCs was directed by the site manager. In particular, if the site manager has the discretion to direct LOSCs in what to do, how to do it, when to do it and where the work was to be done, then the relationship is akin to that of a master and servant. This relationship then imposes many obligations upon the main contractor in respect of safety provisions for, despite the professional independence of LOSC, the main contractor must provide a site which is a safe place for all who work upon it.

SAFETY AND LOSC

It has often been argued that LOSCs are not as concerned about safety as directly employed workers and that management is not as concerned about the safety of LOSC gangs as with direct labour. The available evidence is attitudinal and not quantified, but the Phelps-Brown report noted that site agents considered the safety record of LOSCs to be equal to that of direct workers. Langford[16] countered this view with a survey to directly employed operatives which demonstrated that some 70% of the sample felt that there was a sharp difference in the safety consciousness of directly employed workers, which was thought to be greater than that shown by LOSC workers.

Periodically the HM Factory Inspectorate intervenes in the argument and in a report published in 1973 noted that 'variable employment practices' did not encourage safe working.

More recently the condemnatory document Blackspot 1988[22] from the Health & Safety Executive noted the lack of control of sub-contractors was leading to a deterioration of safety standards in the industry.

Labour-only sub-contracting

In Ireland some further evidence was gained by a survey of 140 sites by two factory inspectors during 1966. The survey, reported by O'Rourke[6], found that site tidiness was a major factor in accident prevention and that there was a connection between untidy sites and the use of LOSC. On the other hand the survey noted that LOSC were seldom involved in accidents. Two conclusions may be drawn from the available evidence:

(a) the preconditions exist for a poor safety record by LOSC exist;
(b) the preconditions do not necessarily result in higher accident rates in LOSC.

How can these two conditions co-exist? LOSC accident rates may be depressed because of the type of contracts which use their services - repetitive housing sites with traditional materials and limited machinery - are being used. Additionally, social factors, so important in accident performance, may lead to lower accident rates. LOSC may be more cautious because of the limited ongoing benefits available to them and because of this there may be a level of under-reporting. Thirdly, a younger age group, more agile, may be employed as LOSC.

Little research has been done to establish whether there are measurable differences between accident rates for LOSC and directly employed workers and the reasons for any differences where they occur.

TRAINING

Much concern has been expressed about the provision of training in the construction industry and it has been suggested that LOSC make little contribution to the formal training of the industry's future labour force. The Phelps-Brown report argued that there was no clear relationship between training and LOSC. This was confirmed by an NFBTE[17] London Region Manpower study 1972 which observed 'that no positive relationship could be demonstrated between operative to apprentice ratios and the intensity of use of labour-only sub-contractors' employment'.

The figures appearing in the report break down the respondent firms by the proportion of labour employed by LOSC. It is noticeable that where LOSC accounts for more than 50% of the labour force the operative/apprentice ratio is much less than in firms where no LOSC is used.

Size of firm by employeee	Operative/apprentice ratio where no LOSC used	Operative/apprentices ratio where 50% labour is LOSC	Difference
25- 59	9.9	20.0	+10.1
60-114	14.2	21.0	+ 6.8
115-299	10.9	28.7	+17.8
300-599	no data	no data	avg. diff. 11.6

From these data it would seem that where LOSC is used as a component of the employment policy then apprentice training is not given as much attention as in firms where LOSC is not used. This would suggest that the use of LOSC has to reach a critical figure within a company before training is badly hit. From this evidence it may be argued that where firms seek to supplant direct labour with LOSC then training programmes within these firms are reduced.

Moreover, the available funding for training may be depressed by the extent of LOSC. The CITB in its evidence to the Phelps-Brown Inquiry noted that it had attempted to collect training levy from sub-contractors. The CITB compiled a list of 45,000 firms who had carried out work for main contractors and local authorities. It then sent out forms for training levy collection to all 45,000 - the returns were as follows:

Forms issued		45794
Returned by dead letter office	15444	
No reply	16024	31468
	Forms returned	14326
Too small for levy	6667	
Not leviable (out of scope)	7012	13679
Those leviable	647	

If some, say 40%, of the labour force is involved in LOSC and the levy return from this sector to the CITB is so small then obviously the resources available for training for the future are likely to be insufficient for the purposes of the industry.

QUALITY OF WORKMANSHIP

Architects and clerks of works have often expessed concern about the quality of workmanship when LOSC were used[16]. Counter to this view was the observation by Phelps-Brown that the quality of work completed by LOSC was no worse than that of directly employed workers since the temptation to 'scamp' was offset by the LOSC sense of responsibility to main contractors, or more particularly to their trade or craft. O'Rourke[10] recognises the difficulties in monitoring quality but suggests that the incentive of repeat business is sufficient to maintain the required quality standards. Nonetheless, management must be vigilant in setting and monitoring wormanship against agreed standards. McVicar[18] reverses the argument by noting that, in the South at least, LOSC employs craftsmen of good reputation. Yet this view was not borne out by a survey by Langford[16] which sought to evaluate operative and management attitudes to LOSC. Here it was found that 87% of the operative sample and 91% of the management sample felt that LOSC workmanship was inferior to that completed by directly employed labour. Hillebrandt[19] also recorded strong feelings about the quality of LOSC in her evidence to the Bolton inquiry into small firms. She noted that small firms had stronger feelings than large firms in respect of LOSC quality - more small firms thinking it better or worse than direct labour than the larger firms who used LOSC. Even the NFBTE[17] accepted some difficulties in respect of quality when it noted that time rather than quality was the focus of attention. Perhaps this is too stark - control of time will inevitably be a major part of control for site management but the concern for quality is moving apace and if quality assurance is one of the directions of the future then greater emphasis may need to be placed on the quality control of LOSCs' work.

SITE MANAGEMENT CONTROL

One of the central reasons for main contractors to use LOSC is to devolve risk to sub-contractors. But with the devolution come certain drawbacks, for not only may risk be passed on but overall control may be diluted. Many would argue that management control of LOSC is exercised through the lump rate but this is only one of the aspects of management - co-ordination of various labour inputs may be critical. The NFBTE report[17] acknowledges

Labour-only sub-contracting

this point by noting that 'the work of different LOSC is difficult to co-ordinate; the builder tends therefore to lose control of the programme'; in effect the LOSC may be independent of control by the main contractor. This view concurs with the RIBA's stated position quoted by NALGO[20]. 'The system (LOSC) can upset construction programming for three reasons:

(1) that it does not give the general contractor the flexibility he needs in the disposal and movement of his labour force around the project which may vary from day to day to meet particular pressures or in changes of operations necessitated by weather and other delays;

(2) the uncertainty when the gang will work - they dictate and choose their own time;

(3) the bad work done by LOSC who have left the site has to be rectified by the general contractor and so imblances the proper planned use of his permanent labour force.'

The degree of control which needs to be exercised may be a function of the type of project and the skills of the manager himself. As has been noted LOSC predominate in traditional 14 trades in relatively simple projects. Here a high degree of control may not be necessary nor even desirable, for trained workers may be left to carry on working on their own because the work is not subject to technological change and tasks do not have to be specified in detail or be kept under constant supervison. Hillebrandt[19] found that the use of labour-only was greater in larger firms and that small firms (with presumably less access to skilled and professional managerial personnel) expressed less satisfaction with LOSC in terms of supervision, productivity and site tidiness. Thus, it may be hypothesised that the level of managerial control able to be exercised may be a function of the size of the main contract and the numbers and expertise of personnel employed in such contracts.

THE FUTURE

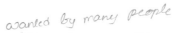

This review of the practice of LOSC has sought to explore some of the macro-effects that may influence the shape of the labour force in the construction industry. To some the continued and extended use of LOSC will be 'a short sighted expedient solution to labour problems' and that 'this form of labour is extensive and expensive'[21]. Others see the use of LOSC as a pattern of employment which mirrors the economic condition of the industry.

The construction industry with its uneven and often very regional workload, finds difficulty in recruiting and training enough skilled labour in times of boom, whilst in recessions labour is shed. This condition means that in times of boom a local labour market is responsive to conditions and able to bargain advantageously for itself, whilst in time of scarcity of work contractors are reluctant to hire direct labour with its moral and financial attendances. Thus, the cyclic pattern of workload may be an influential handmaiden to LOSC. A more even and more controlled pattern of work could provide the structure for a stabilised labour force which could do much to improve the status and image of the construction industry.

REFERENCES

1. MARX K (1867) Das Kapital. New World
2. ROYAL COMMISSION ON LABOUR (1895) HMSO
3. DEPARTMENT OF THE ENVIRONMENT (1973) Report of the working party on self-employment and other issues. HMSO
4. UNIVERSITY COLLEGE LONDON (1983)

5. CONSTRUCTION INDUSTRY MANPOWER BOARD (1976) Second Report to the Minister for Housing and Construction. HMSO

6. CONSTRUCTION INDUSTRY MANPOWER BOARD (1977) Second Report to the Minister for Housing and Construction. HMSO

7. LEOPOLD E (1982) Where have all the workers gone? *Building* October 22

8. LEOPOLD E and LEONARD S (1983) Reorganised labour. *Building* July 8

9. MINISTRY OF PUBLIC BUILDINGS & WORKS (1968) Report of the Committee of Inquiry under Prof E H Phelps-Brown into certain matters concerning labour in building and civil engineering. HMSO

10. O'ROURKE C (1981) The case for labour-only sub-contracting. CIB

11. ibid

12. TRADE UNION CONGRESS (1967) Evidence to the Phelps-Brown Inquiry

13. ibid

14. NATIONAL FEDERATION OF BUILDING TRADES EMPLOYERS (1973)

15. NATIONAL ECONOMIC DEVELOPMENT OFFICE (1973)

16. LANGFORD D (1976) The effects of labour-only sub-contracting upon the construction industry. Unpublished MSc thesis

17. NATIONAL FEDERATION OF BUILDING TRADES EMPLOYERS (1972) Manpower study

18. McVICAR S (1984) Sub-contracting - the pros and cons. *Building* June 22

19. HILLEBRANDT P (1981) Small firms in the construction industry - evidence to the Bolton Report on small businesses. HMSO

20. NALGO (1974) Lump labour

21. DEAN J (1973) *New Civil Engineer* 11 January

22. HEALTH & SAFETY EXECUTIVE (1988) Accident Blackspot. HMSO

VARIATION ORDERS, SITE WORKS ORDERS AND SITE RECORDS

by D F Turner, BA(Hons), FRICS, MCIOB, ACIArb

INTRODUCTION

This paper sets out to clarify for site managers and foremen the meaning and interpretation of the terms variation orders, site works orders and daywork records.

Reference is made only to the clause numbering of the Joint Contracts Tribunal (JCT) Standard Form of Building Contract 1980 in the 'with quantities' edition. Other editions of the Standard Form and also the Intermediate Form of Contract (IFC) produced by the JCT embody essentially the same approach to the matters discussed, but with differences of detail over the use of approximate quantities or of drawings and specification without quantities. These differences are ignored for present purposes of describing the main effects. Central government and civil engineering contract forms are also *broadly* similar in their effects.

The term 'the quantity surveyor' is used for the person acting as such under the contract, while 'the site surveyor' is used for the contractor's surveyor dealing with the project.

DEFINITIONS

Architect's instructions and variation orders

The architect has power under the contract to issue various instructions, falling mainly into four broad groups:

(a) those necessary for the progress, such as about expending provisional and prime cost sums (see below). The contractor cannot act on the matters in question without them - for instance he cannot select his own nominated sub-contractors;

(b) those correcting the contractor, including those relating to defective work;

(c) those introducing some physical change into the works as already designed;

(d) those introducing some change in the conditions of working, if any, imposed by the employer or client in the contract bills over elements such as access and hours of working.

These last two groups are usually known as variation orders and are the concern of this paper.

More fully, the term 'variation' as used in JCT clause 13 means the alteration or modification of the design, quality or quantity of the works as shown upon the contract drawings and described by or referred to in the contract bills. It includes the addition, omission or substitution of any work, the alteration of the kind or standard of any of the materials or goods to be used in the works, and the removal from the site of any work, materials or goods executed or brought thereon by the contractor for the purposes of the works other than work, materials or goods which are not in accordance with the contract. This is the traditional use of the term variation.

But the term variation in the contract also means the addition, alteration or omission of elements as mentioned in (d), these being termed obligations and restrictions and listed as 'access to the site or use of any specific parts of the site, limitations of working space, limitations of working hours and the execution of completion of the work in any specific order'. This is a more difficult aspect of variations, as outlined below. There are also a few cases in the contract conditions where something is declared to be deemed a variation if it occurs, ie it does not need a written instruction to cover it. These are less frequent and are not discussed further, although they are listed in Table 1.

Table 1. Architect's formal instructions

Clause	Subject
2.3	Discrepancies and divergences
6.1.3.	Statutory obligations
7	Setting out works
8.3	Opening up and testing work etc
8.4	Removal of defective work etc
8.5	Exclusion of persons
13.2	Variations
13.3	Expenditure of provisional sums
17.3	Defects during defects liability period
23.2	Postponement of work
30.6.1.1	Documents to quantity surveyor
32.3	Protective work during hostilities
33.1.2	War damage
34.2	Antiquities

There are also various instructions under clauses 35 and 36 relating to nominated work, not listed, which would usually be dealt with by the contractor's main office.

'Deemed variations'

Clause	Subject
2.2.2.2	Correction of contract bills
6.1.4.3	Statutory obligations
22B..2.2	Restoration after damage
22C.2.3.3	Ditto
33.1.4	Work after war damage

Not only must the site manager obey variation orders, but he must not change the design of the works on his own initiative. If he does, his firm may be liable at law for the consequences, even though the site manager might consider that the architect or the clerk of works has condoned his action by silence.

Site works orders

Clause 12 allows for the provision of a clerk of works 'to act solely as inspector'. If he does issue any 'directions' these must be confirmed in writing by the architect as instructions within two working days or they will be invalid, even if obeyed. In practice the clerk of works often has a delegated authority to issue such directions and these are normally referred to as site works orders. The contractor should check in any important case the extent of this delegation, since there will be an intended limit on it (which may not have been made clear). Confirmation will often be by architect's instructions listing the reference numbers of site works orders, although not all will introduce variations.

DEALING WITH ARCHITECT'S INSTRUCTIONS

A difficulty facing the site manager is trying to decide whether or not the work he is doing is adequately covered by the original contract documents or whether it needs a variation order. Even if one is issued it must not be so major as to change the scope of the contract. Such an invalid instruction would be one increasing the scope from three to four blocks of flats, or changing the type of building, say from a fire station to a swimming pool.

The site manager, or his head office, have a right under clause 4 to query whether an architect's instruction is within the scope of the contract. Should it not be, the contractor is not obliged to carry out the work. Obviously, some care is needed here in making such decisions and this will be a matter for head office. Broadly, the contractor need not accept any order changing the nature of the works (says from a swimming pool to flats) or changing its scale (say from three blocks to four).

If the site manager knows exactly what has to be done to conform with the contract documents, any alteration or addition will constitute a variation and must be considered under clause 13. However, the contract documents may not give an exact statement of all the work while including sums of money set to cover the cost. The best examples of this are the use of 'provisional sums' in respect of which clause 13 requires the architect to issue instructions before any expenditure is incurred.

Provisional sums

This term is used for amounts allowed within the bills for work, the exact nature of which cannot be accurately determined at the time the bills are produced. Provisional sums usually occur from one of two main causes. Firstly, the nature of the work is such that exact measurements cannot be made at the billing stage, for example - foundations where underpinning is concerned, or in the joining up of new work to old. In this case provisional, rather than firm, quantities may be included for the work in question and these will need to be dealt with on the same basis as provisional sums of money. The second cause for provisional sums to be included in the bills is that insufficient information was given to the quantity surveyor by the architect. As a result the quantity surveyor has included provisional sums to get the bills out to tender.

It should be normal procedure for site managers to list all works shown in the bills as provisional sums and to see that adequate instructions are received from the architect to cover them. As soon as the architect's instructions are received it will be necessary for the quantity surveyor and the site surveyor to agree as to how the work is to be measured and valued. It is essential that the site manager knows how the work is to be valued before it is started, as this will enable him to keep adequate records, especially of work which will be inaccessible later. If, however, there is any doubt over the method of valuation it is in the interest of all concerned that daywork records should be kept as a precaution.

If the bills are based upon the latest, Seventh Edition of the Standard Method of Measurement of Building Works, there is a distinction made between provisional sums for defined work and for undefined work. These should be identified separately in the bills. The former type is to have outline information about the work covered by the sum, so that the tender can allow for planning and programming effects and extra costs of preliminaries (given at the front of the bills for such elements as supervision of the project, temporary accommodation, services and facilities and also major plant). The latter type omits the information and the tender is held not to include for these effects and costs.

When instructions are issued to expand sums for undefined work, the site manager should liaise closely with the site surveyor over the records to be kept, as the subject is too complex to cover in detail here. These may relate to expenditure where the work is performed and also in the overall site organisation. Records may also be needed of amendments to the programme, possibly leading to prolongation of the contract period or to disruption and prolongation costs.

Prime cost items

It is important to see that the architect issues adequate instructions, not only concerning the expenditure of prime cost sums but also concerning related items, particularly where wide variations exist between the items described in the bills and those actually installed. For example, labours and fixing to a fitment actually costing £500 may be very different from those allowed in relation to PC sum of £100 in the bill for its supply. It is, therefore, essential to see that adequate architect's instructions are received and understood for all items listed in the bills as prime cost sums.

It is usual to find that PC sums for sub-contractors have associated items for attendance listed with them in the contract bills. One of these will be for 'general attendance', which is defined separately in the Standard Method of Measurement to which the bills should conform, as 'the use of the contractor's temporary roads, pavings and paths, standing scaffolding, not required to be altered or retained, standing power operated hoisting plant, the provision of temporary lighting and water supplies, clearing away rubbish, provision of space for the sub-contractor's own offices and the storage of his plant and materials and the use of mess-rooms, sanitary accommodation and welfare facilities provided by the contractor for his own use'.

Any other attendance requirements are to be listed in the bills, so that they may be priced. As they are likely to vary between sub-contractors, there is the possibility that, when a sub-contract tender is passed by the architect for the contractor to accept, there may be different attendance items from what were listed. For instance, general attendance covers only the contractor's own standing scaffolding and not scaffolding erected early, kept late or altered for the sub-contractor's purposes - quite apart from special scaffolding solely for the sub-contractor. These can be costly differences which should be covered by a variation instruction, so that payment is adequate. While they should be noticed when the quotation is accepted and so before the details reach the site, the site manager should check the bills against what he has to do in practice to ensure that there is no gap.

Handling instructions

The site manager must keep track of all architect's instructions and site works orders issued and be able to refer to them easily. This means careful filing and reference systems must be maintained and copies of all variation orders must be sent to sub-contractors concerned at the earliest time possible. For this to happen, the contractor and the architect must agree on a method for issuing and dealing with architect's instructions. The simplest method of ensuring that nothing is missed is for the architect to issue all his instructions as numbered variation orders through the main contractor's office. This enables the contractor to establish a procedure for handling the distribution of information to those concerned. Confusion, however, exists when the architect issues some instructions as formal variation orders and others as letters to the contractor or in the form of amended drawings, without covering these by confirming orders.

Variation orders, site works orders and site records

Probably the most important check required is to see that instructions all follow the same route. The architect should be discouraged from sending some instructions to the site and others direct to the contractor's head office. In any case the architect has no right to give instructions direct to the contractor's suppliers and sub-contractors (whether nominated or not) and the contractor should insist that all instructions are channelled through him to avoid possible chaos. It is a wise precaution to ensure that copies of all instructions are sent to the main office, if not initially then as confirmation of decisions made on the site. This allows special problems about ordering and claims for loss and expense, for instance, to be identified quickly and action to be taken. It is the main contractor's responsibility to ensure that the site manager receives adequate instructions from the site surveyor as to how these instructions are to be treated. This will enable the site manager to know, before the start of the work, whether or not daywork vouchers are necessary. It must be remembered that all architect's instructions must be dated - the date and time at which they are received on site may be of vital importance for adequate payment, or even payment at all. This is discussed under the next heading.

A frequent cause of dispute is the oral instruction which was never confirmed, leading to doubt over exactly what was said, when it was said, or even whether it was said. Clause 4 permits the contractor to confirm such instructions back to the architect in writing within seven days and they then become binding within a further seven days if the architect does not react. Until then the contractor need not, and in safety should not, take any action. His written version of the instruction should be complete and unambiguous in his own interests.

Figure 1 outlines the various clauses dealing with confirmation of instructions and the way in which they take effect. The IFC84 contract does not allow for any confirmation system. It may be noted here that variation instructions do not always lead to 'extras', as there may be omission of work. Even here, records are needed, while the omission may omit too much if priced strictly at the bill prices, as noted under the next heading.

Valuing variations

Where a variation under the contract has been made, clause 13 gives several alternative rules for measuring and valuing extra work and these are to be used in the following order of priority:

(a) where the work to be executed as a variation is both of a similar nature and carried out under similar conditions to work given in the contract bills, the prices given in the bills shall be used;

(b) where either of these criteria does not apply, then any prices given shall be used as a basis for agreeing new prices. The results are often termed 'pro rata rates' and there are a number of subsidiary provisions about when different prices may apply to varied work, whether the actual measured quantity of work has been changed or not. The following are particularly important:

   (i) when there is change of character, which includes a change in specification of the materials or method of application or installation;

   (ii) when the conditions under which the work is performed change. This may mean work which is displaced to a different time in the programme, so that plant etc. is retained or changed to a less suitable item which then happens to be available. Other

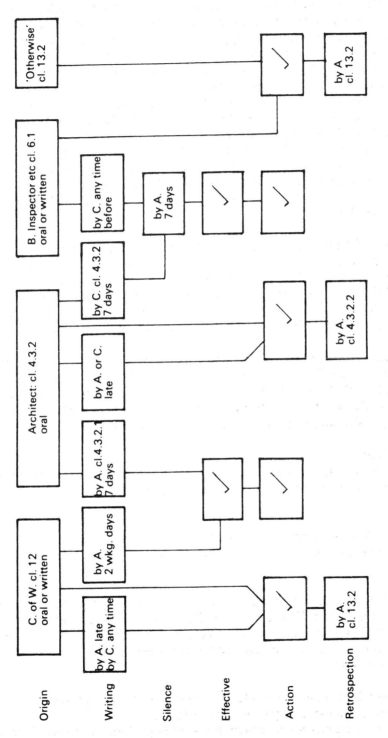

Figure 1 Ratification of variations

example conditions are more restricted access, or working above or below other trades, needing greater care;

(iii) when there is a drastic change in the quantity of work in a particular item in the bills, so that utilisation of resources become more or less economical than was allowed for in the original bill prices;

(c) where work is not comparable with that given in the contract bills but can be measured and priced, a 'fair valuation' shall be agreed. The results here are often termed 'star rates';

(d) where work cannot be measured and valued, daywork shall be used. This method affects the site manager and is considered below.

Where items are omitted, the prices in the bills determine the value of these items; if the omission of any item varies the conditions under which any remaining items are carried out, the prices of these shall be revised as under (b) above. This could be the case if, for instance, most of the work in one trade was omitted and the plant already on site became uneconomical in use or labour was employed discontinuously. It may be necessary to increase the prices for work *not* omitted to allow for the share of costs in work which *is* omitted. The site manager needs to be alert to warn of such occurrences, which he is well placed to see.

Probably the most difficult aspect of this work is in catering for the effect of work which must be completed out of sequence, due to the timing of instructions. It may happen that, while the valuing of the varied work is satisfactorily covered by the rules already outlined, the resulting site conditions affect other work (which is not itself varied) in such a way that there is a change in sequence, or piecemeal working is adopted. If extra expense is involved then it is essential that the reason for the change in sequence, the timing and the resulting effects are clearly established in writing at the time the work is done. It is very difficult at some later stage to recall exactly why a particular course of events was necessary and, to avoid unnecessary argument, the cause must be established at the time.

There will often be cases in which sub-contractors incur extra costs and the contractor will be responsible for raising any variation orders with the architect on their behalf. Otherwise, the contractor may find himself having to pay the sub-contractors and then not recoving the excess from the building client. There may also be cases in which the contractor and sub-contractors will have charges against each other for work done or expense caused that is not recoverable from the client. Separate records will be needed here to avoid confusion.

The more difficult type of variation is that related to obligations and restrictions, such as imposed working hours. Here the effect usually does not show up as something which can be priced in a few adjusted prices for a restricted range of work. It may affect the cost of many items or it may affect costs which can be isolated only within the site organisation costs rather than be related to any particular measured items. A change in working hours will mean that the fixed costs of travelling time and daily transport of workpeople are borne by a different number of productive hours each day. Extra handling due to restriction of access or storage space can have repercussions of an extensive nature. Anywhere the site manager sees a change in his basic running costs, even without any change in the finished building, he should draw attention to it and also keep whatever records are possible in the circumstances. He should not be put off by the fact that these cannot be as clear-cut as daywork sheets, but may necessarily be quite sketchy, even impressionistic. Some evidence is better than none, every time.

LOSS AND EXPENSE

Sometimes the effect of instructions is not just to put work out of sequence so that it is still performed in an efficient and orderly manner, although at greater cost than expected when tendering. The effect may to be create conditions of disturbance, so that proper control by the site manager is undermined by the timing or nature of the instructions and there is disruption or prolongation of activity; there is inefficient working due to the disturbance. Work may become piecemeal, productivity may drop or standing time may occur, among other things. An extra gang may need familiarisation time.

This situation cannot be dealt with entirely by using the variations provisions in clause 13, especially when unvaried work is affected by a ripple or knock-on effect, but needs the provisions of clause 26 about reimbursement of loss and expense. These are complex and beyond the scope of this paper. They require the contractor to give notice when the situation is likely to develop (as the variations provisions do not), so that the site manager must be alert for anything likely to be troublesome - and give warning to the site surveyor who will give the formal notice required. They also use a different basis of payment, perhaps even involving head office costs, while still requiring suitable records, and are not always easy to distinguish from the type of variation which involves change of conditions.

DAYWORK

Clause 13 gives further rules where the nature of the work makes accurate measurement and particularly valuation, impossible. These rules cover daywork in accordance with *Definition of prime cost of daywork carried out under a building contract* agreed between the RICS and the BEC, with percentage additions given in the contract bills. There are similar rules for specialised work under the appropriate definition.

These daywork rules can apply, however, only if 'vouchers' specifying the time spent each day are submitted to the architect not later than the end of the week following that in which the work has been executed. These vouchers are commonly called daywork sheets, but it must be remembered that they only represent a record of work done which constitutes a variation to the contract and do not indicate that the work will necessarily be valued at daywork rates in preference to using measured rates, which may give a higher or lower valuation.

The preparation of daywork records

Remembering that these vouchers are only a record of work done and do not automatically establish a claim for payment, the following points will prevent some of the more common faults occurring when daywork sheets are produced and rejected:

(a) each daywork sheet must relate to a specific architect's instruction, preferably to a variation order or site works order, and the reference number of this must be clearly stated;

(b) daywork vouchers should be submitted by the contractor to the clerk of works or architect for signature. Sub-contractors carrying out daywork should first submit their vouchers to the contractor, who will then include his own attendance and submit fresh vouchers to the clerk of works;

(c) daywork vouchers must be submitted for verification not later than the end of the week following that in which the work has been done. This means that daywork records must be prepared weekly. Sub-contractors must be informed of the procedure and of the day on which their own vouchers must be submitted to the contractor;

(d) daywork vouchers should describe the work done exactly. Many daywork sheets are prepared in such a way that minimum space is allowed for the description of the work. Since the daywork account may not be valued for some months, exact, clear and concise descriptions of the work are essential, even to the extent of supplementing the record with sketches or photographs of the work done. There may also be other work proceeding at the same time and accounted for on a measured basis, for which it will be necessary to establish a boundary separating it from the daywork;

(e) time must be allowed for setting out, covering up finished work, cleaning away and making good. These and other costs may be higher than normal, depending upon the time at which the variation order is issued and on whether overtime is worked;

(f) the architect may request not just the record of hours spent by men on daywork but also their names. This means that sub-contractors must be instructed to produce their day-work accounts in a prescribed manner. Rates for labourers, craftsmen, and foremen differ and, therefore, the status of the man must be shown as well as his name;

(g) plant and materials must be recorded as well as labour. It is, therefore, important to record the desciption and type of plant used, the cost of any small quantities of materials specially ordered and any extra delivery costs, as well as the labour required. When plant available on site is being used uneconomically instead of bringing the more appropriate plant specially on to the site, this must be specifically agreed to by the architect, as higher rates may apply. In accounting for these items it is necessary to observe the provisions of the *Definition of Prime Cost of Daywork* referred to earlier;

(h) it is always easy to forget the obvious when preparing daywork sheets. In particular, one should remember such items as scaffolding, shoring, special sections of formwork with only one possible use, extra long barrow runs, protection of finished work adjacent to the job, access problems, difficulties in working in confined spaces, sharpening tools for breaking concrete etc. Disorganisation to following trades may arise and should be dealt with separately as has been mentioned in relation to loss and expense;

(i) daywork vouchers must be produced at least in triplicate, one copy going to the clerk of works, one to the site surveyor and one to the site manager. These should be carefully filed in sequence, together with the instructions relating to them;

(j) once the voucher is signed as a true record of work done, this is not the end of the story and drawings, sketches, instructions and delivery notes appertaining to the work must be kept, or passed to the site surveyor. These items will be particularly important if eventual valuation is not on a daywork basis.

CONCLUSIONS

It is not possible in a paper such as this nor is it necessary to consider the valuation side of the measured and daywork account. It is, however, essential that the closest liaison exists between the quantity surveyor, the clerk of works, the site surveyor and the site manager. It must be remembered that the quantity surveyor may not see the work done, therefore, the site manager must give the fullest information. Too much information is always better than too little and in final accounts, which can hang on for years after the completion of the work, it is often impossible to remember what has been done without records.

BIBLIOGRAPHY
1. ROYAL INSTITUTION OF CHARTERED SURVEYORS AND THE BUILDING EMPLOYERS CONFEDERATION. Definition of prime cost of daywork carried out under a building contract.
2. JOINT CONTRACTS TRIBUNAL. Standard form of building contract 1980, private or local authorities edition with quantities (amendments are issued at intervals)
3. JOINT CONTRACTS TRIBUNAL. Intermediate form of building contract 1984 (amendments are issued at intervals).
4. TURNER D F. Building contracts : a practical guide. Longman.
5. TURNER D F. Building contract disputes. Longman.
6. WOOD R D. Building and civil engineering claims. Estates Gazette.
7. WOOD R D. (1988) Contractors claims under JCT80. The Chartered Institute of Building.
8. DICKASON I. (1985) JCT80 and the builder. The Chartered Institute of Building
9. POWELL-SMITH V. and SIMS J. Contract documentation for contractors. Blackwell.

MODULAR BUILDING SYSTEMS AND THE ROLE OF CONSTRUCTION AUTOMATION SYSTEMS AND ROBOTICS

by Brian L Atkin, Peter Atkinson, Colin E Bridgewater, Javier Ibanez-Guzman and John Bayes

SUMMARY

The automation of traditional construction methods has proven to be a difficult task. This is due mainly to limitations arising from the adaptation of conventional robotic manipulators to suit the construction environment. Robotic manipulators and construction plant and equipment have had different development paths and there are significant differences in the requirements of each. This paper discusses these differences and examines the underlying philosophy of construction automation systems and robotic tools. Several examples are given of prospective robotic tools for construction and one of these, a cladding fixing robot, is described in some detail.

INTRODUCTION

This paper describes research which is aimed at the development of integrated construction automation systems and robotic tools, based on modular building systems. The starting-point has been the definition of a 'parts-set' of components equipped with standard joints and fixings[1,2]. It is intended that the components will be manipulated by on-site robotic tools, enabling buildings to be completed quicker and more safely than by traditional construction methods. Several tasks that might benefit from the use of such robotic tools have been considered and include column positioning, floor laying, beam placing, floor grouting, partition placing, floor cage laying, ceiling cage placing and cladding fixing. Four of these are outlined in this paper and one of them - a cladding fixing system - is described in detail.

UNDERLYING PHILOSOPHY

Robotic manipulators and construction plant and equipment have evolved from different traditions, having equally different objectives[3,4]. Despite increasing concern in the construction industry over productivity and costs, technological change has been taking place at a much slower pace than, for instance, in manufacturing. The fundamental difference rests in the nature of the end-product delivered by both industries. The products of construction are more durable and are used over relatively longer periods of time[5], whilst the products of manufacturing either become obsolete or worn out over shorter periods. On-site working conditions are also different: in construction the work place is continuously changing, resulting in an ill-structured and dynamic environment. The effect of this is to impose considerable limitations on the technology and on the degree of acceptance of automation systems on-site. By comparison in manufacturing, a static and well-structured work environment is not unusual. Work objects are generally much smaller and tend to be transferred from work cell to work-cell, thereby encouraging automation.

Another basic difference lies in the origins of robotic manipulators, devices which began life in the machine tool industry. For them, component stiffness is an all important property in obtaining accuracy and repeatability in machining and assembly. In construction, equipment such as cranes are generally rugged but not particularly stiff. In manufacturing, robots have a low payload to weight ratio (approximately 1:20); the ratio is much lower for a crane. Consequently, the nature of loads manipulated in both industries is reflected in the different types of plant and equipment employed.

Differences are also apparent through the need for mobility involving, amongst other things, sensing and obstacle avoidance. Clearly there are myriad problems which must be solved before construction robots become commonplace on site.

One argument against attempting to solve problems specifically for construction is that if one waits long enough they will be overcome by researchers in other fields and can thus be applied to construction at a far lower cost. However, the problem of construction automation is an urgent one which can be solved, at least partially, by analysing the differences between conventional robotic manipulators and construction plant and equipment. It would then be possible to make progress by marrying together the useful properties of each. While it is still not possible to effect a well-structured, factory-like environment, it is possible to create a 'factory in microcosm' or work-cell in which a specialised robot manipulator can work within a confined area before being moved on to its next work position. This could be achieved either by building a sophisticated work-cell that includes several robotic tools, as in the case of the steel erection system[6] being developed in Japan; or by adapting currently-available construction plant and equipment, as in the case of the cladding fixing robot described later in this paper. A review of the applications[3] highlighted that current construction plant and equipment is just beginning to make use of modern high-tech devices such as sensors, radio controls and signal processing engines which are well proven elsewhere and relatively inexpensive. The use of these devices can transform mechanical construction plant and equipment into something approaching automation.

Another objective of the research was to design robotic tools rather than fully automated solutions using sophisticated sensor-based control systems. The tools are simple, dedicated machines that are easier to use than more generalised machines; that is, they are not aimed at solving a multitude of problems at once. The following sections describe four such robotic tools.

COLUMN POSITIONING SYSTEM

In conventional steel frame erection processes, columns are positioned first, followed by beams. Columns are lifted by a crane and held in the vicinity of a foundation pad. Fine positioning of the column is accomplished by operatives pulling with ropes. Finally, an operative will climb a ladder or an improvised scaffold tower and release the crane's hook from the column end. The operative is very often in an unbalanced position whilst performing this task.

The proposed column positioning system has been designed to remove the operative from the vicinity of such dangerous work and to improve the efficiency of steel erection. The system consists of an automatic, remote-controlled clasp and inclinometers and a column positioning robot (CPR) working in conjunction with a mobile crane.

For automation purposes, it is proposed to use a clasp attached to the crane's hook and a pair of inclinometers mounted on the hook's pulley. The clasp would be attached to the top of the column by an operative. The release mechanism would be remotely-controlled and would probably be electrically operated. This automated device would thus consist of two locking mechanisms (clasps), a radio receiver, other remote control units and power supply units. A similar device has been developed by Shimizu Corporation (Japan), namely the Mighty Shackle Ace.

Modular building systems

FLOOR LAYING ROBOT

In conventional, in situ, reinforced concrete floor construction, floors are cast with beams on timber shuttering and left until the concrete has sufficient strength to support itself. The amount of material and labour costs in this approach led to the development of precast centering systems such as hollow beam, and plank and beam. Precast components do not require temporary support, but are cumbersome to erect in one operation.

The proposed floor laying robot has been designed to facilitate the handling of planks made of reinforced, aerated concrete or materials with similar load to weight ratios. The system consists of the floor laying robot and a mobile crane with an operative.

It is envisaged that the mobile robot would have a simple end-effector which would enable the planks to be lifted and lowered into position. In order to reduce costs, the robot arm would have vertical motion only. The extra degrees of freedom required for final plank positioning would be provided by manoeuvring the vehicle which would have steerable wheels at both ends, thus allowing omnidirectional motion.

FLOOR CAGE PLACING ROBOT

Traditionally, services have tended to be placed in ceilings where it is difficult to work, instead of putting them within the floor void where access might be easier. This conventional approach may also mean that several specialist trades are working at the same time in confined spaces and in undesirable postures. Operatives must raise large, heavy components over their heads for fixing within the ceiling void, requiring access from temporary scaffolding or platforms.

A key element in the modular approach to the construction of high-tech buildings is the use of floor cages containing all the services that are likely to be required. The raised floor system thus proposed would be made of prefabricated cages to provide a void for all the services that the floor is expected to carry[1,2].

The proposed floor cage placing robot has been designed to manipulate floor cages, removing the operative from handling a heavy load in an uncomfortable position and performing a monotonous task. The system would consist of a loading mechanism, an external work platform and the floor cage placing robot. Operatives would be needed for fixing the cages in their final locations, for positioning the work platform and also controlling the robot.

In order to deliver floor cages to the particular level on which they are required, it is proposed to use an external access work platform. This platform would be a modified version of a commercially-available product. The standard unit would require a ramp to be attached to it in order to unload the cages on to the working floor. For placing the floor cages on to the floor, it is proposed to use a radio-controlled vehicle.

The floor cage placing robot is thus very similar to the floor laying robot and it is possible that the same vehicle could be used for both operations by simply changing end-effectors and certain control modules.

CLADDING FIXING SYSTEM

Traditionally, external walls were constructed as loadbearing elements, using basic materials such as masonry and bricks. Today, non-loadbearing external walls are commonplace with loads transferred to the ground by means of a frame. Because external walls need no longer be loadbearing, it is possible to enclose buildings in a variety of modern materials.

In a detailed study of the cladding process[7], it was found that a cladding system could be conveniently divided into three areas:

— cladding panels;
— joints between panels;
— fixing systems.

For each of these, current building practice was examined and any ramifications for an automated system were noted. The end result of the study was a detailed description of panel sizes and weights; joint profiles and jointing techniques; fixing methods and components required; automation tools and aids necessary to carry out operations; and an erection sequence for a robotic cladding system. Similar studies will be necessary as other areas of the parts-set are developed in greater detail.

Cladding panels can be of any material as long as they obey a set of rules which facilitates their manipulation by robot. In this way, the designer is given the freedom to clad the building in almost any style using large-panel methods. Provision has been made to allow the building to be clad in more traditional materials such as brick or render. Panels can be of many shapes and sizes, constrained only by their total weight and certain key dimensions. In other words, a panel must be of such a weight and size that it can be lifted safely by the robot.

Joints between the panels are based on adaptations of current building practice and are applied from within the building rather than from the outside. This is made possible by installing a joint treatment on the panels in the factory and then finishing the joint after it has been placed.

Additional to the material considerations of the cladding panels is the need for a set of purpose-designed power tools to aid in the jointing process. Many of these comments also apply to the fixing process, in that parts can be fixed to the panels and to the structural members off-site, thereby speeding up the on-site processes. At the same time, the use of power tools to aid in the installation of fixings would mean a reduction in the time that it takes to place a panel and thus a reduction in the length of time that the robot is likely to be idle.

As for the 'mechanics' of fixing cladding, it is not unusual for operatives to carry panels up ladders and then to manhandle them into final position. Naturally, this involves working in uncomfortable positions which are also potentially dangerous. It is proposed to automate the placing and fixing of the cladding panels by means of the cladding fixing system, thereby removing the physical effort required by the human operator in this dangerous process and improving the efficiency of the external wall cladding process. The system consists of an external access work plaform, a cladding robot and a rough-terrain, fork-lift truck.

The cladding robot has been designed to work from a modified commercially-available mobile work platform as illustrated in Figures 1-6. Sets of cladding panels are delivered to the stock-cradle attached to one side of the work platform using a rough-terrain, fork-lift truck. The cladding robot is rail-mounted on top of the work platform to provide horizontal

Modular building systems

movement parallel to the cladding rails. The robot consists of a manipulator arm with two degrees of freedom; lateral movement is provided via a pair of lazy tongs and angled, swivel motion. Its end-effector has pneumatically-operated suction cups which are designed to hold a single panel. The manipulator arm would be used to place and hold the cladding panels in their final position while an operator fixes them. The stock-cradle in the work platform is designed to accommodate up to four cladding panels at one time. The panels may be up to 1.75 m x 2.00 m in size with a maximum weight of 100 kg.

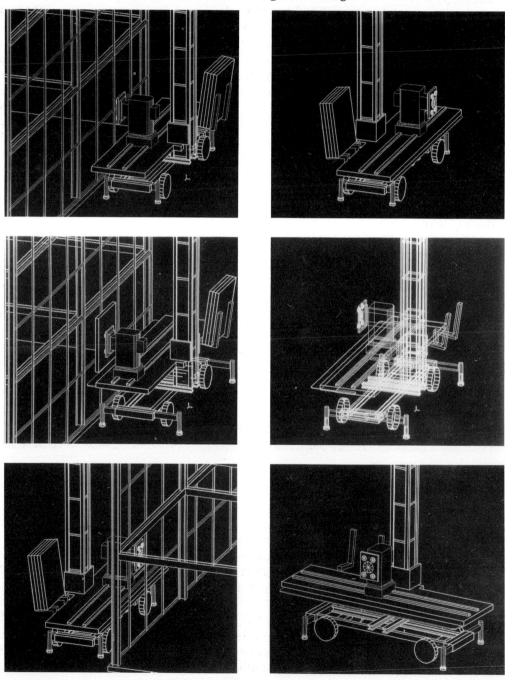

Once the work platform has been loaded it is then raised to the required level under operator control. The robot is now ready to commence its task, part of which can be fully automated to create a work cell. The operator would signal the robot to locate itself in front of the cladding panels stored in the stock-cradle. The manipulator arm of the robot would move forward until the end-effectors were in close contact with the cladding panel. Detection of contact would activate the suction cups to pick up the panels. The manipulator arm would move backwards and the angled, swivel motion would move the panel from its previously angled position to a vertical position facing the cladding rails. The lazy tongs of the manipulator arm would now push the panel towards its final position. The operator would then take over command using a pendant control which would command the robot to move the panel into its final position. The operator would be standing on a floor within the building in close proximity to the panel, but in safety. When the desired position of the panel has been reached, the robot would hold it in place while the operator secures its temporary fixing. The operator would signal the robot to release the sunction cups and the robot would then automatically proceed to pick and place the next panel. This would, of course, involve some motion of the robot along the cladding rails. Such an operation would be performed under automatic preprogrammed control. Cladding fixing would continue along the length of the platform which itself can either be raised to the next level or moved along the building to its next reference position.

CONCLUSIONS

By redefining certain basic methods of construction it is possible to develop a family of robotic tools which are matched to those methods. In proposing robotic tools, it is important to identify the essential differences between conventional, robotic manipulators and construction plant and equipment and then to marry together the most desirable features of each. By such means it is possible to create a 'factory in microcosm' or work cell in which simple robots, under a combination of operator and programmed control, can speed up, simplify and make safe otherwise laborious and dangerous tasks. The robotic tools described in this paper are simple, purpose-built devices that employ existing technology. It is the intention to develop the cladding robot in more detail using simulation techniques in order to achieve an optimal configuration prior to detailed design.

ACKNOWLEDGEMENT

The authors gratefully acknowledge the financial support of the Science and Engineering Research Council for the project 'Opportunities for Robotics in the Construction of Buildings for Industry' (GR/E57788).

REFERENCES

1. ATKIN B L, ATKINSON P, BRIDGEWATER C E & IBANEZ-GUZMAN J. (1988) A new direction in automating construction, Proc. 6th Intl. Symp. on Robotics and Automation in Construction, Construction Industry Institute, San Francisco, pp119-126.
2. ATKIN B L, ATKINSON P, BRIDGEWATER C E & IBANEZ-GUZMAN J. (1989) Parts-set and criteria for modular building systems utilising robotic tools, Occasional Paper 21, Department of Construction Management, University of Reading.
3. ATKIN B L, BRIDGEWATER C E & IBANEZ-GUZMAN J. (1989) Investigation of construction robotics research in the United States, A Report to the SERC, Department of Construction Management, University of Reading, September.
4. ANDEEN G B. (1988) Design differences between robotic manipulators and construction equipment, Proc. 6th Intl. Symp. on Robotics and Automation in Construction, op.cit., pp253-259.

Modular building systems

5. SKIBNIEWSKI M J. (1988) Robotics in civil engineering, Van Nostrand Reinhold.
6. HASEGAWA Y, MATSUSHITA Y, NAKAMURA T, MIYAZIMA T, TERAO M, NISHIMURA M, MIZOUE Y, and TAIRA J. (1988) Development of a new steel erection system - Wascor Research Project (Part 3), Proc. 6th Intl. Symp. on Robotics and Automation in Construction, op.cit., pp515-522.
7. BRIDGEWATER C E, IBANEZ-GUZMAN J, ATKIN B L & ATKINSON P. (1990) Cladding components in the parts-set , Occasional Paper 23, Department of Construction Management, University of Reading.

TRANSPORTING AND PLACING OF CONCRETE

by J R Illingworth, BSc(Eng), FCIOB

INTRODUCTION

Concrete production, reduced to its simplest terms, involves three discrete phases: mix, transport and place within the forms. Great attention is paid to the design of the mix to be used to achieve the specified strength, while the requirements for placing are also spelled out in some detail in concrete specifications. Surprisingly, little or no mention is made of haulage and distribution (transport). Such matters seem to be regarded as the contractor's concern and nothing to do with with the structural engineer. In fact, if these items are improperly considered, unnecessary additions to the client's final cost will usually arise. Concrete quality may also suffer and the contractor's image with the client become a poor one.

Every contract, or part of a contract, will have a unique set of conditions requiring a solution and it follows that site management and those who carry out the original planning at the tender stage, need to be well versed in the equipment available and in the comparative cost of each type in relation to situations for which each is best employed.

INFLUENCE OF THE SITE

Before examining the choice of plant for the transporting and placing of concrete, it has to be remembered that all sites are unique and will have characteristics which affect plant choice. It is prudent, therefore, to be fully acquainted with such factors before examining the possible plant choices open.

Character of the site

What effect does the site and its geography have on choice?

(a) What are the boundary conditions? Will there by any objection to cranes swinging over adjacent property?

(b) Do adjacent buildings or trees limit the choice of plant?

(c) Will noise create problems with adjoining owners?

(d) Does the contour of the land, and access on it, restrict the choice of certain types of plant?

(e) Are ground conditions such that temporary roads or other forms of access will be essential, particularly in the winter?

One or more of the above features may have a significant impact on the final cost of plant being used. For example, the site conditions may not allow the most efficient method to be used.

Effect of the permanent works

The choice and siting of plant is often affected, and sometimes dictated, by what has to be built. For example, it may be necessary to leave out, temporarily, parts of the permanent

Transporting and placing of concrete

structure so that the most efficient handling equipment can be used for the majority of the work. Alternatively, the selection of plant may be dictated by the relation of a building to the site and its boundaries, with a resultant limitation of choice.

Whatever the choice, it must be remembered that, while plant may be easy to establish on site before construction begins, removal may become very difficult once the permanent structure is in place. It follows that the siting of whatever plant has been chosen must relate to the ability to dismantle rather than to erect.

Once all the above matters have been adequately examined, consideration of the possible methods of handling the concrete can begin.

AVAILABLE METHODS

Those involved in concrete construction are fortunate in having a wide range of equipment available from which to make a choice. On the other hand, it can equally well be argued that such a wide range makes choice that much more difficult.

To simplify the problem it is useful to categorise plant into one of three groupings:

(a) linear;

(b) two-dimensional;

(c) three-dimensional.

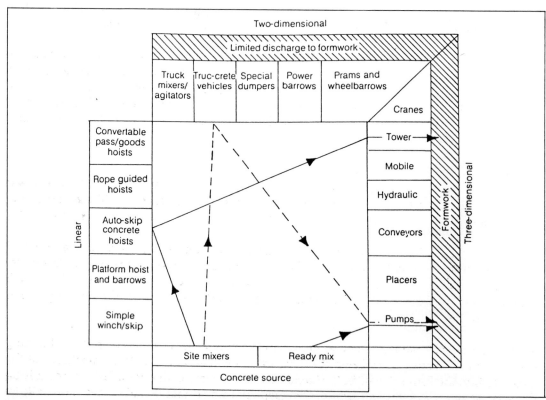

Figure 1 Concrete plant. Relationship between categories.

Figure 1 illustrates not only the divisions into the three categories, but the sub-grouping that also arises. In addition, it must be recognised that some plant has a general purpose role, within which it can handle concrete, while other items are specifically designed to handle concrete.

Further study of Figure 1 indicates that linear methods cannot generally distribute concrete directly into the forms; two-dimensional methods have only a limited capability in this direction and only the three-dimensional plant has a direct discharge capability, limited only by the geometry of the equipment in question.

Notwithstanding the above comments, it will most often be the case that more than one item of transporting plant will be needed to provide the most efficient chain of events - a point that will become clear at a later stage.

With the Figure 1 in mind, the various dimensional categories can be examined in more detail.

(a) Linear methods

Linear methods always operate in the vertical plane and take the form of hoists, ranging from a simple rope and pulley to highly sophisticated models capable of operating to great heights. In general, they have to be associated with other transporting methods to get concrete into the forms.

Hoists are relatively cheap to hire and operate, though it must be remembered that the requirements of the Construction (Lifting Operations) Regulations[1] will add costs to their use, over and above the normal erection and dismantling and service costs.

In the handling of concrete, the real value of a hoist is to provide a cheap and/or fast means of elevating concrete to the working level from whence it can be transported to its final resting place by a tower crane, barrows or whatever provides the best cost solution. This approach, in the case of the tower crane in particular, allows the major item of plant to carry out an additional workload, while the slow hoisting operation is carried out by a much cheaper item of plant.

(b) Two-dimensional methods

Two-dimensional methods, in relation to concrete handling, relate to all types of wheeled vehicles. While classed as two-dimensional, they can, of course, travel up hill and down dale. What is being indicated here is the ability to travel along a plane, even if it is a twisted one.

Wheeled vehicles play a major part in the transport of concrete. Today, the ready mixed concrete vehicle is virtually the universal method of providing concrete to the building site. As such, it provides the high capacity end of concrete vehicles - usually 6m³. Unless it is economic to provide a major batching plant on site - and there is room - the small site mixer is a thing of the past.

With ready mixed concrete, there are two cost elements - the cost of materials and mixing - and the cost of haulage (transportation) to a given location on site. If the discharge of the concrete into the forms is within the capability of the vehicle (into foundations, moulds on the ground or in-situ ground beams, for example) no further ongoing transport will be needed.

Transporting and placing of concrete

The efficient use of ready mixed concrete demands a number of management actions:

(a) ordering in good time - asking for a 50m³ over the phone at the last minute is a recipe for trouble;

(b) ordering in full loads. Part loads attract a premium charge;

(c) giving delivery rate required to the supplier and start and stop times;

(d) giving clear instructions as to location and date required;

(e) specifying the type of concrete in the clearest terms.

In addition, site management must remember that good access is needed for heavy vehicles and that the trucks need to be washed down before leaving the site. While the vehicle may carry its own water for this purpose, a suitable washdown area needs to be provided by the contractor.

Truc-crete vehicles

Often known as beetle trucks because of their shape, the Truc-crete vehicle provides the middle range of wheeled concrete transport (Figure 2). Its capacity is usually in the 2-3m³ range. The transporter body is designed to fit any suitable chassis and is designed to have good flow characteristics when discharging any type of concrete mix. Discharge is on the same principle as that for tipper lorries; hydraulically operated re-mix blades are located across the mouth of the discharge.

Figure 2 Truc-crete vehicle. (Benford Ltd).

Vehicles of this type are most economic on sites with sufficiently large amounts of concrete to justify central batching plants of 20m³ per hour output or more and where haulage distances to the point of placing are not too great.

They are capable of feeding a concrete pump, high speed concrete hoists or concrete conveyors. In other words, any onward distribution system that can cope with the wheeled transport's delivery rate. As with ready mixed concrete trucks, direct discharge is possible to the forms only in ground level, or nearly so, situations (Figure 2).

Dumpers

Dumpers are truly the maids of all work on the building site. As such, they are frequently called upon to handle concrete. The natural result is that special dumper systems have emerged specifically to handle concrete. They incorporate hydraulic tipping; hoppers to provide proper flow characteristics for concrete, even if stiff or very lean mix, and the facility of the hopper to be swivelled to provide front or side discharge (Figure 3).

Figure 3 Rotaplacer dumper. (Johnson Machinery Ltd).

One disadvantage is that, having such a short wheelbase, dumpers require stiffer mixes than usual to stop the concrete spilling over while in transit - unless the access is suitable.

Transporting and placing of concrete

Wheelbarrows and 'prams'

In spite of all the advances in the mechanisation of concrete handling plant, the humble wheelbarrow still has a role to play. Its association with a simple hoist may well still be the cheapest solution for concreting the upper floors of a small structure.

In the United States, the use of motorised prams, or power buggies, is a common sight. Yet in the UK, attempts to introduce them have been a failure. The reason, almost certainly, is that ready mixed concrete was used much earlier in the US and rapid discharge was needed. For tall structures, this led to high speed hoists being used to elevate the concrete to the operational floor. Once there, the conventional barrow was too slow to keep pace. This fact, combined with the high cost of labour, led to the development of the mechanised wheelbarrow. In the UK, by contrast, construction was wedded to the tower crane and slower rates of pour from site mixers. Even today, high speed hoists feeding tower cranes at the working level, or the use of concrete pumps, are seen as more cost effective.

(c) Three-dimensional methods

Three-dimensional methods for handling concrete cover the whole range of crane types, conveyors and the movement of concrete by pipeline (either by placers (pneumatic) or hydraulically operated pumps). Of these methods, it should be noted that cranes are general purpose plant, while conveyors, placers and pumps are specifically designed for handling concrete. All are capable of direct discharge into the forms, within the limits of the geometry of their design.

Cranes

Cranes cover a wide variety of design and three dimensional capability. As such they deserve individual consideration by type.

Tower cranes

Of all crane types, the tower crane possesses the greatest three-dimensional capability, both in terms of height and reach. Operationally, the hoisting speed is slow compared with slewing. Its use in handling concrete is best suited to limited raising and lowering, combined with the horizontal distribution element only. This is readily achieved by feeding the crane by a high speed hoist, via a floor hopper, at the operating level. Hoists, by comparison, are cheap and the arrangement frees the crane for other work related to the concrete operation, viz handling formwork and falsework, together with the reinforcement (Figure 4).

In economic terms, the characteristics of the tower crane are that:

(a) erection and dismantling costs are high;

(b) weekly operating costs are low in relation to what it will do.

Therefore, it follows that the tower crane, in its most economic role, is in a long term situation. The long term is necessary to amortise the high fixed charges involved and to reduce the weekly cost element of such charges.

Lorry mounted strut jib cranes

The lorry mounted strut jib types are slowly dying away in the face of the continuing dominance of the telescopic hydraulic models. While still seen in relation to the handling of heavy pre-cast elements, even in this field, the hydraulic crane is becoming available in greater and greater capacities.

Figure 4 35 storey building with tower cranes and associated twin high speed concrete hoists in the foreground.
(George Wimpey)

In cost terms, the disadvantage is the requirement to erect and reeve the jib after the base machine arrives on site. The reverse applies on completion of lifting; all of which costs money and time.

Hydraulic, telescopic cranes

This crane type has been probably the most significant crane developed since that of the tower crane. It is immediately available for use on arrival on site and because of its characteristics it can operate in very restricted locations, in or out of structures, where other types are unable to go. With high mobility and ready availability on hire, it must be regarded as a short term handling option. In other words, while expensive to hire, it can, with proper planning, be brought on to site, used fully for a given time and then removed.

Excavator conversions

Cranes converted from an excavator have a valuable function for the contractor. Being tracked they have a rough terrain facility, are very stable in such conditions and cheap to operate. As such they have a cost effective multi-purpose role on site. This is particularly true where ground conditions are difficult - for example, pouring concrete to oversites on housing estates where ground conditions and access are difficult.

Conveyors

Conveyors possess unique characteristics when used for the transport of concrete. A narrow belt runs continuously with a relatively small quantity of concrete per unit length. The load to be carried by any section is, therefore, small and the corresponding power requirements

Transporting and placing of concrete

are also small. By running the belt at a fast speed, high outputs can be achieved at the discharge end. Where concrete is the material involved, the conveyor system must be specifically designed. Concrete is abrasive and wet and all parts of the equipment need to be designed to resist the penetration of water and grout, while all moving parts have to be grit proof and 'sealed for life' lubricated.

The leading maker of concrete conveyors in the world makes a big feature of the fact that conveyors can handle any type of concrete that can be discharged from a ready mixed concrete truck - something that a concrete pump cannot do, and with which it competes.

In view of this situation, it is surprising to many people that conveyors only find limited use in the concrete handling role in this country. By contrast, they are still used extensively in the US. The reasons for the difference of attitudes lies in the historical development of the transport of concrete in the US and Europe as a whole. Ready mixed concrete was being widely used in the US while Europe was still wedded to site mixers. To empty truck mixers rapidly, a fast means of transport was needed and, at that time, the conveyor was the best means available. In Europe, concrete pumps were developing and fitted in well with the site plant for mixing concept. As pumps continued to develop and increase their capacity, the use of ready mixed concrete was, at the same time, taking over from the site mixer.

In the UK today, the use of conveyors is limited to short length portable units, rather than chains of linked conveyors, which are frequently seen in the US. More recently, fold away lengths of conveyor have become available, mounted on ready mixed concrete trucks to provide an enhanced discharge reach. Companies providing this service charge a premium on the basic concrete price. This covers the saving of other equipment being provided by the contractor and the necessary reduction in truck load capacity to offset the conveyor weight and maintain the legal axle load limit. A full comparative cost analysis is required to assess the value of this service (Figure 5).

Figure 5 Conveyor unit mounted on ready mixed concrete truck. (Leibherr Building Machinery Ltd)

Concrete pumps

The modern concrete pump, twin cylinder and hydraulically operated, and capable of outputs up to $110m^3$ per hour in the larger varieties, first began to make an impact in the early sixties. Beginning with trailer mounted models, they rapidly progressed to lorry mounted varieties, followed by adding folding hydraulically operated booms. Such pumps are now universally fed by ready mixed concrete, except on major civil engineering work, with high quantities of concrete needed to be placed at a fast rate, where a site batching plant of appropriate size can perform the same role cheaper.

A more recent pump development, designed to further enhance use potential, is the mounting of a hydraulic boom pump on a ready mixed concrete truck. While truck concrete capacity is reduced for the reasons given above, under Conveyors, much smaller, but awkward to reach pours, can be pumped with one piece of equipment hired to order.

The concrete pump, as a three-dimensional piece of equipment, has a very wide range of operation. Its range, in broad terms, is a function of the power available at the pump. It will, however, also be affected by the type of aggregate in use. Rounded material will pump further than angular or crushed stone. It is a highly sophisticated piece of plant, yet it is the only item of all plant that can handle concrete that is limited by the concrete mix. To pump, the mix has to suit the pumping operation.

In spite of the apparent disadvantages of the concrete pump, it has the capability to transform the concept of planning concrete handling on site. Given that the specified mix can be made pumpable — and very few structural concrete mixes cannot — the pump needs to operate at a high output level to be profitable. For utmost efficiency, therefore, concrete placing needs to be delayed until the amount of concrete available is adequate to achieve cost effectiveness of the pump.

Formwork and reinforcement are labour intensive activities, whereas the placing of concrete can be highly mechanised. With pumping, the intermittent pour principle needs to be used, ie, formwork and steel fixing are programmed as efficiently as possible and run continuously. With the high outputs needed for pump efficiency, the concreting time will be much less than the corresponding time for the other two items. With the availability of ready mixed concrete and a ready capability to hire-in pumps, they can be brought on to site for the limited time needed.

Planning in this way has further advantages. Site management is only too well aware that if the formwork and steelfixing are not ready at the appropriate time, the pump may have to be paid for for a much longer period than is necessary, while standing time may arise for ready mixed concrete vehicles. Site managers are much more likely to be ready with hired-in items than if they are using their own plant; a useful incentive to be efficient.

A more detailed examination of this philosophy may be found in Site handling equipment[2].

The concrete pump can also be used for low output but will be more expensive. If concrete is needed in very restricted locations, a concrete pump may well be the only option open to the planner, as the pipeline takes up little room. It may also save much money on access where a site is difficult and, as a result, become cost effective for low outputs (Figure 6).

Figure 6 Value of concrete pump on difficult site. (Burlington Engineers Ltd agents for Fdk W Schwing Gmbh)

Concrete placers

The concept of transporting concrete by compressed air along a pipeline enjoyed a vogue in the sixties and seventies, but has now virtually disappeared. For details of this technique reference may be made to *Movement and distribution of concrete*[3].

Elephant trunking and plastic tubing

The use of interlinked tubes, to provide a means of moving concrete by gravity, can perform a useful role in appropriate circumstances. Such a method is to be preferred to the use of chutes. Various types of plastic tubing can also be adapted to provide economical gravity distribution systems for concrete. Situations suitable for their use are fully described elsewhere[3].

METHOD SELECTION

With all the methods available to handle concrete, one may be forgiven for wondering how to make the right choice. Indeed, selection needs to be a properly disciplined analysis. It should proceed on the following lines:

(a) examine all the extraneous factors that may impact on plant choice — character and geography of the site. Estimate influence of what has to be built; boundary conditions and permanent obstructions etc;

(b) examine problems in relation to access for transporting plant. Study how they may limit choice;

(c) list all apparent available alternative methods — once listed, reduce to minimum by rejecting those that, by inspection, will not be suitable after more adequate analysis;

(d) carry out a full cost comparison of the remaining alternatives. This should involve:
    i) comparison of alternative methods;
    ii) evaluation of the influence of choice, on a cost basis, on surrounding activities;
    iii) effect of intangibles, ie, what will be the cost consequences if unforseen delays arise when a particular method is used? Are they more of a risk with one method more than another?

Ideally, these tasks should have been carried out at the tender stage to enable accurate pricing. Site managers should acquaint themselves with the situation at the first opportunity. If the estimator has not worked in association with the head office planners, the site manager may need to carry out the analysis and cost comparisons himself. Only then will he be sure that the concrete operation is being run in the most cost effective way, or that adequate money was allowed in the tender.

Direct comparison of alternatives

When making decisions on handling plant and evaluating cost effectiveness, two scenarios are likely:
- where handling equipment is required to perform a mixture of roles, eg, lifting and placing a variety of constructional materials in 'unloading to store' or to the required final location;
- where only one material is handled on a productive output basis. Obvious examples are in-situ concrete and precast concrete; either structural or as cladding.

In the first case, direct comparative costing, for the sake of simplicity, is best done in cost per week terms. The handling plant in this case is regarded as a service machine or machines to all trades and operations involved.

In the second case, more detail is possible. The plant operating cost can be linked to output and graphs of unit cost related to output established. Such graphs can be of great advantage in the early planning stages, where output adjustments may have to be made several times in order to balance with work loads elsewhere. They allow several alternative solutions to be plotted on the same graph so that the cheapest solution can be read off at will for any particular output required.

Comparative costing graphs and their preparation

Before any cost comparison is attempted, it is crucial that competing methods are compared on precisely equal terms. For example, if concrete handling methods are to be considered, the comparison should cover the entire mix, haul and place sequence. If ready mixed concrete is involved in one method and site mixed concrete in another, the cost of materials should also be included.

Transporting and placing of concrete

Cost elements

The cost of all handling plant is compounded from two different elements — weekly running costs and fixed charges.

The first relates directly to output per week, while the second, in its contribution to unit cost, depends on the total production of the item in question.

Weekly running costs

Components are:
(a) weekly hire rate for all plant involved;
(b) maintenance allowances;
(c) fuel and lubricating costs (including electricity in appropriate cases);
(d) true cost (to contractor) of all labour involved.

With contractor owned plant, these figures will be put together separately. With hired-in plant an all-in figure will be given in some cases, while in others, the driver may be a separate item.

Fixed charges

Components are:
(a) haulage charges for all plant involved, to and from site;
(b) erect/dismantle costs — some plant may involve very high erect/dismantle charges;
(c) cost of all temporary works associated with the plant in question — eg, access roads, foundations, drainage etc;
(d) cost of installing the necessary services — power, water, etc;
(e) all labour charges associated with the above at a true cost to the contractor.

Weekly running costs (a), fixed charges (b) and the total output required (c) can all be considered as constants. While this statement is not strictly true, it gives an acceptable degree of accuracy for the cost/output graph. Variable factors in this situation will be output/week and unit cost. Let these variables be x and y. These variables and constants have a linking relationship, which can be expressed by

$$y = \frac{a}{x} + \frac{b}{c}$$

This equation is a particular form of hyperbola and all graphs plotted will have this basic form.

By varying x over the probable output range required during the contract, a series of values for y will be obtained from which the cost/output graph can be established. If alternative approaches are plotted on the same sheet, the output parameters where each method will be cheapest can be read off. Figure 7 illustrates the situation in practice — where an examination was carried out of five alternative mix, haul and place methods.

This method of comparative costing is particularly valuable at the planning stage as graphs can be produced in advance of detailed planning. Once the output required is established, a direct read off from the graphs will show which handling approach is the cheapest.

Such graphs can also be used as a management tool during the operation duration. If output per week is less than planned an immediate idea of losses arising can be read off.

Having examined the direct comparison of alternative solutions, attention must now be paid to the second and third stages of the complete process as already put forward.

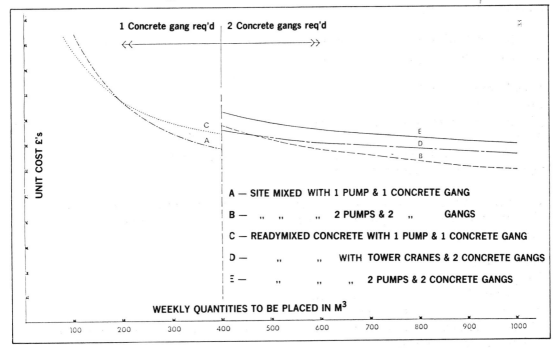

Figure 7 Cost comparison graphs for five alternative methods of mix, distribute and place concrete

Related items

Any method decision will usually have repercussions on related items. For example, a method of concrete supply may well affect the steel fixing and formwork. In other words, it is quite feasible that a method of mix, haul and place concrete may involve greater or less quantities of formwork than an alternative approach.

If the comparison is to be complete, the changes in cost of the formwork must also be taken into account and the increase/decrease set against savings/increases from the direct comparison. Examples from actual site experience indicate that there are occasions when the cheapest concrete method may not be the cheapest in overall terms as the increase in formwork cost more than offsets the apparent saving.

Method can also affect the contract as a whole. Preliminaries and overheads now represent a very significant part of the contract sum. Any method, even if not the cheapest, which effects a considerable saving in contract time can produce financial savings to the contract as a whole which may prove to be the deciding factor in method approach.

Intangible factors

Finally, attention must be paid to the intangible factors. These are those matters to which cost cannot readily be allocated, since they may or may not arise and at the time of the exercise this cannot be forecast.

Transporting and placing of concrete

Examples of intangible items include:

(a) vulnerability of method to delay by lack of labour availability;
(b) consequences resulting from inability to continue work in inclement weather;
(c) risks due to plant break down.

With such items, the assessment of risk and attendant cost will be to some extent a gamble or, looked at another way, an insurance premium. Decisions in this field have to be largely a matter of experience related to the probable circumstances relating to a particular contract. Nonetheless, intangible factors can be very real and if two alternatives show near enough the same cost in the direct comparison, decisions made may depend entirely on an adequate assessment of intangible factors.

PLACING CONCRETE

The final act in the concrete chain — placing the concrete in the forms — needs to take account of the method of distribution used. The first batch of concrete transported will inevitably adhere to skips, mixer interiors, conveyor belts and the inside of pump pipelines. It is good practice for the initial batch handled to be richer in cement and sand than the specified mix, so that the various elements in the handling process will be coated with cement paste, while the mix does not become leaner than specified. In the case of concrete pumps, it is essential that the pipeline is coated with cement grout before any concrete is introduced.

It follows that the same action needs taking in relation to the forms and the reinforcement. Initially, grout will coat these items causing some leanness to the mix unless the initial batch is deliberately made richer.

Vibration

The strength and durability of concrete is greatly affected by the density achieved when placing. For each 1% of air left in, the strength loss is approximately 5%. For this reason, concrete is mechanically vibrated to remove as much air as possible. The poker type immersion vibrator is the most favoured for such work, but external vibrators can be clamped to the formwork. If external methods are specified, siting and fixing calls for expert attention. It must also be remembered that for this method the formwork needs to be much more robust to counter the effects of the vibration on the form construction.

Poker vibrators can be petrol, air or electrically operated. External clamp-on vibrators are electrically motivated in most cases.

CONCLUSION

The transportation and placing of concrete is a key element in the total concrete process. While this paper deals primarily with the plant involved and its economic selection, the labour element must not be neglected. Knowledge of the appropriate gang sizes for particular methods is important. Those involved in the finishing and vibration of concrete require adequate training in the use of the equipment — vibrators, power floats and vacuum mats etc. The ultimate quality often will rest with the skill of the operative.

Finally, it should be clear that the selection of method and plant is not something that just happens. It is an important part of construction planning, itself, in turn, a key service to construction management.

REFERENCES
1. Construction (Lifting Operations) Regulations (1961).
2. ILLINGWORTH J R (1972) Movement and distribution of concrete. McGraw-Hill. pp239
3. ILLINGWORTH J R (1982) Site handling equipment. ICE Works construction guide. Thomas Telford. pp67

BIBLIOGRAPHY
1. ILLINGWORTH J R (1977) The influence of ready mixed concrete on site distribution methods. Proc. ERMCO 77, 5th International Congress, Stockholm
2. ANSON M, ASTON D E and COOK T H (1986) The pumping of concrete — a comparison between the UK and West Germany. Research Report. University of Lancaster and the Polytechnic of Wales. pp34
3. BRITISH CONCRETE PUMPING ASSOCIATION (1988) The manual and advisory safety code of practice for concrete pumping. pp103
4. DEWAR J D and ANDERSON R (1988) Manual of ready mixed concrete. Blackie.

GOOD PRACTICE IN LAYING FLOOR SCREEDS

by E G Mold CEng, MICE, FIHE revised by K F Endean BSc, MCIOB

DEFINITION

A floor screed is a layer of substantial thickness, usually of mortar made up of Ordinary Portland cement and clean sharp concreting sand, or occasionally of small aggregate concrete, laid and thoroughly compacted on to a prepared base slab and brought to a designated level to receive other flooring. The screed may be floated or trowelled smooth to provide a suitable surface for the specified flooring or other finish.

As traditionally laid a screed is not intended to act as a wearing surface.

Occasionally, with screeds over 75mm thickness, a concrete containing small aggregate (10mm) may be specified. If this is laid at a low moisture content, as for traditional screeds, the fine trowelled finish required for thin sheet and tile flooring may be difficult to obtain directly. However, the finish should be acceptable to receive thicker tiles or in situ flooring. Alternatively, the concrete screed may be laid as for normal concrete workability (eg, with 50mm slump) rather than for cement and sand mixes. There will be some delay in floating and trowelling operations allowing the water in the mix to evaporate between the several stages of finishes. Accurately set side forms will be required from which the concrete is levelled. These finishing operations are described in *Concrete ground floor construction for the man on site. Part 2. For the floorlayer*[1].

MATERIALS

The cement should be Ordinary Portland cement to British Standard BS 12. It should be stored under cover and kept dry. The sand (or fine aggregate) should be clean sharp concreting sand to British Standard BS 882.

The sand should be tested at intervals to make sure it remains consistently well-graded as described for the relevant grading in BS 882. Bricklaying sand, very find sand (eg, to BS 1199) or sea dredged sand which can be seen to contain a large quantity of flat shells should not be used. Fine sand can cause increased shrinkage and shells laying in the screed surface can cause finishing problems.

The ease of thorough mixing, compaction and fine finishing will be greatly affected by the sand's general particle size, ie, whether the sand is predominantly coarse or fine and also its evenness of grading. A well-graded sand within grade M of BS 882 is recommended where thin sheet and tile flooring is to be applied. Coarser sands, including those within grade C of BS 882 will be acceptable for thicker tile in situ flooring.

Screeds made with coarser sands may need additional attention in production, eg, control of water content and finishing to achieve thorough compaction and a surface to receive thin flooring (eg, up to 3mm thickness). A more open textured screed surface often obtained with coarser sands will be satisfactory for thicker tile or in situ flooring, providing that the screed surface is sufficiently flat and strong. The final choice of sand must depend on experience of and expertise in the use of local aggregates.

WATER CONTENT

The water content of a mix must be judged to suit the floorlayer's compaction and laying technique. Sufficient water must be used to enable the cement and sand to be mixed well and to allow the screed to be compacted thoroughly over its full depth.

All screeds must be thoroughly compacted. The use of a screed material which has been laid too dry, badly mixed and poorly compacted is at the root of most soundness problems. On balance, a mix slightly wetter than has been commonly used in the recent past is to be preferred.

The term 'semi-dry' screed is meaningless as it can cover a range of possible water contents, depending on individual interpretation. A screed which is too dry cannot be fully compacted by hand methods. If the moisture content of a screed is too high it will be difficult to control screed levels accurately. Skill and experience are necessary to judge water content correctly.

Judging water content

To judge water content a ball of the mixed screed material should be pressed in the hand. On opening the screed material should be moist enough to hold the ball together but not too dry to let it crumble apart.

On a large job, where the costs of testing can be justified, trials can be made before the contract starts to determine the optimum moisture content of the particular mix to be used. On site, a 'speedy' moisture-testing instrument can then be used to check the screed moisture content and to help maintain it at the determined optimum. This may vary between 7 to 9%, depending on the cement content and sand grading.

THE MIX

The mix for cement and sand screed will generally be specified, with typical proportions as shown in Table 1.

Table 1 Concrete mix proportions

Category of loading and use of flooring type	Mix by weight of screed in situ		Mix by volume (for some small building jobs)		
	cement:	dry sand	cement:	dry sand	or damp sand
Normal light to medium loading conditions, eg, in domestic buildings, offices, laboratories, schools etc. and for screeds under thick and rigid flooring (eg. concrete or quarry tiles)	1:4 $\pm$ 1		1 bag	0.13m^3 (130 litres)	or 0.16m^3 (160 litres)
Above normal loading conditions and where the risk of screed crushing failure would be unacceptable because of very difficult access after building takeover, eg. hospital operating theatre floors trafficked with loaded trolleys, sterile environment research laboratories etc.	1:3 $\pm$ 0		1 bag	0.1m^3 (100 litres)	or 0.12m^3 (120 litres)

Good practice in laying floor screeds

Note 1

The volumes shown in column 3 are based on the figure given for weight-batching, and have been calculated assuming sand with a bulk density of 1500 kg/m3; for damp sand, bulking of 20% has been assumed. For sands with different bulk density and bulking characteristics, volumes should be adjusted.

Note 2

The mix proportion of the screed as placed may be specified by weight with permitted tolerances as shown in column 2. The contractor should satisfy himself by preliminary and occasional tests in accordance with BS 4551, that the mix proportions specified are being achieved in the in situ screed. Where this level of precise control is not required, eg, in some small building jobs, the equivalent proportions by volume in column 3 may be satisfactory.

For jobs where more precise control of mix proportions is required and volume batching is used, the contractor should satisfy himself by preliminary and occasional tests in accordance with BS 4551, that the volume equivalents in column 3 of Table 1 will achieve the specified mix by weight in the in situ screed.

The contractor should remember that the accuracy of the screed mix in situ will be affected by the following:
(a) batching accuracy;
(b) efficiency of the mixer and mixing process;
(c) moisture content of sand varying 'bulking', (relevant where volume batching is used);
(d) bulk density of sand (where volume batching is used);
(e) quality of sand and its grading, influencing variability of mix.

For greatest accuracy sand should be batched by weight. However, if equipment for weigh-batching is not available on site, volume batching may be used, by taking whole bags (or multiples of whole bags) of cement, together with the corresponding measured volume of sand required as shown in Table 1. It should be noted that a full one-bag mix may exceed the recommended capacity of some smaller mixers available to builders. With such mixers it will be necessary to use a half-bag mix, ie, half of the cement from a bag mixed with half of the sand volume stated in column 3 of Table 1.

If volume batching is used, the sand can be measured in a suitable container having a known volume (eg, a box, bucket or wheelbarrow) to maintain a consistent and accurate mix.

The tolerance on mix proportions of screed material delivered to site ready-mixed can be obtained from the supplier.

MIXING THE SCREED

The sand, cement and water should be mixed throughly in a forced action mechanical mixer which is clean and well maintained. Sand should be added to the mixer first, then the cement, then sufficient water to give the required moisture content.

For a screed to gain full strength all the same particles must be fully coated with cement paste during the mixing process.

Screed mixing by hand is not recommended for commercial practice as it is very slow and may not give a consistent end result. Hand-mixed materials need to be mixed dry at least three times to a uniform colour otherwise mixing will not be adequate.

Forced-action mixers, eg, trough or pan type are the most efficient machines for mixing low moisture content screed material. They should not be used to mix more material in a batch than that recommended by the mixer manufacturer.

Tilting-drum mixers are not recommended for mixing traditional screed mixes because they do not distribute cement efficiently in low moisture content materials. 'Cement balling' and build-up of cement paste on the mixer drum can lead to severely reduced cement contents in parts of the in situ screed, which may then be too weak to withstand service loads.

If a proprietary screed pump is used, its forced action mixer should provide thorough mixing of the screed materials. The sand and cement should be mixed for at least three minutes before being pumped to the laying area. Some screed pumps are said to be incapable of pumping cement-rich screed (eg, richer than about 1:5 mix proportions). However, most problems of pumping screed material occur because of the use of poorly graded sands.

Ready-mixed screed
Ready-mixed screed material is available from several major suppliers. It is accurately batched and well mixed under factory controlled conditions, so is particularly useful on a confined site where there is little room for mixing. It can be supplied specially retarded for use over a 6 to 24 hour period.

Screed material should be used soon after mixing and certainly within one hour after adding water - sooner in dry weather - unless a chemical retarder admixture is allowed and used in the mix.

Ready-mixed screed material must be covered during transportation and on site with plastic sheeting to prevent water loss. Where possible a stockpile should be shaded from the sun.

During hot weather, the addition of an extra small amount of water to screed material which contains a retarder may be acceptable. Unused non-retarded material that has stiffened must never be re-mixed with additional water then used in a screed. This practice, called 're-tempering' could produce a weak screed.

MAIN WAYS OF USING SCREEDS
The method of using the screeed will be specified but the main ones are as follows:

Type 1 - Bonded screed
A bonded screed should preferably have a thickness in the range of 25 to 40mm. It is laid on a previously hardened concrete base slab which has been mechanically roughened (eg, scabbled - or shot blasted) and thoroughly cleaned of all debris and dust; the slab is water soaked to reduce suction, all excess water removed and a cement grout brushed well into the surface. Immediately after grouting, the screed is thoroughly compacted on to the roughened, water-soaked and grouted slab.

Bonded screeds should not be thicker than about 40mm as this increases the risk of loss of adhesion. The thicker the bonded screed, the more likely it is to lose adhesion and become

Good practice in laying floor screeds

hollow. If thick screeds must be used, the risk of curling can be reduced by making them much thicker, eg, 100mm or greater, to hold themselves down by their own weight. At this thickness, a screed or overslab will have to be constructed in concrete.

Bonding is the most satisfactory way to lay a relatively thin screed and reduce the risk of curling, hollowness and cracking. But even with good workmanship some risk of slight hollowness cannot be ruled out. A proprietory bonding agent may be used instead of cement grout but this does not replace any of the other requirements for proper preparation of the base slab. The instructions of the manufacturer of the bonding agent must be followed.

Where a screed must be bonded to a base of rigid precast concrete units, mechanical scabbing may not be allowed as it might crack some units with thin walls. However, if additional cleaning and roughening is required, shot blasting machines can be used to provide an effective bonding surface. Alternatively, units may be given a sufficiently roughened surface during production; but this surface may need cleaning (eg, by water pressure jet and vigorous brushing) just before grouting and screed laying. If water is used in large quantities for cleaning, it may penetrate to floors below; building work on lower floors must, therefore, be programmed so that there is no risk of damage if water penetration occurs.

The practice of laying a screed 'partially bonded' (eg, a 40mm thick screed laid on a concrete slab with only a tamped texture surface), is not recommended and should be avoided for large areas. This practice leads to a high risk of hollowness and subsequent disputes over acceptability of screed. 'Partial bonding' may be acceptable in lightly loaded buildings with small rooms, eg, housing.

Type 2 - Unbonded screed
Unbonded screed should have a minimum thickness of 50mm (preferably greater) if it is laid on a damp-proof membrane or on concrete which will not provide a good bond, eg, an old oil -contaminated concrete, or a smooth high-strength concrete which cannot for some reason be roughened, or a base concrete containing some waterproofing admixtures.

Laying a screed unbonded will create the serious risk of curling and hollowness, as the damp-proof membrane or poorly prepared base prevents holding down by bonding. However, the risk of curling of an unbonded screed can be reduced by increasing its thickness. By making an unbonded screed or overslab 100mm thick or greater, and of concrete, the risk of curling is reduced to a minimum.

Type 3 - Floating screed
A screed should have a minimum thickness of 65mm if it is laid on a resilient layer of insulating material. Floating screeds in more heavily loaded buildings, eg, parts of hospitals and offices, may be specified as 75mm minimum thickness.

The greatest care must be taken in constructing a floating screed, as the weak support of the resilient insulating materials will enhance any weakness in the screed. It is also difficult to compact a screed on a resilient material so extra care is needed. Compaction on an insulation quilt is even more difficult than on rigid board insulation.

It is almost certain that some differential curling will occur at joints in a floating screed. The use of a concrete overslab of at least 100mm thickness will minimise this risk.

Light structural mesh reinforcement, eg, D49 to BS 4483, may be used to control any shrinkage cracking which may occur but it will not prevent the curling of screeds. If the reinforcement is continuous through joints, this will help to avoid steps caused by differential curling. Any reinforcement should be placed approximately central in the screed depth.

A separating layer of building paper, laid above the insulation, will prevent screed materials penetrating joints between insulation boards.

BAY SIZES AND CONSTRUCTION AREAS

Screeds should be laid in areas as large as possible, consistent with obtaining good levels and to minimise the number of formed butt joints. If screeds contain electrical heating cables, strip construction may be used with the strips divided into bays with plain butt joints to suit the heating mat layout.

Joints

Vertical butt joints are used at daywork joints in large screed areas and between bays containing electrical heating elements.

Whatever the total area of screed to be laid in one day, separate strips 3 to 4m wide should be constructed to obtain good levels.

Long strips or large areas of screed may unavoidably crack at intervals as they dry and shrink. Small bay construction prevents ramdom cracking, but bay edges are likely to curl. When thin flooring is to be applied, steps at formed joints are likely to be more objectional than random cracks which can be repaired.

It will be preferable, where practical by reasons of size, that screeds in rooms should be completed in one operation, so avoiding formed butt joints.

SCREED LEVELS

For levelling purposes, rectangular timber screed battens, steel angles or levelling strips of screed (in the past called 'wet screeds') are used with tops set accurately to the specified level.

A strip of screed is often more convenient to use to set levels between screed bays which are to be joined together to form one room area.

The setting of levelling strips of screed requires considerable skill if final screed levels are to be accurate.

'Dots' or strips of hardened screed (small sloping mounds or strips of screed sometimes used to support level battens or used as level references) must not be allowed to remain in the final screed. Feather-edged screed laid on the sloping sides of hardened screed 'dots' or strips fail in service.

Levelling strips of screed must be well compacted and laid only a short time before adjacent screed bays so that the strips and bay are properly joined together. A daywork joint must be left vertical with a thoroughly compacted screed edge.

Good practice in laying floor screeds

LAYING THE SCREED

Type 1 - Bonded screed

(a) The base concrete for bonding is prepared by mechanical scabbling or shot blasting to expose the coarse aggregate cleanly.

(b) Screed battens are fixed firmly to good level.

(c) The roughened concrete is cleaned thoroughly, preferably with an industrial vacuum cleaner to remove all debris and dust. Shot blasting machines have their own dust collection unit. The practice of using a compressed-air line to blow dust from a floor area should be viewed with caution. It can be bad practice as it may result in dust settling on previously cleaned areas, may cause a health hazard in confined spaces and the compressed air may contaminate the concrete with oil droplets.

(d) The base concrete surface is soaked with water for several hours, preferably overnight. When laying a new screed area, it is good practice to soak the edges of the adjacent existing screed to prevent suction of water from the newly laid material.

(e) Any excess water is removed just before starting to lay the screed.

(f) A cement grout, the consistency of cream, is brushed into the roughened base concrete surface. A grouted area must not be left uncovered for more than about 20 minutes (10 minutes or less in hot weather or in heated buildings) or the grout will set and lose its bonding power.

Grouting should not be carried out too far ahead of screed laying.

If proprietary bonding agents, such as SRB (styrene-butadiene rubber) or acrylic compounds, are specified instead of plain cement grout, study the manufacturer's instructions *before* use. The base slab must be cleaned and in some cases soaked with water before using any bonding agent.

(g) Levelling strips of screed are laid if they are to be used for levels as an alternative to level battens.

(h) The mixed material is spread over the grouted area to a level about 10mm above the batten or levelling strip of screed top, ie, giving a surcharge - the drier the material, the greater will be the surcharge required. Then the screed is tamped down heavily to the level to give full compaction. This part of the laying is very important if the screed is to be strong and bonded to the grouted base. It may be difficult to compact screeds fully by hand without quite hard work. Compaction of drier mixes can be made easier by using a handrammer or a roller or a plate vibrator. Before adding more screed material on to a heavily compacted surface, the surface should be lightly raked to avoid risk of later delamination at this level.

(i) Any excess compacted screed material still above batten or levelling strips of screed is struck off with the rule or straightedge. Any slightly low areas are filled, re-compacted and ruled out.

(j) The section of the screed just laid is finished by wood float or by wood float and steel trowel, depending on the specification. Work is then continued on down the strip, first grouting, then spreading, compacting, levelling and finishing the screed. Adjacent strips which are cast at one time are completed and joined together and levelled from screed level strips.

(k) The screed is finally cured and protected.

Types 2 and 3 - Unbonded and floating screed

The construction and finishing of screed of these types follow basically the steps (b) or (g) and (h) to (i) outlined above for bonded screed (there is no requirement for roughening and grouting, or for applying bonding agent to the base slab). Care must be taken not to damage damp-proof membranes or insulation layers, particularly when using temporary level rails or screeding battens. The use of levelling strips of screed avoids the problem of pinning down battens.

CURING AND PROTECTION

Screeds are cured by keeping the mixing water in the screed. This can be done easily by covering with plastics or similar sheeting in close contact with the screed as soon as possible after laying. The sheets are kept well lapped and held down in place for at least 7 days. The sheets will also help to give some protection from foot traffic during the early life of a screed.

It is normally not advisable to use spray-on curing membranes, as these may interfere with the adhesion of floor finishes.

Curing is very important and should never be dispensed with or curtailed.

Wind must not be allowed to blow under the sheets used for curing. If the screed is laid early in a contract it should be protected on heavily trafficked routes with hardboard or similar covering. Proprietary self-smoothing compounds applied to a screed surface as an under-layment will also require protection from extended trafficking.

REMINDER: Neither screeds as traditionally laid nor with added smoothing compounds are intended to provide wearing surfaces.

Screeds should be allowed to dry out after curing as naturally and as slowly as possible. Rapid artificial drying-out will increase the risk of screeds cracking and curling.

Screeds should not be laid when frost is expected overnight, unless the area of building in which they are laid can be kept above freezing point until the screed has gained sufficient strength to avoid damage.

PERIMETER SKIRTING

The screed level at the perimeter of a room must be especially accurate to allow easy fixing of timber skirting. Where skirting heating or trunking is to be used, back plates, or temporary battens are fixed to give the level of skirting on perimeter walls before laying the screed.

TESTING THE SCREED IN SITU

It is possible to assess the quality of a screed laid on a concrete base in terms of its ability to withstand the likely imposed loads and traffic in service by using the BRE screed tester.

Good practice in laying floor screeds

Tests can be made with this device from 14 days after laying the screed, and provided that the test limits are not exceeded, it should ensure that the screed gives trouble-free service. The BRE screed tester only confirms a screed's soundness, ie, its quality controlled by those factors which affect the strength and integrity of the screed, such as mix proportions, water content, sand grading, mixing efficiency, compaction, curing etc. The test does not confirm any other properties, eg, the adequacy of bond or screed surface finish which must be assessed separately.

The screed test and testing device are included in BS 8204. The testing device may be obtained from Wexham Developments, Wexham Springs, Slough.

COMMON PROBLEMS IN CEMENT/SAND SCREEDS
Unsatisfactory preparation of base concrete
Screeds laid 25 to 40mm in thickness will curl due to thermal and moisture shrinkage unless they are properly bonded to the base concrete. Proprietary bonding aids are often used instead of satisfactory preparation but these aids are best used to assist bonding on to prepared base concrete.

Screeds laid 50mm thick with partial bond are a high risk because they are thick enough to suffer high moisture shrinkage stresses but too thin to be able to withstand them. With partial bond the shrinkage stress will often overcome the bond stress and curling may become excessive.

Poor compaction
Screeds have to be laid at a satisfactory moisture content to suit the screed layer and methods of laying. Lower moisture content with corresponding low workability coupled with low compactive effort will often cause failure.

Selection of sand for screeds
Fine sand or plastering sand is sometimes used in screeds by layers who consider that by using them they can more easily achieve a closed fine surface. The water demand for workability and hence compaction of these sands is high and, therefore, moisture shrinkage stresses are greatly increased. A grade M or C concreting sand to BS 882 will have sufficient fines in the grading to produce a closed surface with a much lower water demand. Fine sand is a common cause of failure.

Use of free-fall tilting drum mixers on site
Even when the materials are accurately batched the use of free-fall tilting drum mixers cause wide discrepancies in the cement content in portions of the same mix. In a 1:3 mix, variation can be as much as from 1:1 to 1:8. This is due mainly to cement balling at the lower moisture contents required in screed mixes. If the batching accuracy is poor then variations can only increase. Areas laid with low cement content may fail under impact loads and are especially vulnerable if covered with a thin flooring material with low load distribution capacity.

Cover to small pipes or cables
It is better if at the planning stage all pipes and cables can be routed into properly constructed ducts but if a pipe is laid in the thickness of the screed it must be properly bedded and the cover over the pipe should be 25mm minimum. Cracks will often appear in the screed over the top of the pipes and these can be controlled by covering the pipes with a close mesh

reinforcement. Care has to be taken to ensure that the screed material is properly compacted around the pipe and the reinforcement. The danger is that the crack will ultimately reflect up through the flooring and that the edges of the crack will then crumble and fail.

Curling

When a screed is allowed to dry out, it will tend to shrink. If this shrinkage is allowed to occur before the screed has hardened, there is an increased risk of cracking. Drying and shrinkage will first affect the upper, exposed face of the screed, thereby encouraging small bays of screed (or areas of screed between cracks) to curl upwards at their edges.

To discourage early shrinkage and cracking, the screed must be kept damp, until the cement matrix has hardened, by curing under polythene.

To further discourage curling, large areas of screed are usually laid in large bays, with a minimum number of day-joints.

Inadequate drying out

Notwithstanding the requirements for curing, it may be necessary to allow a screed to dry out before applying some adhered or moisture-sensitive floor finishes (eg, vinyl or carpets). It may also be necessary to allow a screed to dry and shrink before applying rigid tile finishes. If this is not done, subsequent long-term screed shrinkage may buckle the tile flooring. The construction programme must, therefore, allow sufficient time both for the initial curing and the subsequent drying phase.

Dampness affecting walls

Floor screeds are usually laid after wall plastering. If the plaster has not been terminated at skirting level, dampness may spread from the screed to the plaster, causing a 'tide mark'.

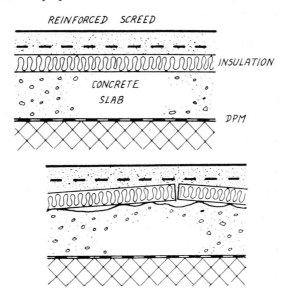

Figure 1 Top: Typical details of insulated floor screed. The concrete slab should have a floated finish
Bottom: Consequences of laying the insulation on a rough - tamped slab

Insulated screeds

Following the revision (1990) of Part 1 of the Building Regulations, it is likely that many more buildings will incorporate ground floor screeds on thermal insulation. It is already known

Good practice in laying floor screeds

that such floating screeds (see above) are difficult to compact properly. It must also be recognised that, traditionally, floor screeds have been designed on an empirical basis, with little or no guidance available to relate specifications to given loading conditions.

Under heavy loadings, it may be appropriate to design such 'screeds' as thin, structural concrete slabs. It is also advisable to provide movement joints at perimeters and at changes in support (ie, where a screed extends from an insulation base on to a solid base).

Heated screeds

It is customary to construct heated screeds in small bays, with soft expansion joints to accommodate thermal movements. However carefully the screed is laid and cured, some 'curling' may occur when the heating cables are switched on. This may be minimised by raising the temperature gradually and it is likely that any 'hollowness' will be less noticeable if the screed is supported on a slightly resilient insulation.

Great care must be taken to protect heating cables. These normally have a flexible coating to absorb movements at joints.

Figure 2 Floor screed with heating cables, being laid in small bays, chequer-board fashion.
(Note: polythene curing membrane was replaced after the photograph!)

Movement joints

It is sometimes necessary to form movement joints directly below joints in rigid tiling.
If the joints are to be formed during the laying of the screed, they must be set out precisely, so that they correspond with the lines of the tile joints. Alternatively, they may be cut in the

hardened screed: this simplifies setting out and also minimises the risk of curling. It is necessary to use a constant-diameter, diamond-edged cutting disc, to ensure a full-depth cut, and cutting cannot be used where joint lines cross any conduits which are laid within the screed.

Tolerances

BS 8204 contains guidance on tolerances for the screed itself. However, difficulties may be encountered if the screed design does not take account of dimensional variations in the base, or if the base construction is unacceptably inaccurate. Factors such as residual camber in precast floor units, or in situ concrete with a tamped finish which is too high, may result in a screed which is too thin.

Supervision

Screed layers, if unsupervised, may follow some of the bad habits or try to 'cut corners' that they may have been allowed to pursue on previous contracts. It is up to supervisors to ensure that the specification written for a particular job is adhered to so that more modern construction practices can be successful.

REFERENCE
1. BARNBROOK G. Concrete ground floor construction for the man on site. Part 2 for the floorlayer. Cement and Concrete Association.

BIBLIOGRAPHY
BRITISH STANDARDS INSTITUTION
(1989) BS 12. Specification for Portland cements
(1983) BS 882. Specification for aggregates from natural sources for concrete
(1985) BS 4483. Specification for steel fabric for the reinforcement of concrete
(1980) BS 4551. Methods of testing mortars, screeds and plasters
(1987) BS 8204 Part 1. Code of Practice for concrete bases and screeds to receive in situ floorings
(1989) BS 5385 Part 3. Wall and floor tiling. Code of Practice for design and installation of ceramic floor tiles and mosaics.
(1986) Part 4. Code of Pratice for ceramic tiling and mosaics in Specific Conditions
(1972) CP 202. Code of Practice for tile flooring and slab flooring. (NB. CP 202 will be superseded by BS 5385, with Parts 3, 4 and 5 of that Standard applying to floors. Part 5, covering terrazzo tile and slab, natural stone and composition block floorings, has yet to be issued)
(1970) CP 204, Part 2. In situ floor finishes
BARNBROOK G. Construction guide. Laying floor screeds. British Cement Association
ROBERTS R.F. Recommendations for testing of sand and cement screeds. Cement and Concrete Association

BUILDING RESEARCH ESTABLISHMENT publications
(1972) Current Paper 78. Method of assessing the soundness of some dense floor screeds
(1984) Information Paper 11. BRE screed tester : classification of screeds, sampling and acceptance limits
Digest 104. Floor screeds
Defect Action Sheet 51. Floors : cement-based screeds - specification
Defect Action Sheet 52. Floors : cement-based screeds - mixing and laying
Defect Action Sheets 120 and 121. Solid floors : water and heating pipes in screeds
(1959) Thermal insulation : avoiding risks.

DEMOLITION AND THE MAIN CONTRACTOR

by K C Kelsall, revised by the Institute of Demolition Engineers

INTRODUCTION

Many redevelopment projects, both in the public and private sectors involve the clearance of reinforced and mass concrete buildings and structures using demolition techniques. Contract completion times are foremost in developers' minds and consequently new techniques are being used to give the main contractor a cleared site as soon as possible.

The following information is designed to help the client and contractor, through the site manager, to carry out demolition and site clearance in a more efficient and safer manner.

PRELIMINARY INVESTIGATIONS

Demolition is a high risk process in terms of damage to life and property. Therefore, it is of the greatest importance before placing a contract to ensure that:

(a) the demolition contractor has experience of the type of work being offered;
(b) fully comprehensive insurance is taken out against all risks and that these policies are currently in force;
(c) an experienced supervisor is continuously in charge of the works;
(d) the contract is priced to include all safety precautions, as specified by Local Authority Byelaws BS 6187 and the Health and Safety at Work etc Act 1974;
(e) the completion date is realistic and not dependent on risks being taken to adhere to it;
(f) the client advises on the previous use of the buildings/premises;
(g) a method statement is submitted and approved.

DEMOLITION — BRITISH STANDARD CODE OF PRACTICE BS 6187

The Code of Practice was first published in 1971 with the help of members of the National Federation of Demolition Contractors and representatives from a cross section of the construction industries and Government departments.

To site managers dealing with demolition works, this document is invaluable as a reference to recommend methods of safety on site. It must be stressed, however, that the contents are only recommendations and are not legal requirements. An outline of the main contents of the Code is given as Appendix 1.

Reference should also be mnade to the Health and Safety Executive Guidance Notes GS 28, points 1-4.

PRELIMINARY ARRANGEMENTS PRIOR TO START OF CONTRACT

Notification should be given to local authorities and service authorities six weeks prior to the commencement of works under the Demolition Works Notice Building Act 1984, sections 80-83.

It is also a requirement to notify adjoining property owners/occupiers under the above mentioned section.

It is imperative, prior to commencement of the work, that the following meetings are organised by the demolition contractor and site manager:

(a) Local authority highways engineer

Agreement will need to be reached regarding the extent of scaffolding and protection to pavement and roads; scaffold and pavement crossing permits should be applied for at this meeting. Depending on the type of job being undertaken, it is at this meeting that permission can be received to close footpaths completely and to encroach into the road with hoardings. (See the Highways Acts of 1959 and 1971).

(b) Police

Whatever has been arranged with the highways engineer must be agreed with the police, so that traffic and pedestrians can be controlled, thereby maintaining a smooth flow in the area.

(c) Service authorities

The following service authorities should be notified and asked to bring plans of services in and around the are of the demolition:

electricity;

gas;

water;

telephones;

private services (radio relays etc).

At each meeting with the representatives of statutory authorities careful consideration must be given to:

(i) the need for all services entering the property to be cut off and sealed outside the site;

(ii) the position and depth of services alongside the site, serving adjacent properties and agreement as to the protection required from heavy machines and vibration. It should be confirmed that the contractor cannot take responsibility for services not notified to him at the meeting.

(d) Adjoining property owners

Meetings should be arranged with the client's surveyor and a surveyor representing the adjacent property owner to agree a schedule of conditions, or as sometimes termed, a schedule of dilapidations, so that the exact condition of adjoining property can be agreed before demolition takes place. (Note should be taken of section 29 of the Public Health Act 1961).

(e) Residents close to the demolition works

To maintain good public relations it is advisable to contact all surrounding residents and to notify them that demolitions are about to take place. They should be asked to co-operate by keeping windows closed wherever possible to minimise the dust nuisance.

(f) It is now a legal requirement to notify the Health and Safety Executive on its form F10 if a demolition contract is to be longer than six weeks. This is usually submitted by the demolition contractor.

(g) If any plant within the demolition contract is lagged with blue asbestos, the Health and Safety Executive requires 28 days notice in writing before it is demolished.

ASBESTOS REMOVAL
A detailed analyst's report is required along with the proposed method statement which has to be considered and approved. All work is to be carried out by a licensed contractor in accordance with the Control of Asbestos at Work Regulations and Health and Safety Executive approved code of practice and guidance notes.

SAFE DEMOLITION METHODS
The following provides an outline of the basic techniques of demolition.

Hand demolition
This implies demolition of structures using hand tools only. Lifting blocks may be used for lifting and lowering members once they are released. Buildings should be demolished in the reverse order to their erection, storey by storey, allowing for the different types of construction.

Debris should only be allowed to fall internally or externally to the ground where a horizontal distance of not less than 6m exists between the point of fall and public ways, footpaths etc. Alternatively, the distance should be half the distance of the drop, whichever is the greater. In other cases skips or chutes should be used. Holes should be made in the floor when rubble is being dropped internally and these holes should be large enough to eliminate the possibility of rubble being deflected. The build-up of debris on walls or floors should be avoided, particularly when demolishing chimneys or lift shafts.

Pusher arm

This technique involves the progressive demolition of a structure, generally a wall, using a machine fitted with a pusher arm exerting horizontal thrust, the building being first brought down to the required level by hand demolition. The pusher arm must be of steel, no other type being acceptable, and the machine must be located on firm level ground and in no circumstances overloaded.

In the past this type of equipment was usually a home-made arrangement. There are now several purpose-designed and built machines which are capable of reaching as high as 21 metres and of pulling as well as pushing.

Deliberate collapse

This method involves the removal of key members of the structure to effect collapse. Expert engineering advice must be sought before this method is used, and then only on isolated, reasonably level sites and where the whole structure is to be demolished. Sections of a structure should not be pulled down in separate operations if instability of the remaining structure results.

Balling

Balling is the method of progressively demolishing a building by the swinging of a weight suspended from a crane or similar plant. Plant for balling work needs to be robust and is generally of the tracked type. It should be stood on firm ground outside the building to be demolished and the cab must be strong, with shatter-proof glass wire mesh windows.

An anti-spin device should be fitted to the hoist rope which should be inspected at least twice daily. The jib should be lowered (but never more than 3m below the top of the structure) as the demolition proceeds. It should never be over-stressed and the length of rope should never be such as to allow the swinging ball to reach adjoining or neighbouring buildings. Care should be taken not to get the ball trapped while dropping arches, floor slabs etc. If this does occur, release should not be effected by tugging as this can overload the machine. Holes in floors should be made to allow debris to fall through.

Buildings of any height may be demolished using this method, providing the crane is of adequate size and design. It is advisable to check with the manufacturer of the crane when using a demolition ball at heights in excess of 30m, to ensure that the equipment is capable of carrying out the works.

Wire rope method

Only steel wire ropes over 38mm diameter should be used and the length of the rope should be twice that of the height of the building being pulled down. This method should not be used on buildings over 21m high.

If more than one pull is needed, the ropes for each section should be set up in advance, in order that no return to the building is necessary by any workers. The ropes should then be attached to a well-secured winch or track vehilcle, which should only tug in the direction the section of the building is facing.

The driver of the machine must be well protected from flying debris and ropes.

Demolition by the use of explosives

The ever-increasing use of mass, pre-stressed and post-stressed concrete in building and civil engineering works means that the work of the demolition contractor is now more difficult

Demolition and the main contractor

and exacting. By using explosives, the most back breaking section of the demolition can be carried out swiftly and precisely. The scope of works which can be undertaken by the use of explosives is increasing rapidly, as it is constantly being tested and tried. Explosives can be used on virtually all demolition contracts being let at the present time. Many such contracts are being carried out in close proximity to other properties which are to remain standing. Even so, by the use of controlled blasting, demolition work can be carried out quickly, safely and efficiently. It must be stressed that any demolition work using explosives should be carried out by experienced and authorised contractors. Any demolition contract in which explosives are to be used must be done with the full co-operation and consent of the local authority and with its guidance.

It is advisable to use firms who are members of the Institute of Explosive Engineers wherever possible.

Other aids to demolition

(a) Thermic lance cutting

This method is used mainly in the cutting of heavy concrete and is an additional aid to any of the previous demolition methods. The equipment consists of a simple lance tightly packed with steel rods; the tip is pre-heated to ignition and then fed with oxygen to provide an intense heat source. One advantage is that it can be used in a confined spaced.

(b) Diamond saws and drills

These machines are a further aid to the demolition contractor where fine limits in cutting concrete and steel have to be employed.

(c) Hydraulic burster

This is a device which is placed into a hole or series of holes and is then expanded using hydraulic force. This causes cracking but with no vibration and no risk of fly-out.

(d) Hydraulic hammer

This method involves attaching a heavy hydraulic powered hammer to the boom of a 360° or 180° excavator. It enables concrete to be broken fairly accurately and the resultant debris is usually very small. Although a little noisy under certain conditions, it can be an economical method for breaking concrete.

(e) Scrap shears

These are an attachment to a 360° excavator to enable steel buildings to be demolished by machine rather than exposing operatives to working at height with oxy/propane equipment. They also allow scrap to be processed quickly ready for removal from site.

(f) Concrete shears nibblers

Again these are attachments to a 360° excavator. This piece of equipment can be used to demolish concrete framed structure in areas of buildings where it is not possible to use a conventional method of demolition with a demolition ball. This attachment not only demolishes concrete buildings but processes the concrete to enable the salvage of reinforcement bars.

SITE FACTORS

Plant and equipment

All plant and equipment should be serviced and checked frequently. Only skilled operators should be in charge of machines, which must be of the required type, size etc for the job in hand. All plant operators are required to be certificated to the type of machine they are required to operate.

Protective clothing

Special precautions should be taken for work on buildings where chemicals have been stored, or where asbestos, lead paint, dust or fumes may be encountered. Working with these materials may require respirators. For general protection helmets, goggles, footweat and gloves must be provided.

Projecting nails

Care must be taken and a careful watch kept for projecting nails, sharp pieces of metal and splintered wood etc.

Shoring and underpinning

When removing sections of buildings which could leave other parts unsafe, temporary support positions must be predetermined in order to allow shoring etc. Drawings should be produced and verified by a structural engineer.

Working areas

Areas where work is to be carried out should be well signposted and a clear warning given that demolition works are in progress. This can include illumination where necessary.

Overloading with debris

Care must be taken to avoid overloading sections of buildings with debris, either against party walls or on suspended floors.

Weather conditions

Where possible, adverse weather conditions should be noted, eg, snow which may drift against unsafe walls and suspended floors and impose similar dangers as experienced with debris.

Flooding

Care must be taken to ensure that there is no hazardous build-up of water.

Overhead cables

Crane heights should be checked against any surrounding overhead cables to ensure no damage can be caused to either.

Skips

Where builders skips are used these should meet the requirements of the Highways Act 1971.

THIRD PARTY PROTECTION
Scaffolding and hoarding

These should be erected and illuminated to the local authority's specification to give protection generally.

Security

Any building which is partially demolished and the surrounding site should be properly secured and closed against entry when demolition operations are not in progress.

Demolition and the main contractor

Dust
This should be kept to a minimum by dousing with water whenever necessary.

Noise
The use of suppressors and silencers can keep noise to a minimum.

LEGAL REQUIREMENTS
The Construction (General Provisions) Regulations 1961, Statutory Instrument No.1580 apply to demolition and consider the operation under four sections:

(a) the application of the Regulations;
(b) the supervision of demolition work;
(c) fire and flooding;
(d) precautions to be observed in connection with demolition.

Application of the Regulations
The requirements of the Regulations apply to the demolition of 'The whole or any substantial part of a building or other structure'. No exact demarcation line exists and the site manager is advised when in doubt, to consider that the Regulations apply.

Supervision of demolition work
Under the 1961 Regulations special precautions were taken to define the responsibilities of 'self-employed' demolition workers. They are called 'individual contractors' and within the

meaning of the Act, each is a contractor who 'personally performs the demolition operations without employing anyone else thereon'. Regarding these individual contractors, the Regulations state that they shall not begin any operation except after consultation with any other individual contractor undertaking the same work, and any 'competent person' who might be appointed to supervise the works. It is not only necessary that these consultations shall take place, but also, that they are to include the 'method by which, and the time at which, the operation is to be carried out'.

It is the duty of all contractors other than 'individual contractors' to elect a 'competent person' to supervise the demolition work. Where more than one demolition contractor takes part in such work they may either:

(a) elect a common 'competent person, or

(b) make arrangements to ensure that no operation is undertaken by their workmen except after consultation between all involved, to ascertain the method of doing the work and when the work is to be done.

The term 'competent person' has no legally defined meaning and it is for each employer to decide for himself whether a person is competent to carry out the work in question. In the event of legal proceedings the employer may be called upon to satisfy the Court that any such person chosen is, in fact, competent.

Fire and flooding
Before demolition begins and during the progress of the work 'all practical steps' must be taken to safeguard those employed from the RISK of fire or explosion from leaking gas mains, or from the RISK of flooding.

Precautions to be observed in connection with demolition
The Regulations draw particular attention to the following points:

(a) that no part of the structure shall be overloaded with debris or material and so be unsafe to those employed;

(b) that all 'practical precautions' must be taken to avoid danger from the collapse of any part of the structure or building;

(c) that precautions must, where necessary, be taken to prevent the accidental collapse of any part of the structure or building;

(d) that certain work may only be carried out, by either workmen experienced in the work and under the direction of a competent person, or by people under the immediate supervision of a competent person. This work includes any part of the work where a risk of collapse exists, and the cutting of reinforced concrete, steelwork or iron work joining part of the structure or building.

Under Section XI of the same regulations — 'Miscellaneous' the following points of special interest are included and should be considered in connection with the demolition work:

Demolition and the main contractor

Electricity — the danger to persons employed.
Protection from falling material.
Protecting nails and loose materials.
Construction of temporary structures.
Avoidance of danger from the collapse of a structure.
Protection of the eyes.
Keeping records.

BIBLIOGRAPHY

ACKNOWLEGEMENT

The photographs used in this contribution were made available by the Controlled Demolition Group of Leeds.

APPENDIX

DEMOLITION, BRITISH STANDARD CODE OF PRACTICE BS 5187
1. General. Scope. Definitions

2. Preliminary procedures. Survey. Contract. Miscellaneous

3. Protective precautions. General Safety of personnel on site. Safety and convenience of third parties. Protection of property. Special items

4. Methods of demolition
   These include:
   Hand demolition
   Mechanical demolition by pusher arm
   Mechanical demolition by deliberate collapse
   Mechanical demolition by demolition ball
   Mechanical demolition by wire rope pulling
   Demolition by explosives
   Other methods of demolition

5. Recommendations for demolition of various types and elements of structure
   Small and medium sized two-storey dwellings (not exceeding 10m in height)
   Large buildings with loadbearing walls (three storeys and over)
   Framed structures
   Cantilevers (not part of a framed structure)
   Bridges. Masonry and brickwork arches
   Independent chimneys
   Spires
   Pylons and masts
   Petroleum tanks in the ground
   Chemical works, gas works and similar establishments
   Basements
   Special structures

SETTING OUT

by E M Baker FCIOB, Revised by B M Sadgrove

INTRODUCTION

Considerable time and effort has been expended in studying and analysing most of the processes that contribute to the completion of a building project. For example, mixer cycle times, crane utilization charts and critical path programmes are legion, but one would have to search hard and long to discover a study of the time and cost of setting out.

How is this important function carried out? How much does it cost if done badly, or save if it is done expertly? Can it be improved and how does management arrange and keep in contact with this task? It is the intention of this paper to provide some answers to these questions.

Setting out in its most elementary form ensures at the very least that the building faces the right way in the correct field; at its most complicated it can dictate the whole progress of a project.

Management could profitably consider in greater depth the basic principles involved and question whether they are 'setting out in the right direction'.

PRELIMINARY SETTING OUT

In the rush to get work started on site, the investigation and checking and more importantly, the recording of site boundaries, party walls and bench marks can often be neglected. There are several reasons for this, including a shortage of staff and/or information, or the reluctance of management to spend money on permanent setting out stations or other aids to planned setting out.

Site personnel frequently change during the course of a contract, so the responsibility of management at this early stage is to have a document or drawing prepared, indicating very clearly — and preferably from some permanent building or feature beyond the site perimeter — the source of this information on boundaries, party wall agreements and all related information that is used initially to establish the position, line and level of the future structure. It is not sufficient just to obtain information from the statutory suppliers of gas, electricity etc — one recent site, for example, involved eleven different authorities — the information they give will only be approximate, so site investigation and plotting on the preliminary site layout is an essential task in the interest of safety and also of future building progress.

MAIN GRID LINES

When first positioning main grid lines is the time to consider their future use, maintenance and accuracy throughout the contract. If the building is going to climb and the site is reasonably clear, check points of main lines can often be established at roof level of nearby buildings, to be used later when the initial ground profiles are no longer usable.

The appreciation that setting-out pegs and profiles do get knocked, dug out and disturbed, is one step towards avoiding the waste of effort and recrimination that is all too common. As a preliminary task the site manager must decide the routes of his supply vehicles, diggers and drain excavations, and have these plotted. Given this information, the person delegated to provide setting out facilities has the opportunity to position his profiles and pegs to avoid these trouble spots. It can be argued that site conditions frequently do not lend themselves to this desirable situation but if most of the facts are known, then the management's engineering services may have to choose different, more robust or remote methods of providing this vital facility. The cost of doing this has to be considered against the cost of constant replacement of disturbed markings with the consequent risk of inaccuracy, the dissatisfaction of the site staff and delay.

As contracts get larger and faster and sites become more confined, major building organisations are adopting a method called 'co-ordinated setting out'. This method depends on the co-operation of the architect and engineer to produce a plan which indicates the proposed buildings, roads and all other obtainable information on services and site boundaries etc. On this drawing the setting-out surveyor can plot his setting out stations in the 'green belt' area (areas most likely to be undisturbed) and plot readings between these stations, which will permit at least one of them to be available at all times to allow the setting out of principal grid lines. With the co-operation of all levels in these early stages, coupled with the issuing of a drawing showing the agreed lines and stations, the necessary importance and authority is given to this vital task. It has even been possible to negotiate conditions with sub-contractors and site users to deduct costs if they cause damage or obstruction to these agreed stations.

The cheapest way to discover that one cannot maintain a permanent reliable profile is on a drawing board or model before operations commence. At this stage it may be revealed that a few extra bags of ready-mixed concrete, some tubular guard rails, plenty of paint or warning signs are all that are needed. By comparison, on one job, piles were driven on which were indicated the main setting-out lines. The piles were surrounded by a 1.8m metal fence (with lockable gate) to ensure that they were undisturbed throughout the contract.

If the problem cannot be solved by this method, at least it has been highlighted, and the solution may mean renting, or obtaining use of ground outside the main site, or the erection of towers. It could imply that different, or more expensive, instruments are required, or it may have to be accepted that pegs, profiles etc will have to be transferred as work proceeds, thereby necessitating more staff. At this stage the make up of the bills of quantities might well be examined to see if provision has been made in the estimate for the difficult situation that has been revealed. Very often, engineering services are allocated to management on a job cash value basis which has no relationship to the ease or difficulty of varying sites. A £¼ million job may be complicated and need constant attendance, whilst a £5 million job may only need initial simple plotting.

DELEGATION OF DUTIES AND RESPONSIBILITIES

It is of utmost importance at the early stage of the job to establish, very clearly, who is responsible for providing and maintaining line and level, and also the limitations of their responsibility and authority. One way to clear up any misunderstanding is to create a hypothetical situation, for example, a row of columns 1m out of position and check through the processes and staff who would be associated with this error. By carrying out this exercise and tying up any loose arrangements, similar situations like this can be avoided.

THE CONTROL OF ACCURACY

For those concerned with accuracy, BS 5606 : 1978, Code of accuracy in building, is recommended reading. This publication is intended to give assistance to builders in avoiding problems which arise if the effects of inaccuracy are not anticipated. It defines some of the terms used, gives tolerances and sources of inaccuracy and gives permissible deviations of both instruments and materials etc used in setting out processes.

As it is becoming customary for new contracts to refer to this document, site managers need to be familar with its recommendations. On the subject of specifying permissible deviations, the Standard states 'Absolute accuracy exists only in theory; all construction and manufacturing processes are subject to inaccuracy'. It goes on to state 'The overall object in specifying permissible deviations is to define standards of dimensional acceptability to enable components and construction to be fitted together to achieve a satisfactory building'.

Having established the degree of accuracy required and tolerances permitted one can select the tools and the men to achieve them. Greater accuracy and final tolerances call for increased supervision and better training of both workmen and supervisors, and often at greater cost. The selection of tools and instruments is also based on permitted tolerances.

A modest breeze can produce considerable sideways bow to twine and so cause serious errors. Synthetic lines are stronger and, therefore, can be thinner and still reduce the errors due to wind. Plastics and steel reinforced plastic tapes can be surprisingly inaccurate; use steel tapes to BS 4484 for accurate setting out. Put aside one tape as a 'standard' against which working tapes should be checked at regular intervals. Standard tension should be applied for critical measurements and corrections made for slope and temperature, and for sag if using catenary taping.

Water levels can give trouble if some parts of the tubes are at different temperatures, or if air is present. Even with every precaution one should not expect a greater degree of accuracy beyond ± 5mm.

The limitations and expected accuracy of levels, theodolites and spirit levels etc are given in the Standard. For example, one should not expect a greater accuracy than ± 5mm in 5m when using a theodolite for vertical plumbing.

The site manager must not only ensure that the correct instruments and tools are chosen for the specific purpose, but also that the persons concerned with setting out are aware of the permissible deviations or manufacturing tolerances of all the component parts to be used in the construction. Bricks, particularly hand-made facings chosen for aesthetic purposes, are sometimes outside the British Standard and those within the BS can vary by $\pm$ 15mm in 5m. Precast concrete and other specialist manufactured goods should either conform to an appropriate standard, or have a tolerance agreed and made known to all those responsible for their positioning and final performance.

Special job conditions, adjacent buildings or scaffolding, may influence the required tolerances and must be considered, agreed, understood and publicised to the interested parties.

The important point to remember is that 'special' as compared to 'normal' accuracy will in all probability cost more, and management must decide whether it is necessary.

AIDS TO MAINTAIN PROGRESS OR PREVENT DELAY
Establish a routine
When delegating the setting-out function and control it is necessary to plan the sequence of events, for example, periodic checking of main grids and instruments is imperative.

A routine must be estabished and priorities fixed and clearly understood by those requiring the services of the setting-out organisation. These people cannot function efficiently if the priorities are settled by the chargehand carpenter shouting louder than the ganger needing drain invert levels.

Considerable repetitive work can be avoided by the provision of jigs, gauge rods and templates. Delays can be avoided if jigs or templates can be made of specially manufactured items, or items of long delivery. For example, curtain walling can progress if two identical templates are made of the special glass or infill panel. The manufacturer can proceed to work to one template whilst setting out of brackets etc can proceed by utilising the partner template on site. This method can also be utilised for setting out of precast units, columns, copings etc.

Operatives can often understand and prefer to use a gauge rod from a given datum and this can reduce the likelihood of error.

Multi-storey buildings and grid lines
When constructing tall buildings, a 'secondary' grid is usually required and generally this is provided as soon as the ground floor slab has been cast. These grid lines may need to be offset from columns and walls in order that they can be projected to the upper floors.

These grids should be registered in a permanent form, either by scribing or by centre punch marks on a steel plate, bolts set in the slab, or by masonry pins. Provision must be made to enable these lines to be transferred to the upper floors, so either holes should be cast in the floors, or the precast units, if forming the floor, should have joints wide enough to be able to use an optical plumbing instrument.

A temporary bench mark (TBM) should be transferred, agreed and recorded, to a column or lift shaft wall at ground floor level. This TBM, if possible, should be a standard dimension above floor level, and to avoid mistakes should always be described clearly as '1m above finish floor level', or structural floor level whichever is desired or more practical.

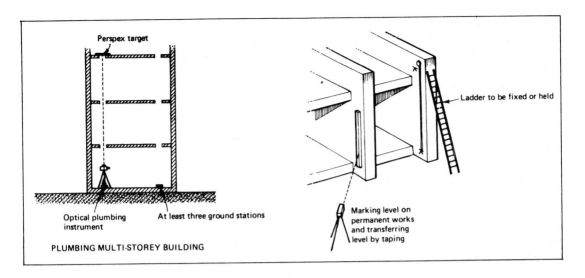

Perspex target

Ladder to be fixed or held

Optical plumbing
instrument

At least three ground stations

Marking level on
permanent works
and transferring
level by taping

PLUMBING MULTI-STOREY BUILDING

As the building progresses, this TBM is transferred to each floor by taping down directly to the ground floor TBM. In this way any settlement of the building will not affect vital dimensions, and creep or gain of cumulative floor to floor measurements are avoided. Once the TBM is established it can be distributed where required on progressive floors by water or optical levels, and should be so marked that they remain to be used throughout the contract by all trades following the frame construction. Before finally covering these marks with wall finishes, it is advisable to record the position of the main bench marks on a drawing. Should any dispute arise at a later date these would be easier to find and with only local damage to the wall finishes.

The laser beam projector and separate sensor have been added to the surveyor's tool list. Weighing about 6kg, robust, waterproof and easily set up, the projector emits a pencil-thin beam of light horizontally or vertically with a range of about 300m. The beam can be made to rotate, thereby providing a plane of light. Powered by rechargeable batteries, the beam of an appropriately low-powered laser is harmless and does not obstruct work in the way that string lines sometimes do. The cost compares well with most conventional instruments.

Set up to give a horizontal plane of light, it is extensively used for level control in ground works, flooring, kerb laying etc. A sensor fitted to the controls of a dozer, say, can automatically grade ground to the required levels and falls. When levelling, the operator uses the sensor on the staff, eliminating the need for two-person levelling and mistakes due to signalling etc between them. The instrument is self-levelling within close tolerances. If it goes out of level, this will be indicated and usually the beam will be cut off automatically.

Used to give line, the laser can be set to give a horizontal line, or a line to a given slope; useful, for example, in drainlaying and pipelaying. The laser beam can also be set perpendicular for plumbing falsework and slipforming. The control of tall buildings is one application, but shaft and tunnel control also use the laser.

SUMMARY
One certain way of improving site setting out would be to find out the cost of bad setting out. Obviously, this is a very difficult task as the effect of one small error can be far reaching.

Setting out

Nevertheless, considerable improvements and cost savings can be achieved if site management takes a closer look at the methods they already employ, to check if they are making the best use of the highly paid men and of the expensive instruments that are often misused or abused on sites.

Instant setting out
Every effort must be made to eliminate the attitude 'Joe is waiting for a line'. This is not a good reason for someone to take a hurried look at a drawing and then go rushing out with the instruments; it is fraught with danger and frequently leads to costly errors.

Setting-out as part of management
To divorce the setting-out processes, or to isolate them from management, by placing these processes as a buffer between management and the construction team is both costly and wasteful. Setting-out processes, and the men that provide them, should be a link and a valuable means of interpreting dimensions and drawings to the site personnel. Once allocated to the site they should be fully under the authority of the site manager, they should work the site hours and use site facilities. To be able to provide the best service to the site, they should be represented at site meetings and have a knowledge of variations and revisions to site details or programmes; in short they should be active members of the management team.

More substantial profiles and more thought in their siting and preservation, will not only raise the status of those engaged in providing them but will encourage greater accuracy. The time saved in not having to constantly replace these can be better spent on quality control, programming and forecasting, so that these men can see and prepare for a future in management.

BIBLIOGRAPHY
1. BUILDING RESEARCH STATION (1970) Accuracy in setting out. Digest 114. HMSO. pp8
2. BRITISH STANDARDS INSTITUTION (1978) Code of practice for accuracy in building
3. INSTITUTE OF BUILDING (1978) Setting out : an annotated bibliography. 2nd edition.
4. SADGROVE B M (1987) Setting-out procedures. CIRIA and Butterworths

ACKNOWLEDGEMENT
The illustrations of plumbing a multi-storey building are reproduced from the author's book 'Setting out procedures'. The co-operation of the publishers CIRIA and Butterworths is gratefully acknowledged.

FORMWORK AND SITE MANAGEMENT

by John G Richardson, FICT, FIIM

Introduction

This paper is based on one originally prepared by E M Baker FCIOB. Ted Baker was renowned for his pragmatic approach to construction and it is hoped that this present contribution reflects the essentially practical nature of the original.

CAPITAL ASPECTS

Management decisions at all stages of formwork operations have a major impact on the eventual success of those activities in terms of performance and economics.

Achievement of specified finishes at the concrete face, optimising on available skills and resources, whilst attaining economic returns from investment in equipment, demand a substantial input of practical and technical knowledge by management.

The site manager who recognises the importance of formwork in the overall context of concrete construction can, by employing basic skills of management, ensure the success of formwork operations. There is immense scope for: planning of technical input, which is of extreme importance; motivation of the workforce, particularly in the adoption of innovation; co-ordination of concrete supply, handling and placement; the careful control of the total operation from form design to striking.

In this paper various aspects of formwork operations are considered with the objective of identifying critical steps in the control of operations, as well as examining current trends in materials and applications.

ESTABLISHING STANDARDS

Early establishment of standards is an important step towards determining the optimum formwork system. All too often the standards required of the constructor are wrapped up in the language of the specification writer. Frequently, it is left to the contractor to interpret specifications for concrete finish, only to meet with rejection of work at site on the basis of a differing interpretation of the same document.

The Formwork Committee of the Institution of Structural Engineers and the Concrete Society, emphasises in its report *Formwork : a guide to good practice* the importance of the establishment of standards by the preparation on site of a pre-production model of the critical parts of a building or structure. The report recommends that a full size section of concrete should be produced incorporating all the detail which will be encountered in the course of construction. Features, finishes, joints, openings, drips, returns and items such as grooves for asphalte should be formed in this model in order that intended standards of accuracy and finish should be demonstrated and agreed upon by the parties to the contract.

It is believed that such a model should also include repair to a corner or some similar detail. This will ensure that should some damage occur, there will be a reference or a 'yardstick' to which such remedial work can be matched.

At this stage in the work, management will be well advised to ensure that the workmanship and skills employed are truly representative of those that will be employed during construction. Opportunity should be taken at this stage to present counter-proposals on feature and detail which experience has proved will improve the buildability of the final structure. The experienced site manager will have much to offer in this area resulting from personal experience gained through a range of different contracts.

Report No.24 of CIB (International Council for Building Research Studies & Documentation) sets out in the form of comparative photographic illustrations, varying degrees of defect and various 'shadation' of concrete surfaces. The intention being that the specifier should stipulate the range in each instance within which the concrete should fall. It is not intended that, as in the case of colour cards for paint, the specification should be absolute, simply that the range should be determined. The report also quantifies blemishes such as blow holes. The use of such a 'tool' in specification will do much to clear up the ambiguities otherwise encountered as a result of traditional specification techniques.

The site manager should avoid the problems inherent in the establishment of standards based upon small samples of work. It is difficult to achieve a match in the case of factory produced precast concrete and almost impossible in the case of site product with all the variables to be encountered. It may be found that sensible standards can be established by reference to a previously constructed building or part of a building. In American practice it is not unusual that some early part of the construction, such as a basement wall, is used as a sample for the work later in the contract.

Increasing use is made of concrete for purposes of interior decorations, these applications demand carefully detailed forms to achieve accuracy and crispness of presentation   Acknowlegement The Metal Box Company Reading

With attainable standards established, consistent with the qualities expressed in the specification, the work of formwork design and construction can begin.

THE FORMWORK DESIGN STAGE
At the design stage the site manager can contribute much towards ensuring that the eventual system meets the objectives of generating the specified finishes in an economic fashion.

Formwork and site management

Formwork design is essentially a practical process. Research has improved knowledge of the pressures on formwork and the ability of various materials to contain those pressures, whilst providing the required reuse. Whilst this information assists in the preparation of formwork design and detail it is of little value unless applied by a designer skilled in the use of materials and the production of detail appropriate to the work in hand. Here the task is to ensure a balanced approach to problems, maintaining a balance between the academic and the practical approach to design. In some instances 'overdesign' may prove more practical and improve the economic outcome of the work.

During the last decade there have been major changes in the attitudes of specifying authorities and their site representatives. Volumes of concrete permitted to be placed in one operation have increased and greater lift heights are almost universally adopted. Recent pours of 2000m^2 of concrete in building foundations and 10m lifts of concrete in abutment walls to motorway bridges for example, have resulted from the new enlightened approach of such authorities, coupled with initiative on the part of contractors.

Whereas safety and the mechanics of the formwork, pressures and forces and the way in which they are contained are important, they do not present significant problems for the qualified design engineer.

Issues of importance are the selection of the appropriate materials to generate the required surface finish, application of concrete technology, knowledge of rheology, the design of concrete mix and detail such as the way in which the forms, be they of traditional materials or proprietary formwork, are framed, handled, treated and stored between uses.

The site manager's knowledge of plant, equipment and systems and his skills in evaluating alternative proposals can all be brought to bear on the form design. Co-operation with plant, equipment and service specialists and discussions with the proprietary supplier of formwork and formwork sub-contractors are all areas where basic management strategies, coupled with an enquiring approach to the topic, can improve performance and ease construction tasks.

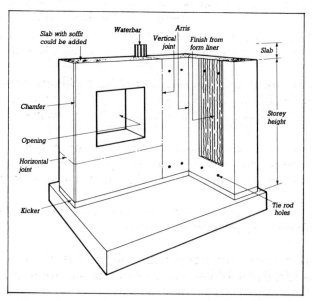

Figure 1 Composite sample or trial panel (hypothetical)

MATERIALS SELECTION

As discussed earlier the selection of the materials for use in facing the forms is a major item in the sequence of design and detail. Although the carcass of the form is important in generating the strength and rigidity required for structural and handling purposes and for containing the concrete mix without allowing leakage, it is the form face that generates the visable result. The visual outcome of the operation may well be in evidence for the 30 to 50 year life of the building. The form face and facing materials determine to a large extent how many uses can be achieved from the equipment and thus are critical to the overall economnics of the work.

Considerable research has been conducted into the properties of facing materials used in formwork and the interface between such materials and the concrete of the structure. The main points emerging from the many sources of information can be usefully summarised and will serve to provide pointers in selection of the facing or sheathing to the formwork:

- all concrete 'weathers' to some degree with age. Unless this is considered in design it is likely that the appearance of the building may suffer;
- concrete cast against smooth surfaces tends to 'craze' and collect dirt, detracting from the visual appearance of the structure;
- feature, particularly vertical striation assists in directing water flow, controls staining, masks joints and generally disguises the different shadation of adjoining lifts and bays of concrete;
- mechanical considerations permitting, a fast rate of placing, consistent with the achievement of good compaction, is likely to produce the most satisfactory face finish;
- form and mould treatments, the use for example of chemical release agents and scientifically prepared oils and coatings, yield additional reuse value as well as providing a degree of consistency of appearance at the finished face;
- mechanical tooling of formed faces is a destructive and time consuming process and can well be replaced by the adoption of the latest generation of surface retarding agents.

With these facts in mind the site manager may well decide to approach the client or his advisors with a view to using one of the flexible form liners currently available. These liners, provided they are of substantial section, consistent with the feature incorporated, will yield 50 or more uses with consistent results, depending on the care exercised in use.

Where plain surfaces are essential to the function of the building then it is likely that ply or timber derived materials will be used as the sheathing to the forms. It is advisable in these cases to discuss the likely visual outcome with the building designer.

Considering the requirements of the Department of Transport Specification for Road and Bridge Works, and that of the Property Services Agency General Specification Concrete Work, which are used as yardsticks for much of concrete construction:

- used ply and particle boards leave a surface that meets the requirements of Class F1 concrete where there is no visual requirement;

- sound used forms, ply or proprietary panels produce Class F2. Some degree of repair is acceptable, imperfections and non-uniformity of colour are normally permitted as is making good;

Formwork and site management

- Classes F3 and F4, high class finish with and without ties, are intended for visible surfaces and, due to the many variables involved, call for extreme care in materials selection, placing, compacting and curing techniques. These finishes carry a real risk that minor blemishes and patchiness will detract from excellence of finish even when extreme care is taken in production.

In the case of the last two finishes, where possible, it is advisable to negotiate for a lightly textured concrete face. This can be achieved by using a lightly textured form, light surface retarder, or by subsequent brush blasting of the surface. Each of these techniques will reduce crazing to an acceptable minimum.

The use of the scientifically prepared coatings and treatments available from the service industry surrounding formwork has increased. No longer does the contractor buy 'mould oil' pure and simple. Hopefully, he takes guidance from the information forthcoming from research, chooses the treatment wisely and then controls the use of the material sensibly at site.

A material, such as a chemical release agent, that costs three times that of a basic oil may at first sight seem expensive. That material, applied using a fine mist spray will coat far more than three times the area of a similar quantity of oil, more effectively and provide greater consistency where appearance matters.

Surface retarders eliminate destructive, tiring and morale destroying tooling processes.

INNOVATIVE MATERIALS

A number of new materials have become established in the formwork field. Notable innovative materials include glass-reinforced cement, pressed steel permanent forms, expanded polystyrene permanent forms and latterly, controlled permeability formwork.

Glass-reinforced cement (GRC) permanent forms may provide the ultimate solution to satisfying particularly demanding specifications. High quality finishes are achieved in the works in course of manufacture of GRC. Moulding techniques allow the production within simple moulds of the most complicated, or traditional details incorporating stone or brick finishes in the permanent formwork. Selection can be exercised prior to installation on site. Additionally, a new generation of stainless steel mesh-reinforced GRC permits substantial spans to be achieved without intermediate support.

Pressed steel permanent forms have revolutionised the formwork industry. Used in conjunction with a steel frame or reinforced concrete they provide instant working area, economy in reinforcement and minimal prop requirement. As with GRC there are no striking operations to delay succeeding work.

In the industrial field, where complicated shapes are to be formed, the proprietary form comprising a moulded expanded polystyrene inner covered with high density plastic sheet as a sheathing provides economy in forming voids, cores and ducting in the concrete mass. With care several reuses can be achieved from a product that costs less than one third of equivalent, traditionally produced forms.

The use of man-made fibre facing in combination with a porous form sheathing material, Controlled Permeability Formwork, as it is termed, was initally pioneered in the UK on the iniative of a major contractor. The system permits excess pore water to evaporate from the concrete contained within the form. Reduction in the water/cement ratio within the 'covercrete' achieved in this manner provides surface strength and reduced permeability, improving the resistance of the cast concrete to freeze/thaw and aggressive attack. Surfaces produced from permeability form faces are lightly textured, providing improved bond for applied finishes. As an added bonus, the resulting reduced formwork pressures enable the carcass of forms to be of a lighter nature. These features combine with a reduction in striking times serve to maintain the economy of formwork operations where such systems are employed.

There are few materials that cannot be used to form concrete when some particular problem has to be solved. Concrete, GRC, plastics and card all have their place and will be used by the enterprising form designer. The site manager should keep alert to trends and examine the viability of materials with a view to using them when some special characteristic lends itself to speed of construction or the solution of some particular construction problem.

STOPENDS AND DAY-JOINTS

The site manager in co-operation with his form designer should make a point of discussing with the client the incorporation and location of day joints and construction joints. These must be sited to allow the optimum reuse to be achieved in the most direct manner suited to the formwork module. This discussion is one of the most important aspects of the formwork operation.

In complicated construction it is likely that up to 40% of the surface area of formwork is in stopends and day joints. In such a contract the forms account for some 50% of the cost of constructing the concrete frame. Day joints and stopends are extremely important items in cost evaluation.

Stopends are required to sustain the same pressures as the form face yet frequently the construction and integrity of this part of the formwork is left to the tradesman, perhaps not the best informed person as regards the stresses and strains involved. Money spent on detailing properly located stopends or the purchase of disposable materials for their formation will be amply repaid in speed of construction and performance. Faces formed by stopends are visible in a great amount of finished work and provide an ongoing commentary on the skills and abilities of those involved in the construction.

Expanded metal lathing and patent assemblies for installation of starter bars have made for increased economy in joint construction. These materials reduce the labour content as well as eliminating the damage to the formwork caused by drilling forms and opening formers for projecting bars. The disciplines of the bar assemblies also provide an assurance that projecting bars are not overlooked when the time comes for their reinstatement into their vital location at the interface between succeeding lifts and bays of concrete.

Traditional specifications calling for 'scabbled surfaces' at stopend and construction joints have historically increased construction costs, whilst possibly detracting from the structural value of the joints so treated. Surface retarders used at stopends and joint positions meet general approval today and improve the economics of this otherwise costly area of work.

Formwork and site management

PROPRIETARY AND SPECIAL FORMWORK

It makes good sense to take advantage of the 'free consultancy' available from the involvement of the proprietary formwork industry.

In the case of fast-track construction, access to the database offered by the supplier can result in savings in labour, time and materials. Early consultancy, whilst details of the building can still be modified, allows the adoption of system equipment with its inherent speed of erection and simplicity of handling operations.

Early consultation is important where, for example, the use of table or flying forms is envisaged as it may be necessary to stiffen slab edges by the incorporation of additional reinforcement, or for the designer to analyse the results of application of some element of construction loading into the structure. Consultation is part of the service offered by both the proprietary formwork supplier and the special form manufacturer. They survive in a competitive area of industry and consultation is best confined to a few regular suppliers with whom the constructor does business.

Special design is necessary where the job is massive or complicated. Practical formwork detail is important to achievement of speed and economy of use.

SUPPORTING WORKS

Considerable emphasis has been placed in recent years on the topic of falsework. For many years this area of construction was the province of the formwork designer but as construction operations grew more complex and the loads on the supporting works increased it became more the province of the Temporary Works Engineer.

Following on the publication of the Bragg Report and committee work by the professional institutions, BS 5975 : 1982 Code of practice for falsework, has been published and accepted throughout the industry. It has resulted in considerable improvements in practice and safety.

The site manager in his consideration of formwork matters should acquaint himself with the requirements of the standard. An hour or so spent with the document will provide a reminder of a number of interesting aspects of construction, as well as serving to underline the importance of the in-process checks that form an essential part of the process of providing support.

In the area of support for formwork there can be few items of equipment that have contributed as much to the safety of the operations as the 'system prop'. The lacing, bracing and tieing facilities integral in prop systems ensure that the dangerous practices highlighted in the Bragg Report and in each publication of the Health and Safety Executive are avoided.

The introduction of alloy and lightweight equipment, primary and secondary beams and quickstrip equipment have also contributed to the efficiency of the industry, allowing as they do speedy handling and frequent reuse to be obtained. The soldier member for supporting wall forms, with the possibility of a full moment joint, allows substantial forms to be constructed and safely handled and erected with minimum expenditure of crane time.

In the special formwork field, advances in the stressed-skin type of form have permitted either elimination of, or a reduction in the amount of through ties needed for a given area as well as reducing the amount of support needed from the ground. In many cases special forms incorporate working platforms and take support from previously constructed mature concrete.

STRIKING TIMES

It is difficult to consider any of the activities of concrete construction in isolation. Equally, it is impossible to consider formwork without considering the technology of the concrete that the forms are designed to contain.

The site manager making sensible decisions regarding formwork method and design needs to take into account the emerging technology of concrete with its impact on, for example, the timing of operations.

Much has been learned of the characteristic of concrete described as its maturity. Maturity for a particular mix of concrete, with particular aggregate, cement content and water content, correlates very closely with the 'strength' of the concrete. Maturity is a value (expressed in degrees centigrade, hours) which is derived from the temperature history of that concrete.

Given similar maturity in the structural element and a test specimen (achieved by storing the specimen in an environment corresponding to that within the concrete in the element) the test will provide a sensible assessment of the 'strength' of the concrete in the structure, its ability to withstand the forces imposed on striking, its ability to sustain loading and its resistance to frost and mechanical damage.

These facts have led to the adoption in BS 8110 of striking times based on the maturity of the concrete in the structure. The result is seen in more realistic striking times which allow the constructor the opportunity of reducing turnround time on forms by accelerated curing techniques, enhanced cement contents and so on.

Concrete technology has evolved in the area of large pour construction. Knowledge has increased of the characteristics of concrete in conditions such as the massive foundations to prestige buildings to the stage where rules have been established that allow the construction of deep slabs and large pours, avoiding problems of cracking and distortion due to differential temperatures across the section due to the heat generated by the mix and by subsequent cooling. As a result contractors are able to cast concrete economically in large pours, secure in the knowledge that systematic insulation or ventilation can maintain the temperatures within acceptable limits thus avoiding damage.

THE DEVELOPING ROLE OF FORMWORK SUBCONTRACTORS

Trends, particularly in 'fast-track' construction are towards increased employment of specialist formwork subcontractors. Whilst such firms have existed for many years, some have been reluctant to employ new technology, new materials or similar innovations, their main resources being a fund of tradesmen practised in traditional and now often outmoded methods of form construction, erection and handling. Formwork activities inevitably feature on the critical path of activities determining hte overall duration of the construction phase of the building process.

Increased speed of construction thus demands enhanced outputs from formwork and the formworker. Enlightened subcontractors now employ techniques of applied design and detailing of forms, use proprietary formwork equipment and accessories, new and innovative materials. Technological inputs such as the use of results of research into the development of early strengths and accelerated curing techniques are combined with accurate strength assessment in determining striking times. In this way the turnround times demanded in fast-tracking are economically achieved, particularly where the subcontractor is appointed early in the course of the contract when his input on the potential of methods and materials can be of most benefit.

Today's specialist subcontractor frequently provides additional services such as detailing, supply and fixing of normal reinforcement and prestressing tendons as well as concrete mix design also handling, placing and compaction using fast and efficient methods which integrate with formwork system; becoming involved in all aspects of the construction of founds, frames and superstructures. In the case of the larger firm the input and expertise they provide will have been gained in the course of constructing the most prestigious building in the UK.

CONCLUSIONS

Formwork is a vast topic. However, it is not by any means an exact science. Consequently, there are few rules. Essentially, formwork activities provide the site manager with the opportunity to innovate, to capitalise on the funds of knowledge available and thereby achieve economies. These benefits cannot be achieved without taking into account the technologies of the materials and the capabilities of available equipment. To be successful, it is important to understand what is required of the finished product and to plan the maximum use of skills, equipment and the specialist knowledge available 'in-company' as well as in the research and commercial fields.

BIBLIOGRAPHY

1. BRITISH STANDARDS INSTITUTION (1982) Code of Practice for falsework BS 5975.
2. CONCRETE SOCIETY (1986) Formwork : A guide to good practice.
3. AMERICAN CONCRETE INSTITUTE (1984) Committee 347. Formwork for concrete.
4. RICHARDSON J G J(1977) Formwork construction and practice. Chapman and Hall.
5. NEVILLE A M (1981) Properties of concrete. Pitman.
6. GAGE M (1974) Guide to exposed concrete finishes. Architectural Press.
7. BRITISH CEMENT ASSOCIATION (1980-on) Appearance matters. A series of guides to concrete practice.
8. RICHARDSON J G (1987) Supervision of concrete construction. Vols 1 & 2. Chapman and Hall.
9. Q I TRAINING LTD (1989 — on) Concrete practice series of video modules. Formwork etc Q I Training Ltd. Swindon.

DETERRING VANDALISM ON BUILDING SITES

by A J Charlett, BA, MPhil, MCIOB

INTRODUCTION

Vandalism is a malaise of society and its current level is causing concern to those responsible for maintaining law and order as well as to its victims.

The building industry, although not a major sufferer, experiences a range of attack from small incidents costing a few pounds to repair, to large scale havoc causing damage running into thousands of pounds. Building contractors, when asked by a Home Office Standing Committee on Crime Prevention[1] to estimate the cost of damage caused by vandalism on site, considered that such damage constituted between 0.2-3% of the contract value. With an annual output of over £40,000 million it can be seen that the cost is not insignificant.

Financial loss may also result from the over-running of contracts, where this can be attributed directly to having to repair damage resulting from vandalism. The cost of vandalism is heavy enough for large contractors, particularly when working on narrow margins, but the effect on small builders can be catastrophic.

The extent of the problem is not clear. The number of cases of criminal damage recorded by the police in 1988 was 593,923 but a great deal is either not reported to the police or is not recorded by them where the damage does not exceed £20 (a minimum limit set by the Criminal Damage Act 1971). It has been estimated that only 7% of cases of 'minor' vandalism are reported to the police whose records, consequently, considerably under-represent the scale of the problem. The reasons for this low report rate are numerous but likely factors[3] include:

(a) the incident not being considered sufficiently serious enough to notify the police;
(b) the incident not being witnessed at first hand;
(c) reporting the incident may seem too late for effective action to be taken;
(d) fear of victimisation;
(e) fear of becoming involved in Court proceedings.

There are no official national statistics on vandalism and its effects on the community and on the economy. Estimates of the scale or cost of the damage can only be conjectural.

A number of different agencies have studied the problem of vandalism in an attempt to devise solutions but a panacea has not been found. The general consensus of opinion points to tackling the problem from two standpoints. Either by preventing people from wilfully damaging their surroundings or by trying to stop them from wanting to do so by improving the environment. The latter solution is a matter for the community as a whole and cannot hope to be tackled by one sector of the community in isolation.

The object of this paper is to analyse the type of problems occurring on building sites and to look at methods of deterring the occurrence of vandalism on sites. It is not intended to offer a cure for the problem.

ANALYSIS OF SITE VANDALISM

In studying vandalism it is useful to build up a mental picture of the types of person involved in this activity and to examine some of their motives. In this way, methods can be evolved for deterring a large proportion of the vandalism occurring at present on unprotected sites.

Vandalism has been characterised as an opportunistic kind of offence. This sets it apart from other security problems on site, such as theft. The vandal will most likely commit damage to property if it is made vulnerable and if the chances of being apprehended are negligible. Thus, unprotected sites, having large amounts of vulnerable equipment and property which are usually in isolated areas, are particularly favoured.

An additional factor, which seems to have a direct bearing on the incidence of vandalism, is that of remoteness of ownership. Often children will vandalise property that does not appear to have any apparent owner to prevent them, or to complain about the subsequent damage. In this way abandoned cars, derelict property and buildings under construction are prime targets. Vandals have also been found[4] to react against a drab environment. Buildings in the course of construction present this drab aspect and create the impression that they are not owned by a specific person. An untidy site can often heighten this illusion of dereliction. Thus, a child damaging equipment and property on such a site does not feel a sense of moral shame, as would probably be the case if the property looked 'cared for' and a sense of ownership could be established.

Vandalism appears to be a crime of the young. It is committed either by young children in the course of unsupervised play, or by older adolescents seeking status among their peers and excitement. Building sites are often seen by these children and adolescents as an adventure playground. Statistics categorising the incidence of vandalism by age and sex can only provide an outline of the general picture, since they only include people convicted or cautioned for committing criminal damage. A much larger proportion of people are never apprehended, due, in part, to anonymity of the offence.

Current statistics[5] suggest that of the 16,100 people convicted or cautioned for criminal damage, approximately 29% are aged between 10 and 17. A large proportion (approximately 35%) of these juveniles are in the 10-14 age group, and the majority of them (approximately 90%) are male.

The motives behind vandalism are compatible with the age groups concerned. In particular, young children see the building site as a playground and their destructive behaviour may go little further than breaking windows and smashing fragile materials during the course of their play. Some psychologists believe that this type of behaviour is regarded by the children concerned as a highly institutionalised form of rule breaking.

The older child poses a more serious threat, since he or she will often trespass on to a site and cause wilful damage in response to a dare made by a member of their peer group. In these circumstances a property which has attempted to be secured now poses a bigger challenge to the offender and is more likely to be the target of attention, because of its high prestige value. The group is an important factor in vandal behaviour, since an act carried out by an individual within a group may be far more serious than by an individual acting alone. The site itself may add to the willingness of an individual to commit an act of destruction. Objects which can be used for inflicting damage on property, such as bricks, stones, metal tubes and

Vandalism on building sites

stout timbers, are readily to hand, and when this is coupled with the site's isolation, ensuring uninterruption, the vandal's bravado does not have to be overwhelming in order to succumb to the temptation.

In a survey[6] of the prevalence of vandalism among 584 adolescent schoolboys aged 11-15 from a northern city having a large vandalism problem, 40% of the boys interviewed admitted to having 'smashed things on a building site' at least once in the previous six months. The same survey highlighted a much higher incidence of vandalism on unoccupied property than on occupied property. 68% of the boys interviewed admitted to having 'broken a window in an empty house', but only 32% admitted to having perpetrated the same act to one which was occupied. Although these figures cannot be applied universally, they do give an indication of the scope of the problem.

Vandalism by adults is largely a means to another end. The motive may be theft from the site, revenge against a former employer, an expression of frustration against the environment, to draw attention to a political campaign or to draw attention to the perpetrator (a method often used by old lags as winter approaches in order to regain admission to prison).

Although each of these different groups will react in different ways to deterrents used on the building site there is one overriding factor. The incidence of vandalism can be reduced by installing effective deterrent measures.

DETERRENT MEASURES

Because of the opportunistic nature of vandalism, a great deal of damage can be prevented from being inflicted by the simple expedient of providing effective deterrent measures around the site.

The effectiveness of most deterrent measures is difficult to evaluate and to compare against cost outlay, since nothing happens when the measures are operating successfully. The benefit of these methods can best be appreciated by a contractor who has suffered a large scale damage through vandalism and who has subsequently had them installed with no further intrusion.

The methods used can be roughly classified into three categories:
- physical barriers, such as fences or hoardings;
- psychological deterrents, such as lights and warning notices;
- fear of apprehension, such as security guards and alarms.

Fencing and hoardings

A good perimeter fence will keep out all but the most ardent of intruders. However, it can only be fully effective if it is continuous, of the correct height and constructed from strong materials. There is a problem on many sites, particularly housing sites, where the widespread nature of the operations makes it uneconomic to provide a proper perimeter fence. In these cases it is advisable to enclose vulnerable equipment and materials within an adequately fenced compound and to protect the houses under construction from intrusion and damage by boarding up doors and windows. This concept will be discussed more fully later.

On more compact sites strong fencing is a feasible proposition. There are many different types of fencing available and it may prove of value to consider their effectiveness as deterrents to the vandal.

Primarily, the fencing system may either be open, as with chain link fences, or solid, as with hoardings. The former has the advantage of enabling perimeter security lighting to be used to aid observation of intruders, whilst the latter prevents potential intruders from reconnoitering the site.

Fencing may also act as a check to prevent unintentional trespass on to the site. The contractor has an obligation to protect the general public from incurring injury on his sites and he should meet this obligation by clearly defining his boundaries with physical barriers.

With open fencing, the contractor has a choice of chain link, the cleft chestnut pale and the weldmesh fence. The fence must be securely attached to its supporting posts in a way that cannot be easily undone or that might serve as a foothold to someone climbing the fence. The height of the fence is also critical. British Standard 1722 suggests a minimum height of 1.80m for effective security fencing, dependant of the type of fence adopted. Anti-intruder chain link fencing seems to be a more effective barrier against the vandal than the standard chain link fencing and is available with cranked posts to which can be affixed three strands of barbed wire. If the posts are cranked at a reflex angle over the boundary of the site unauthorised entry by climbing over the fence is made particularly difficult.

At ground level the chain link fence should be embedded in the ground to at least 150mm depth to discourage tunnelling.

To be really cost effective, the fencing must be reusable so that a contractor may utilise the same fence on a succession of sites. In this respect, chain link fences have their limitations, since they tend to sag having once been strained. The contractor may wish to consider negotiating for such fences to be returned to the supplier in exchange for a cash adjustment on his next purchase, or he may consider using an alternative system.

Systems are now available utilising stout weldmesh fencing welded to a frame of angle iron in modular sections. The whole system is supported by cranked posts carrying barbed wire. Such systems are strong and easy to erect and dismantle, giving a high number of uses.

Solid hoardings used to enclose totally a site are a simple and widely used anti-vandal deterrent. They may be constructed from stout plywood, close boarded timbers, or corrugated iron sheeting. Because the support system is at the rear they present a flat vertical surface to the intruder, which is particularly difficult to assail. The top section of the boarding may be coated with anti-climb paint which is non setting and which, apart from deterring potential intruders, also helps to identify persons who have attempted to gain access to the site by marking their clothes.

The Highways Act 1959, Sections 147 and 148, states that a close boarded hoarding must be erected by a person proposing to carry out work to a building within a street, to separate that building from the street. This may prevent the use of open fencing. Where the contractor is concerned about the screening effect of a hoarding to intruders who have gained access to the site, observation panels may be cut at regular intervals into the hoardings. On restricted sites overhanging gantries are often required to support site accommodation which cannot be placed at site level. These gantries can, in themselves, provide a convenient means of access over the perimeter hoarding and should be similarly treated.

The perimeter barrier must also extend across access points to the site and gates must be of a similar construction to the fencing to which they are attached. It is important to ensure that no protrusions such as padlocks, handles or supporting bars, can be used as a means of climbing the gate. Gate posts must be similarly free from protrusions, particularly at the hinges. Gaps greater than 100mm at the sides or underneath the gates must be prevented and railway sleepers should be embedded below the line of the gate to prevent tunnelling.

The type of lock used to secure the gates is most important. It is useless expending large sums of money on perimeter fencing if the gates are then secured by a cheap padlock which can easily be broken, or cut open by bolt croppers. A suitable padlock should be of the closed shackle design, preventing bolt croppers being applied to it. The gates should be fitted with heavy duty padlock staples welded to the inside of the gate. A cut-out enables the padlock to be locked and unlocked from outside the site, but reduces the attempts to tamper with it.

Security lighting

Most cases of vandalism on building sites occur at night. There are three basic reasons for this. Firstly, the site is unoccupied, so that materials and equipment are left unattended, providing an open invitation to the potential vandal. Secondly, the cloak of darkness covering the site makes satisfactory surveillance of the area extremely difficult, thus lowering the risk of any intruder being apprehended. Thirdly, the evening is normally the time that groups of youths gather together in social groups, and, due to the lack of local facilities, or for adventure, they may decide to embark on a spate of vandalism.

Security lighting serves as a deterrent to intruders. Its primary purpose is to deter thieves, or to assist in catching them before they can steal, but it can also serve effectively against vandals, provided it is used in conjunction with a sound perimeter fence and, in cases where security is of high priority, an efficient patrol or security guard.

The lighting columns must be arranged in such a way that they are either directed towards the perimeter fence, or used to illuminate dark surfaces inside the site. The first technique, known as glare lighting, has the dual attributes of illuminating any intruders approaching the fence, so making them feel vulnerable, whilst simultaneously concealing any guard or watchman who may be keeping surveillance behind the line of lights. For this system to work effectively, an open fence must be used, but the materials of its construction need special attention. Research[7] on the effect of fence luminance on the detection of potential intruders has shown that the luminance value of the fence, in relation to its surroundings, can have marked effect on the visual acuity of an observer. A fence of high luminance value can hinder a watchman observing a potential intruder if perimeter lighting is shone on the fence, but the converse also applies, and a fence of high luminance value can hinder a potential intruder from observing the interior of the site. This would suggest that a fence with a covering of high luminance value on the outside and a low luminance value on the inside, should be used.

The alternative technique of illuminating dark surfaces inside a site, provides a clearly illuminated background against which intruders may be clearly seen. Again, this may act as a deterrent since it makes the intruder feel vulnerable.

Often lighting on building sites is provided primarily to enable work to continue outside daylight hours. This may extend appreciably the working day during winter months and thereby contribute in making the installation and operations of a site lighting system cost effective. The siting of lighting towers is, therefore, heavily dictated by the need to provide adequate illumination in the production areas. These requirements may be in direct conflict with establishing suitable locations for effective security lighting. In these cases, a compromise must be reached between the two demands, or alternatively extra lighting must be provided.

The use of site lighting must be economically viable. In order to keep running costs down an efficient light source must be chosen. Suitable lamps are the long life lamps, such as low pressure sodium, which can give 4000 hours operation. These minimise the risk of failure and reduce maintenance costs. An additional factor which needs to be taken into consideration is that the lamps themselves may act as a target for vandalism. Consequently, they must be located at a convenient height to be efficient, but should not be low enough to be an easy target for the vandal. Breakages must be expected and the cost of replacement must be added to the operating costs when deciding whether site lighting is a cost effective deterrent.

The social costs of lighting must also be considered by the contractor. In built-up areas, a continually lit site may prove to be a nuisance to the surrounding neighbourhood, particularly if the lights have to be generator driven. Good public relations is very important in combating vandalism, as people whose houses border the site are often invaluable for watching the site and for raising the alarm if intruders are observed.

The small builder may be able to derive the benefits of perimeter illumination, without having to go to the expense of installing site lighting. Careful positioning of boundary fences around urban sites can utilise the illumination derived from street lighting during the hours when vandalism is most active. Even this low level of illumination at the perimeter fence can provide an effective deterrent to the intruder. However, gates and access points must be effectively protected, since entry may be attempted at those points if the deterrent effect of the perimeter lighting is successful.

Warning notices
Deterrents do not have to be physical in nature in order to be effective. Often a notice warning potential intruders that they will incur serious penalties if caught trespassing, or causing damage on site, is sufficient to deter the less ardent offenders. Any warning signs used should be clear, simple, unambiguous and large enough to be noticeable. It has been found[8] that a large proportion of children committing acts of vandalism are non-achievers at school and so may experience difficulty in reading. Warning notices may also be usefully employed within the site, particularly at places where an intruder may be placed in danger, such as on uncompleted scaffolding, close to generators and transformers and around excavations. Indeed, many such notices will be in use as warnings to site personnel. A general notice at the perimeter of the site warning of the dangers to young people arising from playing on building sites, can also be effective.

It is important that any threats on warning notices of penalties that will be incurred should intruders or vandals be apprehended, are seen to be implemented. Warning notices are soon made impotent as deterrents if the builder is not prepared to take action against apprehended intruders. In this respect, it is important that the builder works in close liaison with the local police and refers all persons apprehended to them.

The general public can also be involved in helping to stop vandalism on sites. Schemes where a contractor has erected a sign at the site offering rewards for information leading to the apprehension of intruders, have proved very successful. The scheme may be extended to circulating letters, outlining the reward system, to householders in the district. The deterrent nature of these measures is further enhanced if maximum publicity can be obtained for a successful apprehension. Local newspaper or radio stations may well be willing to cover the successful implementation of such a scheme. Less successful schemes seem to be those which request local schools and youth groups to publicise the commencement of work on a site in that area. The idea behind the scheme is to inform the groups most likely to attempt to trespass on to the site that such actions will not be condoned by the builder concerned. However, far from deterring potential trespassers, these warnings tend to advertise the site and substantially more children visit the site, mainly from curiosity.

Watchmen and security patrols

The use of night watchmen on sites was a common feature of the building industry until comparatively recently. The job is lonely, has unsociable hours and uncomfortable conditions, and there is likely to be little activity, leading to boredom. These factors have made the job unattractive and this, coupled with the low rates of pay offered by the employers, has tended to restrict the job to the elderly. Ironically, this very group of people are the least capable of apprehending intruders, since they may be unfit and are certainly not as able to deal with a group of vandals who have ventured on to the site as would be a younger and fitter person.

In addition, police surveillance of sites, particularly those which are more isolated - often the most vulnerable - may only be cursory as their resources become more and more stretched. It is unrealistic for builders to rely on the local police to provide their site security. With rising crime figures and the police thinly spread, this approach cannot be recommended.

With the development of security firms the building industry has begun to rely more heavily on their services for site surveillance. One of three methods are commonly adopted; guard dogs, static guards and patrol guarding. The first technique used specially trained dogs roaming loose within a secure compound. However, since the introduction of the Guard Dog Act 1975, no person can allow the use of a guard dog on his premises without ensuring its control by a suitably qualified handler. This increased the cost of security.

Static guards are particularly beneficial where vandalism problems or site security are of prime importance. Effective static guards are expensive, particularly if they have been fully trained in their job by their security organisation and the builder must decide if the scale of loss caused by poor security justifies their employment. Such guards can be made more effective by using guard dogs and by utilising perimeter lighting as described earlier.

To offset the high costs of static guards, many contractors prefer to use security patrols, which offer surveillance to several neighbouring sites and thus reduce the cost of surveillance for each individual site. This technique, although attractive, has become less favoured of late since the visits of patrols to site were often infrequent and fleeting and the builder could not check on when, if at all, the visits had taken place. To offset these criticisms, major security organisations have introduced systems of clocking patrol visits and using radio communication to supervise their patrols.

One major problem with patrol guarding, as with mobile police surveillance, is that on isolated sites vandals can often detect the approach of a vehicle a long way off. This gives them time to either flee the site, or hide until the patrol has moved on. Consequently, intermittent surveillance is not a totally satisfactory method of combating vandalism.

Alarm systems

Intruder alarm systems have kept pace with technological development so that there are now a wide variety of systems available for varying applications. British Standard 4737 covers the main systems currently available, some of which can be utilised for protecting building sites.

Alarm systems are a particularly cost effective deterrent, since they will operate continuously at a fraction of the cost of other forms of security. The most effective alarm systems operate on site buildings, buildings under construction, or adequately fenced compounds and it is sensible to incorporate a system to protect these areas for the low cost outlay involved. One suitable system is the use of magnetic reed contacts on site hutting (which trigger an alarm when the contact is broken) and volumetric detection in stores (where a change of volume in a sealed area triggers the alarm). However, it is sensible to seek the advice of a company specialising in alarm systems regarding the most suitable system for the site in question.

The builder may consider that the alarm system should be extended to include the perimeter fence or hoarding, since early detection of entry to the site is likely to be the most effective deterrent to the vandal. Systems are available to cover these areas but they have tended, in the past, to be prone to being triggered by slight vibrations or by the wind. Alarms are now available which employ microwaves, ultrasonic, passive infra red waves, or seismic detectors. These are more successful in open locations.

Generally, a builder will enter into a contract with an alarm company who will offer advice on suitable systems for the site, install and maintain the equipment and, on completion of the contract, remove the system and resite it. The protection system should be extended to cover sub-contractors offices, stores and compounds and agreement should be reached on how this should be achieved at the commencement of the contract. The alarm system adopted should include a suitable signal once the alarm has been triggered. Suitable signals may be audible, such as bells or sirens, or visual, such as flashing lights or operation of the main site lighting. Alternatively, the system may alert the police or the security organisation responsible for site security by use of signalling equipment working through the telephone system.

Some contractors may decide to give a foreman the responsibility for maintaining the site security measures. It would be his duty to ensure that the site was secure at the end of each day and that the alarm system was in operation. In such cases, the alarm system may be designed to operate a radio alert call to the foreman who could then check on the cause of the disturbance.

As a keyholder, a responsible foreman could also be utilised as a contact either for the security organisation or the police should they wish to notify the builder of any intrusion or wish to gain access to site in an emergency.

Any alarm system used on a building site must be robust; it is of no use to install a balanced vibration alarm system in flimsy temporary site hutting, since the alarm will be easily triggered.

Effective deterrents against vandalism on site should not end with perimeter detection methods, but should be extended to prevent damage occurring on site should unlawful entry

Vandalism on building sites

be achieved. To this end, security measures should be taken to protect the buildings under construction, the site accommodation and the materials and equipment used in the construction process.

PROTECTING THE SITE FROM WITHIN

Should vandals gain access to a building site then the damage inflicted may be considerable. Buildings in the course of construction are vulnerable to attack, particularly in the later stages of construction when fragile components are being incorporated as part of the finishes. In addition, materials, plant and equipment left insecure on site provide an open invitation to the vandal to tamper with them or damage them. Often, materials and equipment on site are used to inflict damage on the site buildings. Even temporary site hutting is not immune from attack and often valuable equipment such as theodolites and levels, are stored in these huts. Another target for attack is the site canteen, presumably because of the stocks of confectionery and similar items stored inside.

PROTECTION OF BUILDING UNDER CONSTRUCTION

The Scottish Local Authorities Special Housing Group[9] has suggested precautionary measures aimed at protecting housing under construction and these measures may also be adapted for other buildings. Suggestions include the provision of a simple one piece shutter of 9.5mm thick plywood screwed to a 50 x 50mm framing to protect windows at ground level. The framing is wedged into the window opening tightly to prevent a crowbar or lever being inserted. External door openings can be protected by temporary doors or barricades secured by two safety pattern hasps and staples with close shackle type padlocks.

Upper windows are less susceptible to forced entry, providing that there is not easy access by way of ladders and scaffolding. Such windows need only be protected against projectiles and a 6mm thick plywood shutter on 50 x 50mm framing may be considered adequate.

Access to upper storeys can be a major problem and ladders should be removed at night and placed in a locked store or secure compound. Additional deterrents can be applied to scaffolding, such as fitting outriggers to the platform at the first floor level with four strands of barbed wire affixed, to discourage climbing. However, this method can constitute a hazard to operatives working on the scaffold during the day.

Housing sites are particularly vulnerable, since they are unlikely to be adequately fenced around the perimeter. Window shutters and door barricades are particularly effective and may be moved from one house to another as occupation occurs. As the houses become occupied vandalism often decreases since occupants are particularly keen to prevent neighbouring properties from being vandalised. This surveillance is particularly effective since it continues during the time that the site is unattended. Since occupied houses are less vulnerable to vandalism than those which are occupied, some builders hang curtains in properties as they are completed to create the impression of occupation. This strategy only remains effective whilst the house and its garden remain tidy.

PROTECTION OF MATERIALS, PLANT AND EQUIPMENT

A high proportion of the cost of a building project is in the materials used. If materials are not properly stored and protected on site then there can be a significant cost in replacement as well as delay to the project.

To prevent damage occurring through vandalism all loose materials should be stored in a secure compound. The compound should have a strong perimeter fence, at least 2.4m in height, topped with three strands of barbed wire. The gates should be of comparable strength to the fence, with welded staples on the inside face, on which can be affixed a close shackle padlock.

Since materials' deliveries and distribution need to be made within the compound, a perimeter fence of weldmesh panels, topped with barbed wire, is desirable. This type of fence enables panels to be removed as required to facilitate access for delivery and vehicles. A secure compound is feasible on all types of sites and is ideal on housing projects where the size of the site makes perimeter fencing impracticable.

Materials in use, such as bricks, blocks and roofing tiles, are better stacked near the point of use, but even they can be protected from wilful damage by careful stacking and covering with a tarpaulin when not in use. The correct storage and deployment of materials is discussed in a number of publications issued by the Instiute[10]

One of the most attractive aspects of a construction site, from a young vandal's viewpoint, is unattended mechanical plant. The urge to 'joyride' is often overwhelming and may be the sole object of trespassing on to the site.

A great deal of damage can be caused to heavy plant by inexperienced hands. Machinery may be started up and set in motion without knowledge of how it can be stopped. The resulting damage to buildings, materials and plant can be considerable.

It is imperative that plant and machinery should be immobilised, preferably by fitting a simple steering lock device and made secure. The secure compound in which the building materials are placed should be large enough to contain all plant and machinery that can be moved.

Some builders have taken their need for a secure compound one stage further and equipped their sites with a galvanised steel clad warehouse, which can be quickly erected and dismantled to provide a completely enclosed compound. Into these warehouses can be placed all vulnerable materials and equipment. They have the additional advantages of keeping materials dry and allowing the operation of a stock control system. A large enough warehouse can provide shelter and security for items of mechanical plant and may enable repairs to be undertaken when required. Such a compound can merit the money expended upon it from the savings accrued from lost and damged materials.

Within the secure compound should also be placed highly inflammable materials, such as petrol and LPG containers. Arson is the more serious and destructive side of vandalism, but can occur if the opportunity and suitable materials arises. Paints should be locked up in a secure shed, since these are often the targets for vandals.

PROTECTION OF SITE ACCOMMODATION

Site hutting is a prime target for vandals, partly because of its often flimsy nature and ease of access, and partly because of the materials and equipment stored inside. Hutting should be of robust construction, having screened windows with mesh protection and doors secured by good qualitiy padlocks. Often the rim locks that are fitted to site huts depend very much on the strength and fit of the framed opening and may not be relied upon absolutely. Ad-

Vandalism on building sites

ditional security can be attained by siting the huts within the locked compound, but this can sometimes be a disadvantrage where space on site is at a premium.

On restricted sites, huts are often sited around the perimeter, or located on a gantry forming part of the perimiter. High level siting of this type is advantageous, since the offices have a clear view of the site and are also on general view, thus discouraging vandals from forcing entry. This arrangement should be located away from the boundary fence, however, otherwise the suport gantry may be used as a means of gaining access over the fence.

Expensive equipment, such as surveying instruments and calculators, should be locked inside a well secured hut. Where a number of huts are being served by LPG, a single cylinder can be used, sited at a safe distance from the accommodation and surrounded by a strong fence.

Where a canteen is provided on site all goods, particularly confectionery and cigarettes, should be locked in a secure cabinet at the end of the working day. These items should never be left on general display, since they may well encourage vandals to break into the canteen.

Site accommodation should be linked in to the general alarm system serving the site. Systems for protecting accommodation do not need to be highly sophisticated, but they must be designed to operate on temporary structures (which may not be as sturdy as the static equivalent) without being triggered accidentally.

CONCLUSIONS

The largest problem to be overcome in combatting the growth of vandalism on building sites is the attitude of contractors that sites are of a temporary nature. Traditionally, contracts have been of relatively short duration and any accommodation or security precautions used on the site were obtained at low capital outlay, reflecting their temporary status.

This situation has changed in recent years, with many contracts of longer duration and materials no longer cheap. Builders are realising the benefits to site efficiency, operative welfare and overall security in providing sites with better equipped accommodation and security measures.

Often the amount of money allocated for measures to deter vandalism on building sites is alarmingly inadequate and this in part may be ascribed to the difficulties encountered in calculating the cost of vandalism. Very few records are kept of incidents taking place and damage caused. A typical response is for the builder to include a contingency sum in the contract for delays and damage. This attitude reflects poor financial policy, particularly where tenders are very competitive and margins pared to the minimum.

The cost of deterring vandalism by some of the methods outlined in the paper may not result in increased expenditure on the part of the builder. He may well find that the money expended is less than that which could have been spent on repairing damage caused or subsidising time lost, through the effects of vandalism on his sites.

In addition, it is possible for savings to be made in builders insurance premiums if satisfactory precautions are taken against vandalism and theft on site. It is unlikely that an insurer will reduce the premium simply on the results achieved by such measures being instituted on one project, but a consistently satisfactory record on many projects may have a beneficial effect. It is certain, however, that if no satisfactory precautions are made the premiums will increase, and the builder will be asked to accommodate increasingly higher excesses. It is worthwhile for the builder to establish a good working relationship with his insurer so that on the one hand the contractor can show the insurer that effective measures are being taken to combat the risk and, on the other hand, the insurer can show the contractor that he is willing to reward him for his efforts by reducing his premiums and/or excesses on his policies.

Not all precautions mentioned in this paper must be implemented to provide effective protection against vandalism. The requirements differ from one area of the country to another and even between sites in the same locality. A different approach may be needed on urban sites to that used on rural sites. The former may have a high density population merely seeking outlets for adventure and play, whereas the latter may be isolated and not subject to frequent surveillance.

Little more than general guidance can be given to the builder, since sites can differ in so many ways. Builders may find that their own experience, gleaned from past projects, is an invaluable guide to how the problem should be tackled on future projects. Alternatively, where they are moving into a new area, existing contractors in that area may be willing to offer guidance on effective precautions. Certainly, an early interview with the local Police Crime Prevention Officer is essential since he will know the patterns of vandalism occurring in that area and may well be able to suggest effective preventative measures.

Combatting vandalism on building sites must be a policy matter, agreed and endorsed by senior management. They must decide which measures are necessary and their scope. From this, middle management should be able to decide on the specific measures for each individual contact.

It is imperative to establish effective measures to deter vandals from the outside of the project rather than wait for the damage to occur. A site once having been vandalised tends to attract continued attention.

It is not possible to completely eradicate the problem of vandalism on building sites. However, the problem can be minimised by adodpting many of the measures discussed in this paper.

ACKNOWLEDGEMENTS

I wish to express my grateful appreciation to those people who assisted me in the preparation of this paper. In particular Mr E Payton, Director of Construction Security Advisory Service, Mr Holdaway, Public Relations Officer, Group 4 Total Security Ltd and Sgt P Dawson and PC J Cracknell, Crime Prevention Officers of Nottingham Constabulary.

REFERENCES

1. HOME OFFICE STANDING COMMITTEE ON CRIME PREVENTION (1975) Protection against vandalism. HMSO
2. Crime statistics England & Wales 1986 (1987) HMSO
3. STURMAN A (1978) Measuring vandalism in a city suburb. Home Office Research Study No.47. Tackling vandalism. HMSO

4. GRIFFITHS R and SHAPLAND J M (1979) In designing against vandalism. Design Council

5. Annual abstract of statistics England & Wales 1986 (1988) HMSO

6. GLADSTONE F J (1978) Vandalism amongst adolescent school-boys. Home Office Research Study No.47. Tackling vandalism. HMSO

7. BOYCE P R (1979) The effect of fence luminance in the detection of potential intruders. *Lighting Research and Technology* 11 (2)

8. BELSON W A (1975) Juvenile theft : the causal factors. Harper and Row

9. SCOTTISH LOCAL AUTHORITIES SPECIAL HOUSING GROUP. The protection of local authority housing (SLASH Research Paper 78/2) 18

10. CHARTERED INSTITUTE OF BUILDING (1980) Try reducing building waste
CHARTERED INSTITUTE OF BUILDING (1980) Materials control and waste in building
WYATT D P (1978) Materials management Part 1. Occasional Paper No.18. Chartered Institute of Building

BIBLIOGRAPHY

1 UNDERWOOD G (1980) Security in buildings. Building sites. *Architects Journal 172,* December 10, pp1159-1161

2. Secure on site? (1980) *Building 239,* November 28, pp28-30

3. CLARK G (1979) TEC guidenotes - 5. Ensuring protection and safety on site. *Building Trades Journal 180,* July 11, pp23-24, 26

4. HEAYES N (1980) Security - lock up your losses. *Contract Journal 296,* July 3, pp45-47,49,51,55,57-58

5. CLARK G (1980) Defining and protecting limits of sites. *Building Trades Journal 180,* July 25, pp18,21-22

6. DAVIES W and WARRINGTON A (1979) Storage and control of building materials. *Building Trades Journal 178,* October 26, pp32,35-36,38-39

7. GARRATT J (1979) Site under siege. *Building 237,* August 31, pp21-23

8. HEAYES N (1977) Site services. Securing your profits. *Contract Journal 276,* April 28, pp47-49

9. (1977) Site security - should you be alarmed *Constructor 49,* March, pp42-44

10. ASSOCIATED GENERAL CONTRACTORS OF AMERICA (1977) Superintendent's guide to theft and vandalism prevention, 19pp

11. DOBSON R A (1976) Security on site. Paper to IOB Seminar 'Materials control and handling on building sites', London, September, 10pp

12. Theft! (1975) Boom time for crime. *Construction News Magazine 1,* November, pp17,,19,21

13. LONDON A V (1974) How secure is your site? *Building Technology and Management 12,* May, pp4-5

14. (1974) Designing for security, *Building Technology and Management 12* May,pp7-8

15. BURTON R J (1974) Security - the police viewpoint. *Building Technology and Management 12,* May, p6

16. CROMPTON R (1972) Site security and crime prevention. *Civil Engineering and Public Works Review,* February, p174

17. HENDERSON R H (1972) Scaffolding : security on building sites. *National Builder 53,* January, p19

18. (1971) Site services keep out those villains *Construction News,* October 28, p29

19. BUILDING RESEARCH STATION (1970) Security. Digest No.122. HMSO. 8pp

20. HARRIS S J (1970) Thwarting the thief : protecting new works, materials and workmen's tools. *Building 218*, June 12, pp113-114
21. BRIGHT, K and DUNSTAN, M (1991). The secure site — an impossible dream? CIOB Occasional Paper No. 44. pp36.

COMMUNICATION

by R J Biggs, MSc, FCIOB, MBIM, MAPM

INTRODUCTION

Although there may be a lot to say about the tools or means of communicating — particularly with the silicon chip in mind — there can be little new to say about communication as the passage of ideas from one person to another. However, if everything has been said before, observation of communication in practice today suggests that most of us are no better at it than were previous generations. On the global scale we still have 'war war' where 'jaw jaw' has failed and at site level experience shows a continuation of many errors resulting from poor communication.

The objective of this paper is to stimulate a greater awareness of the subject and to highlight the need for everyone to improve his or her own communication skills. Every site manager will be aware of the problems that have arisen due to his own shortcomings in this field; the aim is to show that he can help himself to reduce them.

The next section delves briefly into the theory of transmitting thoughts or ideas and this is followed by some guidelines for practical improvement.

'out of one mind and into another'

THEORY

'The final cause of speech is to get an idea as exactly as possible out of one mind and into another'. I would only substitute 'communication' for the word 'speech' in Sir Ernest Gowers' *Plain Words*. Communication is the conveyance of ideas through the senses and preferably a combination of two or more should always be used. Those most useful are seeing and hearing. Together they can be most effective; hearing alone without visual backup is relatively ineffective, although speech is what comes to mind first when the topic is raised.

'Is there a sound in the forest if a tree crashes and no one is around to hear it?' This old riddle serves to bring home the truth that the most important factor in the passage of ideas from one mind to another is not, as may be commonly believed, what is spoken, but rather what is heard or perceived by the listener. With the help of this clue one can arrive at the right answer to the riddle which is 'no'; the crashing tree produces sound waves but without a receiver, such as an ear, there is no sound. Sound is created by perception. This fact is fundamental and if properly grasped can be a great aid in improving one's communicating.

Once it is appreciated that the listener is the important factor it is easy to understand that he or she has to be tuned in or be sympathetic if the speaker's efforts are not to be lost or wasted. The only evidence of success is if there is a response from the audience. Ideally, the listener will say 'do you mean ..' and then put your idea into his or her own words. We must all have known disputes that were sparked by 'But you said ...' and were exploded by 'That's not what I meant'. Effective communication is a two-way activity and continues to and fro until both parties believe they understand each other. Such understanding is unlikely to come about unless an atmosphere of mutual trust and confidence is developed.

Understanding will also be frustrated unless a common vocabulary is used. For example carpenters' metaphors should be used when talking to carpenters. The danger of introducing jargon or specialist words is illustrated in Figure 1.

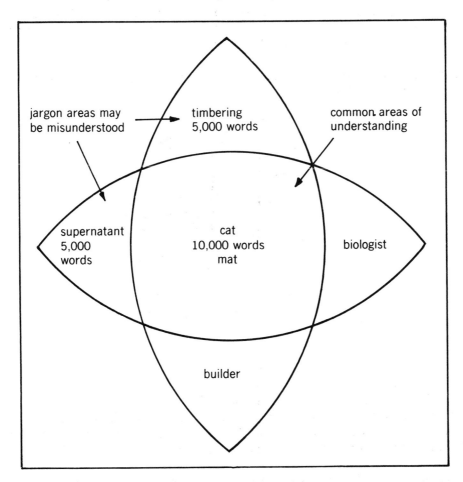

Figure 1

Common experience or vocabulary between communicating parties is essential if they are to understand each other. If jargon is employed it is likely either not to be received at all or to be misunderstood as shown in Figure 2.

Communication

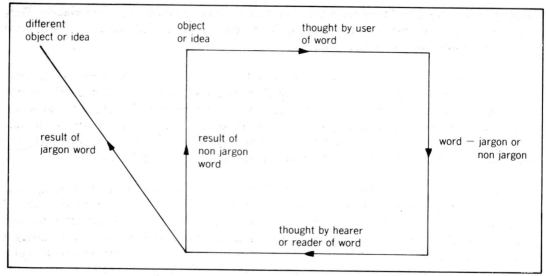

Figure 2

It must be borne in mind that in semantics, eg, language, writing etc, some symbols are arbitrary and some are not. For example the word 'wall' is arbitrary whereas 'cuckoo' conveys something of the bird. Examples of the latter are rare and most symbols are of the arbitrary kind. It is essential then to ensure that the symbols used are understood by the audience. One must always strive to operate within the range of perception of a receiver. The scope for improvement is considerable. It is commonly regarded that only a quarter of the time spent in communicating is effective, the other three quarters being wasted because understanding is absent.

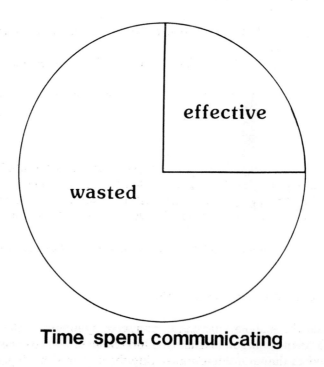

Time spent communicating

Figure 3

Most people believe that they are experts in transmitting and receiving ideas clearly. They also believe that they learnt it in childhood. This belief is unfounded and gives rise to most of the gaffes and bungles which occur.

An important element in communicating is body language or kinesics. Over thirty basic facial expressions have been identified, together with a quarter of a million combinations of gestures. Movements of the body can convey enthusiasm and the hands are most expressive, as are the fingers ticking off a number of points as they are made. In face-to-face situations such animation used effectively can be of great help but it is not to be confused with annoying personal habits which have the effect of distracting the listener.

It is revealing to consider how the emphasis of a particular word in context to its neighbour can alter the meaning of a phrase: eg, '**I** never said he stole money'. If emphasis is placed on a different word each time it will convey a different meaning. 'I never **said** he stole money' would imply that someone else said it whereas 'I never said **he** stole money' would suggest that I said it about someone other than 'he' and so on. This is where enthusiasm can help. The well-motivated speaker will place emphasis on words where it matters.

Electronic data handling allows rapid processing of words and symbols but if it is to be put to good use it must still obey the rules if ideas are to be moved intact from mind to mind.

Communicating is a very deceptive complex subject which is inherently imprecise and ill defined. However, if it is recognised that it is a learned skill it can be accepted that it demands honing to a sharp edge. The next section aims to help put this into practice.

PRACTICAL COMMUNICATION
General
- enthusiasm is a great aid. If you are really trying you are much more likely to succeed in getting ideas across;
- preparation will always be rewarded. Cervantes said 'The man who is prepared has his battle half fought';
- a wide variety of media are available. Take care in choosing what is right for you and your message; it will be well worthwhile.

Speaking
Remember that if you say something that is capable of misinterpretation somebody will misunderstand it and everything that follows. This can be avoided by choosing words that are unambiguous. Avoid 'saying it with flowers', ie, using big words or jargon. Try a descriptive phrase instead or at least define what you mean by the jargon word at the outset.

- Get to the main point immediately and fully. Brevity is a virtue but never sacrifice lucidity for brevity.

- Don't generalise when you can be specific.

- Enlist the aid of the analogy, metaphor, simile or synonym; motoring examples, for instance, since everyone is a road user. Reinforcement of what has been said is essential and this is a way of doing so without insulting the listeners' intelligence.

Communication

- Introduce as much humour as possible since levity ranks with brevity in order of importance.

- Take time to think about what to say; asking questions is one way of gaining space for this.

- Speak as if you were writing. Have an introduction, a body and a summary to the presentation.

- Look at the person or persons being addressed and be as clear as possible. Remember that the meaning lies not only in the words used but also in the manner in which they are presented.

- Since the concentration span of listener is limited, introduce breaks into a long presentation. It is better to have three 15 minute periods with question breaks than to go on for 45 minutes non-stop.

- Feedback from listeners is the only way to ensure they have understood the message. A lot can be gleaned from observing them as you speak; a glazed or nonplussed expression needs little interpretation. Test the audience by asking specifically about what has been said.

'a glazed impression needs little interpretation'

Consider the listeners and try to bear in mind they may 'seize up' over such things as:
- speaking too fast;
- presenting visuals eg, slides or flip-charts, too fast;
- verbal shorthand;
- jargon;
- foreign words;
- too many ideas/minute/hour;
- too many questions;
- too many shocks;
- something not believed without full justification.

Appeal to the visual sense is very important. A variety of visual aids is available and use should be made of one or more to suit the topic/audience.

Listening

As demonstrated previously, listening is an essential part of communicating and demands skills, which for most people are capable of improvement. The following tips may help:

- concentrate on what is being said;

- recognise your prejudices, against the speaker, the message and so on, and make an effort to overcome them. Try to understand the message even if the speaker does not come across in a way you would have liked;

- ask questions if something is unclear but do not interrupt until the individual has finished making his or her point;

- make sure you understand what the speaker wants you to know. If you do not then ask for the key points to be repeated.

The listener should also be aware of the 'traps' given below:

- considering the subject uninteresting;
- evaluating the speaker instead of the speech;
- becoming emotionally involved;
- listening only for facts;
- taking copious notes;
- faking attention;
- becoming distracted;
- avoiding listening when the subject requires detailed attention;
- reacting to loaded words.

Fear can be paralysing, the result being that one does not listen. An example is the person who has asked for directions. The enquirer is afraid of forgetting them and in consequence fails to listen properly as they are given.

Writing

The visual presentation is the most effective form of communication, and spoken messages need to be confirmed in writing. Even better is to draw a picture. The pointers which follow may help:

- keep the number of items in a message as small as possible;
- use sketches to reinforce the message if practicable;
- itemise points and put them in logical sequence;
- underline or highlight important points;
- use associations that will bring the message home to the reader.

One way of improving writing skills is to carry out an audit of written work over say, a two-week period, rating each letter or report against:

- thinking, ie, logic of writing;
- arrangement; of paragraphs, tables, sketches, etc;
- information given;
- policy knowledge; with or against the organisation's rules;
- legal implications ie, undesirable commitments;
- length;
- language;
- courtesy.

Communication

The audit should reveal weaknesses and provide a yardstick against which to measure improvement in subsequent checks.

Reading

Most managers are faced with an ever increasing amount of material to read and there is no doubt that skill in this field can be improved significantly. Books and courses are available to promote increases in speed, comprehension, retention and recall. *Use your head* by Tony Buzan, a BBC publication, is a good starting point.

It is not appropriate to expand this topic further but one interesting point is the suggested use of visual aids in reading such as a ruler, a piece of card or a finger to guide the eyes as they move over the page. This practice has been eliminated for most of us by misguided teachers and parents soon after we learnt to read.

CONCLUSIONS

At the beginning of this paper it was claimed that there was nothing new to say and as such credit must be given to the many writers whose ideas have been drawn upon. It is hoped that an interest has been stimulated, which will encourage the development of personal skills in this vital area of human relations. Its importance in site management and in life generally cannot be over-stressed. One's own level of skill in communicating will certainly affect personal progress in life. Collectively, if we don't improve in transmitting ideas we may not survive. Finally, life is short and so is good communication.

BIBLIOGRAPHY

1. GOWERS E (1973) The complete plain words. 2nd edition. HMSO
2. COOPER B M. Reports and report writing.
3. CIOB. Building management notebook
4. DRUCKER P (1970) *Management Today*. March
5. SAUNDERS W (1979) Chances are you're not communicating. *Supervisory Management*. October
6. CAMPBELL CONNELLY J. Solving tough communications problems
7. BRADEN & TROTTER. Why supervisors don't communicate
8. MELVIN T. Practical psychology in construction management
9. MORTLOCK B (1981) Structure of building communications. *Building Economist* 19 March, pp226-229
10. SHEPHERD P (1972) Communication in construction needs many bridges. *Construction Surveyor* July, p13
11. DEVERALL C S (1972) Communications and control. *Municipal Building Management* 1 pp6-8
12. CURTIS J (1979) Communicating for results.

MODERN SITE CONTROL METHODS SUITABLE FOR THE LARGER PROJECT

by R Oxley, MSc, MCIOB, MBIM and J Poskitt, MCIOB, MBIM

INTRODUCTION

When considering control it is imporant to regard it as part of the overall process of management and, in particular, to appreciate its relationship with planning. Planning and control are interdependent and continuous processes. It is possible to have a plan without control but in this case it will lose most of its value. On the other hand it is impossible to have control without a plan. The stages of control can be set down as follows:

(a) providing a yardstick and the aim to be achieved, ie, setting the objectives, plans, budgets, standards etc;

(b) communicating the aim to those who have to implement it. Often insufficient thought is given in the industry to this stage. Site managers, foremen, sub-contractors etc cannot be expected to achieve objectives of which they are not aware;

(c) monitoring progress with the aid of charts, tables, quantities etc;

(d) comparing progress with the ultimate aim, isolating excessive discrepancies and identifying causes;

(e) taking appropriate action in the light of this information.

Control at site level is concerned with ensuring that project budgets, costs, programmes, supplies and sub-contractors achieve the set targets and that quality requirements meet the standards laid down in the contract document.

For control to be effective the time from initial action to feedback must be as short as possible to allow remedial action to be taken.

Control information should be presented in such a manner that only achievement outside standard tolerances is brought to the attention of managers, ie, the exception principle should be used. Another important point often overlooked is that reports should be condensed to meet the requirements of the particular level of management to whom they are directed.

The control of any project is similar in principle, irrespective of its size. It is control methods and systems which vary with the size of the project. The techniques used for control should be as simple as possible and should present information in a clear and concise manner.

CONTROL OF PROGRESS FOR REPETITIVE CONSTRUCTION

For large scale (defined throughout this paper as in excess of £1,500,000) local authority housing work in traditional, rationalised traditional and industrialised forms of construction, high outputs are required and control techniques have evolved from those used in the manufacturing industries. The methods employed include flow line planning, line of balance and the use of decision rules.

Modern site control methods suitable for the larger project

Flow line and line of balance methods have similar features in that they have very good visual impact and progress can be seen at a glance. Items requiring attention stand out immediately when things go wrong — another advantage of graphical methods. In addition, they indicate trends making it easier to predict the effect of one activity on another. Material deliveries, sub-contractor requirements, plant requirements and information requirements can be ascertained easily from the information provided. For this example the basis of these methods in terms of objectives is the number of dwellings required per week.

Flow line

The single flow lines in Figure 1 represent the programme for each operation. Actual progress can be marked on this chart and early warning is given of differences betwen programme and progress. Remedial action can then be taken. The example shows part of a flow line programme for 120 houses. The completion of several stages of work is shown. Progress has been marked on to the chart up to week 12.

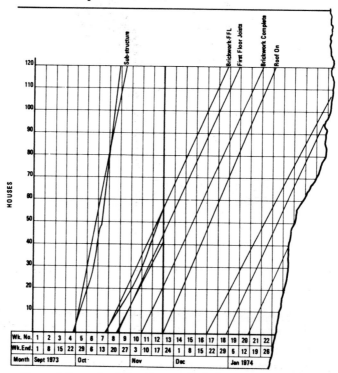

Figure 1 Example of flow line method of programming

Analysis of the progress together with suggested remedial action is as follows:

(a) sub-structure now complete;

(b) brickwork FFL. Started slowly but production is now above that programmed and, therefore, no action is necessary;

(c) first floor joists. Gradually falling behind programme and will hold up progress if no action is taken; suggest extra gangs introduced;

(d) brickwork to completion. Started but production is now much higher than planned. This makes the situation with first floor joists more critical and action on this operation of prime importance.

This method is very useful for head office overall control.

Line of balance

This is similar to flow line but as can be seen from Figure 2, each activity is defined by two parallel lines. The space between these lines are 'buffers' to allow flexibility in the programme.

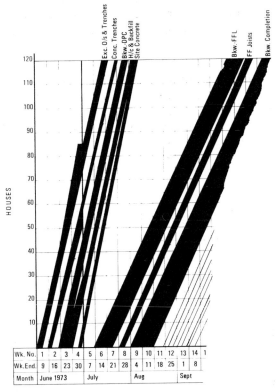

Figure 2

The example given shows part of a line of balance programme for 120 houses required to be produced at a rate of 10 houses per week.

One method of showing progress is to plot the actual start and finish of each operation on each block. This information is best obtained from job cards where this system is used as a basis for bonusing.

Progress has been marked on to the chart up to week 4.

Analysis of progress with suggested remedial action is as follows:

(a) excavate over site and excavate trenches. As site conditions are good allow this operation to proceed at present rate;

(b) concrete trenches. Still falling behind and now four blocks behind programme. Not yet critical, due to following operation also being behind programme, but suggest an extra gang be brought on to pull it back to programme;

(c) brickwork to DPC. Still falling behind and now eight blocks behind programme. Bricklayers in short supply therefore, no extra manpower available. Adjust programme to take account of present situation;

(d) hardcore and backfill. Proceeding on programme but take one gang off this operation to bring production in line with brickwork to DPC;

(e) site concrete. Proceeding on programme. Acceptable for time being.

Schedules based on line of balance quantities

Schedules can be produced based on the line of balance programme and these can be presented in the form of a bar chart consisting of planning cumulative weekly outputs (see Figure 3).

Operation	Month	April 1973				May					June				July				August				
	Wk. Endg.	5	12	19	26	3	10	17	24	31	7	14	21	28	5	12	19	26	2	9	16	23	30
	Wk. No.	1	2	3	4	5	6	7	8	9	10	11	12	13	14	15	16	17	18	19	20	21	22
Exc. over site & trenches		10	30	50	70	90	110	120															
Conc. trenches			15	35	55	75	95	115	120														
Bwk. to D.P.C.				13	33	53	73	93	103	120													
H/c fill & Backfill					21	41	61	81	101	120													
Site Concrete					5	25	45	65	95	105	120												
Bwk. to FFL							5	15	25	35	45	55	65	75	85	95	105	115	120				
FF Joists & Car Roof								5	15	25	35	45	55	65	75	85	95	105	105				
Bwk. Completion									5	15	25	35	45	55	65	75	85	95					
Roof Carcass										10	20	30	40	50	60	70	80	90					
Roof Covering											3	13	23	33	43	53	63	73	83				
Flash & Glaze												8	18	28	38	48	58	68	78				
etc.																							

Figure 3 Schedule based on the line of balance programme

Progress is recorded by shading the bar line representing units as they are completed for each activity. As with traditional bar charts a curser is used. The progress position can be seen at a glance. A time line is used to show the start and end of the operations. Progress has been recorded up to the end of week 12. Analysis of progress with suggested remedial action is as follows:

(a) brickwork to FFL. On programme;
(b) first floor joists are ahead of programme by five blocks and are catching up on brickwork to FFL. Reduce gangs on this operation and transfer to roof carcass;
(c) brickwork to completion. In front by three blocks. No action required at present;
(d) roof carcass. Well behind due to non-delivery of gang nailed trusses. These have now been delivered and the gang from first floor joists will assist in bringing progress up to programme;
(e) roof covering. Not yet started — sub-contractors have agreed to start on Monday next. Delay due to roof carcass operation being behind programme.

CONTROL OF 'ONE-OFF' PROJECTS

Many companies do not underake formal planning and control procedures beyond those carried out at head office. On large, complex projects effective control becomes difficult and often impossible unless planning is done on site. Site planning must be carried out in sufficient detail to incorporate changes and to detect deviations as soon as possible.

Control of progress

The aim here is to keep progress up to the original programme. In order to control progress on large projects short-term planning is essential. This will entail the production of programmes, method statements and other planning information. The breakdown will be in much more detail than that used for the overall programme, thus allowing finer control to be exercised. On large projects this planning would be carried out by a planning engineer under the direction of the site manager. The planning engineer can often offer useful assistance in solving problems.

The short-term programmes used, are in the form of 'stage programmes' covering a stage of the work and/or 'period programmes' to cover a specific period of time. These stage and period programmes incorporate changes caused by variation orders and architects instructions etc and should, when necessary, provide for making up lost time. They should normally be based on operations by individual trades and sometimes groups of operatives. Control of progress on site is based on these short-term programmes and provision should be made to show time and quantity of work done.

Period programmes normally cover a period of from four to six weeks and are up-dated every fourth week, thus providing two weeks overlap. Probably the best time to draw up a programme is after the architect's monthly meeting, thus making it possible to incorporate decisions taken at the meeting.

Provision is often made on these period programmes for controlling plant, materials, subcontractors and information requirements and these will be considered under separate headings.

Short-term progress

Part of a period plan for an office block is shown in Figure 4. The work content shown should be based on expected output and on the same basic information as that used for the incentive scheme. Sub-contractor time periods should be compiled in collaboration with the subcontractor.

The programme shows progress up to the end of week 47.

The bar lines have been filled in to show the proportion completed of each operation and a separate time line is shown. Analysis of the progress on the part of the chart shown is as follows:

(a) electrical carcass, heating carcass and suspended ceilings are completed for blocks A and B;
(b) carpenter first fix Block A is complete. This is one day ahead of programme.

It can be seen that unless carpentry first fix progresses faster, a delay in plastering ceilings is inevitable if this progresses at the present rate. Plastering to walls is also one day behind programme due to a late start by the gang of plastereres.

In the light of this information appropriate action can be taken.

Head office operates overall control which will be based on information received from site. For large projects critical path analysis is often used and it is sometimes considered sufficient for a report of 'key events' to be sent to head office.

Modern site control methods suitable for the larger project

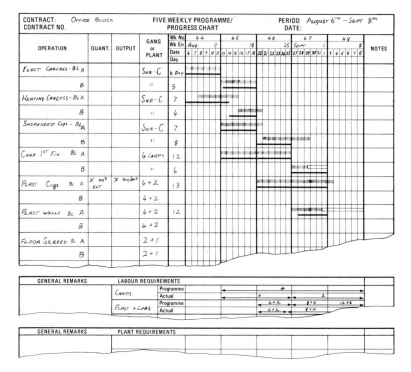

Figure 4 Short-term period programme

Key events are selected prior to the start of the project and are usually on the critical path. The aim of a 'key event' report is to indicate, very briefly, the overall progress of the project. If this is satisfactory no further information is necessary. If variance from programme is excessive, further more detailed reports may be necessary to enable management to take effective action.

Weekly progress

Weekly progress control is based on weekly planning and is concerned with communicating programmes etc to the trade foremen, gangers etc. These can be prepared as detailed bar charts, diagrams, operational instructions, detailed method statement or job card etc. They should tie in with short-term and period programmes.

Control of labour

The labour requirements will have been determined at the overall planning stage when drawing up the programme. If progress is to be maintained it is necessary to ensure that the labour strength is up to requirements. When stage and period programmes are drawn up labour requirements should be summarised as shown in Figure 4. This provides a basis for control. Labour should be recruited in advance and a record kept of actual manpower. If this varies significantly from requirements, remedial action can be carried out by making necessary adjustments.

Figure 4 shows the labour requirements for the contractor's own men for the office block. The actual labour force is shown and from this it can be seen that the contractor is two carpenters under-strength. As no further carpenters appear to be available, despite a recruitment drive, the programme will have to be adjusted and information given to all those affected. In the meantime a concentrated effort must be made to increase the number of carpenters.

Control of materials

(a) Material losses

Materials such as sanitary fittings and door furniture should be kept locked up and be under the control of a material checker or storekeeper. A strict check should be kept on all allocations. Some materials do not get to the site in full loads, examples being concrete, loose bricks, blocks, sand etc. Obviously, regular checking of these materials is out of the question as this would cost more than it saves, on the other hand random spot checks can be very effective and may make potential thieves think twice.

Many items are lost and damaged due to chaotic site conditions. A well thought out site layout with materials and components in pre-determined storage areas wiill reduce this type of waste considerably.

(b) Material deliveries

Schedules will probably have been drawn up for major materials at the overall planning stage. When short-term planning is being carried out, a check should be made on materials to ensure hold-ups due to non-arrival are kept to an absolute minimum. Special attention should be paid to materials required for criitical operations. Delivery times shown on schedules will allow some flexibility, providing storage space is available on site. If materials do not arrive as scheduled, immediate action should be taken.

A record of goods received should be sent to head office weekly to enable a check to be made with orders. This helps prevent over-ordering. On very large sites this is sometimes done on site.

(c) Quality control

Quality control is not within the scope of this paper.

Control of sub-contractors

Information necessary to control the arrival and progress of sub-contractors is initiated at the overall planning stage and it is of great assistance if the sub-contractors are consulted when the overall programme is being drawn up. If sub-contractors have agreed to the programme they are much more likely to co-operate on site. Too many planners are inclined to make demands that may prove unnecessary. This is bad for site relations.

The sub-contractor schedules drawn up earlier should be checked at the short-term planning stage and sub-contractors kept up-to-date with current developments. They should be contacted at least two weeks before they are due on site and their representative should attend planning meetings. Sub-contract programmes can be drawn up based on short-term programmes and handed to the sub-contractors. Obviously the sub-contractors should be involved when drawing up these progammes and should have agreed the time required and call up period.

Figure 5 illustrates part of a programme for plasterers for a college building. Progress is shown up to the end of week 56. The programme can be used as a basis for controlling progress of the sub-contractors and can be discussed at site meetings as a basis for getting the work back on programme.

CONTRACT: COLLEGE BUILDING CONTRACT NO.			SUB-CONTRACTOR: PROGRAMME					TRADE: PLASTERER DATE		
OPERATION	GANG	TIME REQ'D (WKS)	Wk ends SEP 8	15	22	29 OCT	6	13
			Wk No. 54	55	56	57	58	59
PLASTER CEILINGS – GROUND FLOOR BLOCK A	4+2	2	▓▓▓▓					
GROUND FLOOR BLOCK B	4+2	2			▓▓▓			
FIRST FLOOR BLOCK A	4+2	2			▓▓▓			
FIRST FLOOR BLOCK B	4+2	2						
SECOND FLOOR BLOCK A	4+2	2						
SECOND FLOOR BLOCK B	4+2	2						
PLASTER WALLS – FIRST FLOOR BLOCK A	4+2	2				▓▓▓		
GROUND FLOOR BLOCK B	4+2	2						
ETC.								

COMMENTS	SUMMARY OF LABOUR							
	PLASTERERS + LABS	Prog.		4+2	8+4	12+6		
		Act.	4+2	4+2	8+4	12+6		
		Prog.						
		Act.						

Figure 5 Programme for plasterer sub-contractors

Control of information requirements

Schedules should be drawn up at the overall planning stage. They should be checked with the short-term requirements and any other information requirements added. Contacts should be made with architects etc for critical information to ensure it will be available when required. The dates for arrival of information should allow for some flexibility but if information does not arrive on time, chase-up procedures should start immediately.

Control of plant and equipment

The schedule drawn up at the overall planning stage should be checked with stage programme requirements, and plant and equipment required in the period ahead chased up. Unlike materials, plant and equipment cannot be brought on site early to allow flexibility and it is, therefore, vital that checks are made on these items in good time. Contact should be made with the plant department, prior to planned arrival date, to ensure that there will be no delay.

It is also important from the point of cost to see that plant is returned immediately after use. A check should be made monthly and records kept of plant and equipment returned or transferred. Plant breakdowns should also be recorded as a check on plant efficiency and to ensure charges made to the project are accurate.

Cost control

This can be applied in varying degree of detail. A monthly, or perhaps a weekly, summary of the financial position of the project gives a good guide to overall profitability. It does not, however, indicate where gains or losses are occurring. To give this information operational cost control is necessary; it is not normally confined to plant and labour.

Basically, allocation sheets must be completed and checked, giving the hours spent by plant and operatives on each operation. These hours are summarised weekly and the information transferred to a weekly cost sheet which gives actual unit costs and outputs. These outputs can be compared with 'standard costs' or costs from head office which will be based on past records. Appropriate action can be taken where excessive variance occurs.

It is important to realise that the unit cost should not be compared with the labour and plant elements in the bill of quantities (BOQ) rates. Most of these rates are 'average' for the whole job and the rates on the cost sheets relate to a specific situation under specific conditions. Ultimately some comparison should be made with the BOQ rates by accumulating unit costs which make up a BOQ item. This can be used as a guide to future estimating.

USE OF MEETINGS

Communication of information is a very important factor in effective control. Co-ordination of sub-contractors, consultants, suppliers and the contractor's own work force is essential for successful project management and the project manager must accept total responsibility for overall management. Site meetings provide an excellent means of communication and provide an opportunity for effective co-ordination, but they must be properly organised.

Stage and period programmes should be available for discussion on progress, resource requirements etc. One week before a meeting, sub-contractors should fill in a standard form giving information required for discussion at the meeting.

For monthly meetings a chairman should be appointed and a record kept of all decisions. An agenda should be issued prior to the meeting and minutes circulated afterwards. Each decision reached at the meeting should be itemised. Each person and the time allowed for implementing the decision should be recorded in the minutes. The person made responsible for a particular action should report back to the next meeting.

Weekly site meetings are perhaps better if made less formal. These should ilnvolve trades foremen, gangers, sub-contractors foremen etc. Consideration should be given to the previous week's work and action taken to correct any deviations from the plan if possible. The following week's work should be considered and programme decisions taken. Separate programmes will be required for each trade.

CONCLUSIONS

As projects become bigger and more complex formal planning and control procedures become essential tools of management. This paper is meant to give an appreciation of the kinds of control necessary on larger projects.

USE OF OPERATIONS RESEARCH ON SITE

by A T Baxendale BSc(Hons), MPhil, MCIOB

INTRODUCTION

Operations research may be defined as the *design of models for complex problems concerning the optimum allocation of available resources, and the application of mathematical methods to the solution of those problems*[1]. In this paper the possibility of applying operations research methods on site is investigated. The need is also established for data on construction operation times to be obtained if these methods are to be employed. Activity sampling is then examined as a means of collecting distribution time data and an example given of its use. This is followed by a review of the modelling of construction operations by Monte Carlo simulation which has been found to be applicable to many site situations.

Simulation has been defined as the representation of features of behaviour of a physical or abstract system by the behaviour of another system and the Monte Carlo method as a method of obtaining an approximate solution to a numeric problem by the use of random numbers.

The general approach to simulation is considered together with formulation of operation durations and an example is given of the application of simulation to the re-planning of site operations.

The aim of the paper is to encourage those managers in the construction industry, who have to plan the use of resources, to recognise that operations research is a useful tool and has role to play on site.

THE NEED FOR OPERATIONS RESEARCH (OR)

Operations research is not widely used in the building industry and it was with the aim of making greater use of the technique that a study[2] was carried out to establish the shape and degree of resource planning problems experienced by site managers. As a first step it was important to find out if problem areas capable of being solved by OR were recognised. To this end a questionnaire (see Figure 1) was distributed to 37 individual managers who had

Table 1 Recognition of questionnaire simulations

Question no.	OR technique	Recognition	Validated recognition
1	Linear programming	85%	41%
6	Queueing	76%	41%
5	Scheduling	68%	27%
2	Transportation	59%	35%
4	Routing	54%	19%
7	Sequencing	54%	27%
3	Assignment	38%	19%
8	Decision analysis	38%	8%

agreed to respond. The questions asked did not ask the respondents to identify OR techniques but rather the situations which would lend themselves to such techniques. The results are shown in Table 1.

A proportion of the site managers was also interviewed and this allowed some explanation of the terminology. Where the site manager recognised any of the situations described in the questionnaire a brief description was sought and from this a judgement was made as to the validity of the application of the technique. Where validated recognition was evaluated the ranking order changed and the level of recognition dropped significantly (see Table 1).

Please tick the box for YES or NO, if you recognise any of the following situations and give a brief description over the page where this is so

1 Men, machines or materials have to be allocated to a number of tasks.

  1.1 Work is processed through various stages, requiring the throughput to be regulated so as to make the process as efficient as possible.

      YES ☐   NO ☐

  1.2 Selection is made from alternative forms of labour, types of machine or materials in order to make production as efficient as possible.

      YES ☐   NO ☐

2 Resources are required to be transported or distributed from a number of sources to a number of destinations so as to minimise cost.

      YES ☐   NO ☐

3 Resources are assigned to tasks, where the number of men or machines is equal to the number of tasks and only one gang or machine is available and required for each task.

      YES ☐   NO ☐

4 A path is identified to minimise travel time, cost or distance.

  4.1 A shortest path has to be selected for travelling between two places where there are a number of alternative routes.

      YES ☐   NO ☐

  4.1 The most satisfactory circuit to follow in making a number of visits has to be chosen, where there are a number of alternative routes.

      YES ☐   NO ☐

5 There is a fixed sequence of tasks and times can be given to each task so as to decide on the most economical method of performing the work.

      YES ☐   NO ☐

6 Tasks being carried out depend upon the arrival of men, machines or materials which may experience delay or which may arrive together causing some to wait.

      YES ☐   NO ☐

7 Tasks have to be carried out in sequence to minimise the total time.

  7.1 A number of tasks have to be processed by a number of men or machines, the sequence of work of the mean or machines being fixed.

      YES ☐   NO ☐

  7.2 A number of tasks have to be processed by a number of men or machines, the sequence of the tasks being fixed.

      YES ☐   NO ☐

8 A decision has to be made to choose just one set of actions, each action leading to just one of a set of outcomes, there being a payoff associated with each outcome. A rule is devised to assist in the decision making process.

      YES ☐   NO ☐

Figure 1: Questionnaire to site managers on the application of OR.

Use of operations research on site

It could be concluded, therefore, that there is a limited number of situations, recognisable by site managers in which OR techniques may be applicable. These are those shown in Figure 2.

OR Method	Examples of application
1. Linear programming	Choice of sources for materials. Choice of production method.
2. Transportation	Material distribution on and off site. Ready-mixed concrete deliveries.
3. Assignment	Allocation of labour.
4. Routing	Delivery of materials.
5. Scheduling	Planning construction activities.
6. Queueing (Monte Carlo Simulation	Organising excavation plant and disposal of material. Site precasting and erection. Placing concrete. Fixing formwork. Fixing preformed cladding panels.
7. Sequencing	Repetitive production. Allocation of construction plant.
8. Decision Analysis	Selection of construction method. Choice of materials.

Figure 2 Application of OR methods recognised by site managers

It is important to recognise where a standard OR technique may be of use, before looking at the situation to which it is to be applied in more detail. Fixed or deterministic times or costs are usually available for use with an OR method such as transportation, but there may be a problem in obtaining an adequate data base for use with queueing and hence simulation.

DATA COLLECTION BY ACTIVITY SAMPLING
Activity sampling

The Building Research Establishment (BRE) has developed methods of measuring productivity in construction work, designed to provide a basis for estimating the effect of various factors on what is produced and the resources used[3]. Significant factors identified include the scale of work available, the complexity of the building, the employment rate in the region, the time of year and the type of contractor. However, there is too great a variation in these factors to allow reliable estimates of average productivity, whether the study in detailed involving continuous observation, or at a broad operation level using activity sampling methods. Activity sampling on building sites can give estimates for:

(a) labour expenditure for each operation, observations being converted into man-hours depending on the frequency of observations;

(b) who carried out the operation;

(c) how the time was expended and what proportion of time was expended on each activity.

The relationship between the degree of accuracy of activity sampling study and the number of observations is:

$$A = \sqrt{\frac{2\,P\,(1-P)}{N}}$$

where N = total number of observations
  A = degree of accuracy (expressed in decimal form)
  P = percentage occurrence of activity (expressed in decimal form)

The large number of snap observations are made to obtain a desirable degree of accuracy, the BRE requiring an accuracy of ±5% in their studies[4].

The general use of activity sampling is to judge the effectiveness of the utilisation of resources. It is fast and directs attention to idleness and poor productivity. Areas of work not meeting an acceptable standard can be identified at the time the conditions exist, rather than later, as for example in a cost report. Activity sampling methods can vary in format and can be classified[5] as follows:

(a) field-rating (headcounts or activity ratings). All site personnel are covered and classified as 'working' or 'not working';

(b) work sampling (BRE activity sampling);
   A large sample is taken N = $\frac{4P\,(1-P)}{A^2}$
   (from previous formula)

(c) five-minute rating, observations made every five minutes.

SITE DATA COLLECTION

Activity sampling can also be used for collecting data on site operations for use with an OR method. The taking of observations every five minutes can be used to collect initial data from sites, although with a large number of operatives, observations can only be made say, every ten minutes. It is only desirable to make random observations if operations are cyclic over a short period of time. By taking observations at small time intervals an attempt can also be made to obtain a number of observations sufficient for a reasonable degree of accuracy.

An extract from an activity sampling study on the fixing of timber framed housing panels is shown in Figure 3 and the distribution of times extracted from such studies in Figure 4.

CONCLUSIONS

Data will show what has taken place on site as a result of organisation, resource usage, weather conditions and the like. As such it may assist in the planning of resources for another similar contract. However, investigations by the BRE have shown that a confused pattern emerges. The conclusion to a study carried out by McLeish[6] into house building productivity states that considerable variations in man-hours for related operations were discovered, that the number of visits to complete operations were large and that the numbers of different operatives employed in some operations were a factor related to productivity.

Data collected must be of direct assistance to the site manager in producing models for resource planning. Consequently, it is recommended that the five minute rating be used on site where no other information is available and distribution times for carrying out construction operations are required. These distribution times can then be used for queueing or simulation models.

SIMULATING REPLANNING OPERATIONS ON SITE
Monte Carlo Simulation

Queueing theory has been used to provide general solutions for simple situations, but there are few such applications in construction. A better approach is to introduce a dynamic element in a Monte Carlo simulation. The application of a Monte Carlo simulation to construction problems which involve queues of all types is described by Pilcher[7]. A dynamic model of a situation is produced from direct observations or from a similar situation, in order to predict the likely operation of the subject being modelled over a period of time. Analyses of historic data can be translated into predictions of future behaviour, permitting the comparison of alternative courses of action in sufficient time to alter strategic as well as tactical decisions.

The general approach is to simulate a real situation from random samples of activity, relating occurrences or intervals to a distribution curve. An example of a distribution curve from site observations is given in Figure 5. This approach contrasts with the average times and average rates and costs. The times required by the processes that make up construction operations are not constant, they are scattered about a mean value and the probability that they will differ from that mean value by some amount can often be estimated. The factors that influence construction operation times, once the operational technology has been decided, are process design, work content, equipment, labour, working conditions and efficiency of management. The influence of these factors on construction operations also varies between construction sites.

DETERMINISTIC OR PROBABLISTIC DURATIONS

If the outcome of a situation is certain, having selected or knowing all the variables involved, it is known as deterministic. A deterministic duration assigns a specific fixed value to the duration of the operation. Construction operations taking place in a controlled environment may, therefore, use a single value of time duration when the variability is considered small or insignificant. A deterministic analysis of productivity may also be undertaken because it is sufficiently accurate for the purpose of the manager, but the loss of productivity due to the random variation of work cycle durations may be significant enough to justify consideration of probabilistic durations.

Where the value of some of the variables is not known, or an allowance has to be made for the varying probability of certain values, the situation is known as probablistic or stochastic. Where stochastic models are used it is necessary to select values of the variable which are representative of the data available. Then, providing there is some measure of the frequency of occurrence of certain values, random sampling may be used to postulate the likelihood of an event. The use of random sampling ensures that every item in the population has an appropriate chance of being selected. Random numbers can be obtained from statistical tables[8] or can be generated by calculator/computer.

DISTRIBUTION TIMES

Random durations in the movement or flow of resources can delay work cycles and operations by increasing the time that the resources spend idle awaiting release to productive work tasks. Probability distributions are useful in describing observed variations of work task times on site, since they are easily defined mathematically by relatively few parameters. Probablistic functions are then used to define the populations from which random time durations can be taken for the simulation of a system. Random variates can then be generated by entering the cumulative probability function. Table 2 shows an example of the build-up of cumulative distribution from site observations.

Contract: 160 low-rise dwellings		Reference: Timber frame/ATB/79	
Operatives/Plant: A. Carpenter B. Carpenter		Notes: Block of 7,2 bedroom houses	
Time	Operation	Time	Operation
11.30	A. Fix window surround B. Idle	12.20	A.) Fix PW6⅙ B.)
11.35	A.) Fixed unit E8 (third) B.)	12.25	A.) Moving units B.)
11.40	A.) Idle B.) (Unit PW4 2/2 fixed)	12.30	A.) Moving units B.)
11.45	A.) Fix window B.) into external unit	12.35	A.) Fix E9 (second) B.)
11.50	A. Fix unit PW4½ B. Idle (Unit E8 (fourth) fixed)	12.40	A.) Fix E9 B.) Units PW6½ and 2/2 fixed)
11.55	A.) Idle B.)	12.45	A.) Fix window to B.) external unit
12.00	A.) Idle B.)	12.50	A.) Fix PW6½ B.) (Unit E9 (third) fixed)
12.05	A.) Idle B.)	12.55	A.) Idle B.) (Unit PW6½ fixed)
12.10	A.) Moving units B.) (Unit PW4½ fixed)	13.00	A.) Lunch B.)
12.15	A.) Moving units B.)		

Figure 3 Activity sampling study extract. Fixing timber frame panels for housing

Use of operations research on site

Random variates can also be generated from some common types of theoretical continuous probability distributions, such as normal and lognormal. Use of such standard distributions has the advantage that they can be uniquely defined and this also does away with the requirement of specifying the cumulative function interval by interval. A check will have to be made that the distribution of the data collected can be assumed to follow a theoretical distribution. The validity of any simulation is a function or result of how much is known of the statistical distribution of the actual construction operation times. Only adequate sampling or records can establish a true distribution.

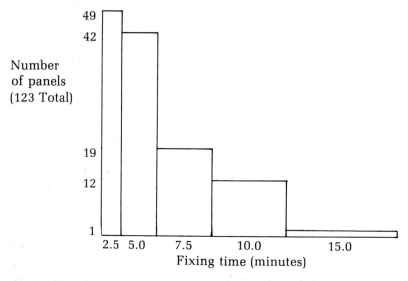

Figure 4 Distribution times. Fixing timber frame panels for housing

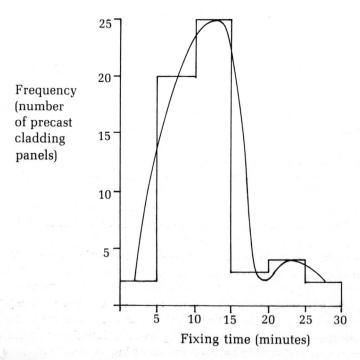

Figure 5 Distribution curve of fixing times for precast cladding panels

Table 2 Site observations of fixing times for precast cladding panels

Interval (minutes)	Number of observations	Relative frequency	Probability density	Cumulative numbers of observations	Cumulative distribution
0 - 4	2	0.04	0.04	2	0.04
5 - 9	20	0.36	0.36	22	0.40
10 - 14	25	0.45	0.45	47	0.85
15 -19	3	0.05	0.05	50	0.90
20 - 24	4	0.06	0.06	54	0.96
25 - 29	2	0.04	0.04	56	1.00
Total	56	1.00			

SITE APPLICATION

The re-planning of operations on site by Monte Carlo simulation has been applied to calculating the time of concrete pours by pump and fixing precast cladding panels[2]. Based on data collected from the first concrete pour, as shown in Table 3, simulation was used to plan the timing of future pours in relation to the volume of concrete that could be placed. Table 4 shows the method of simulating a 100m^3 pour within chosen parameters to give an estimated time.

Alternative resource allocation, for example a varying number of ready-mix trucks, can also be simulated. The data collected from the fixing of precast panels to the first half of a building, as shown in Table 2, were also used in a simulation to re-plan the fixing for the second half of the building. For control purposes comparisons could be made between the original plan, the simulation and the actual carrying out of the work.

FORECASTING PRODUCTIVITY

A contractor has to forecast the results that he would hope to obtain in practice and average historic production feedback is not always an adequate guide to the future performance.

Pilcher[9] said that one of the most important problems in construction work was that of balancing resources, there being two types of resource:

(a) cyclical, usually associated with plant;
(b) that with a random or one-off pattern of behaviour as typified by labour processes.

Construction processes should be measured and the ranges of productivity recorded rather than average readings. The total time for a work cycle can then be found by random sampling from frequency distributions of each part of the cycle. By running such a model many times practice can be realistically simulated. The basic considerations that are required before simulation becomes practical are that the data are adequate and correct, and that the situation is sufficiently complex to make the exercise worthwhile. Simulation should only be regarded as a technique to be used alongside others, such as network analysis, in planning, progressing and resource allocation. The point is also made that for better understanding it is desirable initially or carry out simulation by hand, although simulation could best help in a complex situation and should be used for planning resources at the estimating stage rather than as a site planning method.

Use of operations research on site

Table 3 Frequency distribution data for concreting waffle floor slab by pump

INTER-ARRIVAL TIME (minutes) Two ready-mix concrete trucks		SERVICE TIME (minutes) Concrete pump on trucks	
Midpoint of range	Cumulative percentage frequency	Midpoint of range	Cumulative percentage frequency
3	6	3	2
9	19	5	32
15	54	7	52
21	67	11	71
27	76	13	80
33	82	15	86
39	91	17	91
45	97	19	94
51	100	21	99
			100

Table 4 Simulations for concreting waffle floor slab by pump

Simulation number	Random number	Inter-arrival time	Time of arrival	Random number	Service time	Time at end of service
1	93	45	45	72	11	56
2	81	33	88	02	3	91
3	79	33	121	66	9	130
4	25	15	136	28	5	141
5	64	21	157	58	9	166
6	96	45	202	96	19	221
7	66	21	223	22	5	228
8	02	3	226	32	5	231
9	30	15	241	28	5	246
10	59	21	262	73	11	273
12	24	15	292	80	11	303
13	32	15	307	97	19	326
14	14	9	316	08	5	331
15	53	15	331	32	5	336
16	97	45	376	85	13	389
17	76	27	403	70	9	412
18	64	21	424	72	11	435
19	98	51	475	03	5	480
20	03	3	478	39	7	487

Duff[10] has developed a method of calculation that allows the inclusion of variability in the estimating of activity durations based upon statistical analysis of site feedback. But once such a model, based on existing data, is in use it will need continuous refinement by the collection of more reliable and appropriate data from site. He believes that the analysis of productivity by stochastic modelling should be reduced when a greater understanding of variability is attained. Thus, the area to which deterministic models can be applied will increase, making an improvement in the planning of the construction process. Factors influencing productivity must, therefore, be identified to see what variability or productivity is truly random and how much is a response to factors which can be accurately measured.

CALCULATIONS
As can be seen from the examples of simulation many repetitive calculations are necessary. The disadvantage is that the impact of using the simulation may be lost if time consuming and tedious arithmetic, together with possible errors, has to be carried out. A simulation program can be written in BASIC for use on a micro-computer, to enable simulations to be carried out many times upon the input data[2].

Twenty simulations are required for two ready-mix concrete trucks serving a concrete pump for a $100m^3$ pour and the results of such an exercise are shown in Table 4. The overall time by hand simulation was 487 minutes. A necessary requirement is that a number of simulations have to be carried out and an average time taken. Ten sets of twenty ($100m^3$) simulations were carried out by computer and gave the following times: 414, 428, 444, 600, 478, 412, 434, 456, 464, 472 minutes or an average of 457 minutes. Further simulations can then be made until the average has settled down, the data from this sample giving an average of 432 minutes after 100 simulations and 434 after 1000 simulations for a $100m^3$ pour.

CONCLUSIONS
Parker and Oglesby[11], among others recognised that the attempt to simulate real life situations is limited by the parameters set by the manager. Simulations can be reliable in comparing alternatives — such as the productivity of two proposed systems — but accuracy in predicting actual output or cost for a particular system may be limited by the accuracy of the input data and the length of simulation run. But as concluded by Bjornsson[12] simulation is a very flexible and useful tool which, if used appropriately, can generate a great deal of information. It is also a technique that, without violating the nature of reality too much, is easiest to understand and to adapt to actual situations.

It would appear from the recommended applications of simulation that it may well be best applied to augment the planning that takes place at the estimating or pre-contract stage for large and complex projects. If the construction manager wishes to use simulation on site it may well be in a limited form, within criteria previously set by the estimators and planners.

If data are gathered on a particular site, the frequency distribution so collected will be of particular significance to the unique situation that exists on that site. This aspect of using data collected on site for use in re-planning construction processes is the one shown by the simulation examples given earlier and is recommended to site managers.

REFERENCES
1. BRITISH STANDARDS INSTITUTION (1976) Glossary of terms used in data processing. BS 3527 : Part 1

2. BAXENDALE, A.T. (1982) Construction resource models : planning for physical resources, MPhil Thesis, Bristol Polytechnic
3. (1980) *BRE News* Spring/Summer
4. STEPHENS, A. J. (1969) Activity sampling on building sites. Building Research Station Current Paper, 16/69
5. AHUJA, H.N. (1976) Construction performance control by network. Wiley
6. McLEISH, D. C. A. (1978) House building productivity, PhD Thesis, University of Aston in Birmingham
7. PILCHER, R. (1976) Principles of construction management. McGraw-Hill
8. MURDOCH, J. and BARNES, J. A. (1970) Statistical tables. Macmillan. 2nd edition
9. PILCHER, R. (1977) Simulation as applied to project control, *Proceedings of Institution of Civil Engineers* Part 1, August
10. DUFF, A. R. (1980) A stochastic analysis of activity duration, *Construction Papers* 1 (1)
11. PARKER, H. W. and OGLESBY, C. H. (1972) Methods improvement for construction managers. McGraw-Hill
12. BJORNSSON, H. (1978) The numbers game: quantitative techniques in construction management. Proceedings of the CIB W-65 Second Symposium on organisation and management of construction. Israel

METHODS OF MEASURING PRODUCTION TIMES FOR CONSTRUCTION WORK

by A D F Price, BSc, PhD, MCIOB and F C Harris, BEng, MSc, PhD, CEng, MICE, MCIOB

INTRODUCTION

Work study techniques have gained wide acceptance in the manufacturing industry and are regularly used for estimating and planning, setting financial incentives and in determining levels of production efficiency.

The building industry has been less enthusiastic, seemingly because the changeable conditions and relatively short duration of projects produce excessive variability in output levels. Indeed, the time needed to gather the data has not seemed worth the effort. However, a study[1] into construction productivity has shown that much of the variability can be quickly explained to leave relatively constant levels of output for individual operations, (ie, standard times).

The primary factors in determining output rates were found to be:
- work rate;
- delays and waiting caused by poor management;
- excessive breaks and poor motivation.

The latter two items accounted for more than 50% of the available working time on many sites, whereas work rate varied only slightly. This last finding may be surprising, but the results indicated that when working the effort applied appeared fairly constant the time spent working was largely dependent upon the level of motivation induced through the payment system. Where a combination of good, direct supervision with satisfactory financial incentives was present high levels of motivation were observed. Conversely, low motivation occurred on sites where minimum daywork payments were in force.

AIMS OF THE STUDY

The principal objective of the study was to establish reliable methods of measuring output levels of plant and labour with a view to establishing standard minute values.

MEASURING OUTPUT

Following exhaustive trials, activity sampling combined with cumulative timing proved successful techniques in establishing:
- time standards for major construction operations;
- levels of site efficiency.

Example:

The example used to illustrate the recording methods involves a concrete gang of four men, a crane and skip pouring concrete into shuttering for two columns and two roof slabs forming a section of a framed building (Figure 1). The dimensions and quantities of work are given in Table 1.

This example will now be considered in three sections. The first illustrates the procedure used to collect the data on site; the second describes how this data is translated into standard

Methods of measuring production times for construction work

Figure 1

Table 1 Dimensions and quantities used in example

Roof slab 1 width = 2.0m area = 88.3m^2 Volume of concrete = 29.9 $\left.\vphantom{\begin{array}{c} \\ \\ \end{array}}\right\}$ 33.6m^3

Roof slab 2 width = 2.0m area = 10.6m^2 Volume of concrete = 3.7

Columns   (0.5m x 0.50m x 3.8m) x 2   Volume of concrete = 1.9 $\left.\vphantom{\begin{array}{c} \\ \\ \end{array}}\right\}$ 2.1m^3

Joints   (0.1m x 0.05m x 3.6m) x 12   Volume of concrete = 0.2

Average distance travelled by skip   Vertical   = 10.0m $\left.\vphantom{\begin{array}{c} \\ \\ \end{array}}\right\}$ Total = 38.5m

Horizontal = 37.0m

Capacity of concrete lorries = 5m^3
Capacity of skip   = 1m^3

times for the operations and the final section investigates how these standard times can be used in the synthetic build up of planning times for other operations.

Data collection

The data were collected on a specially adapted Activity Sampling Sheet (Figure 2) made up of the following sections:

(a) Activity sampling

Activity sampling was successful for recording groups of workers engaged in constantly changing work patterns. The procedure adopted required an observation to be taken of each operative every five minutes, with the activity and rate of work being recorded in the first nine columns. The list of codes used in this study are given in Table 2.

Code	Additional Activities	Code	Break points	Comments
31	Concrete joints in PCU	BP1	Skip arrvies at lorry area	Start roof slabs at 1.30pm
10T	Transfer tools	BP2	Skip leaves lorry area	and finish at 4.50pm
16c	Pour concrete to columns	BP3	Skip arrives at pour area	Start columns at 4.50pm
17c	Vibrate columns	BP4	Skip leaves pour area	and finish at 5.30pm
				Weather:- Wet/Cold/Calm
				Date:- 14 March 1983
				REC. Andrew Price

	Man 1		Man 2		Man 3		Man 4			Cumulative Timings (mins)				
TIME	Act	Rate	Act	Rate	Act	Rate	Act	Rate	Comments	PLANT: TOWER CRANE				
1.30	16	100	9	110	9	100	9	110	Delays	Note. Start @ 1.30 (= 0.00)				
35	9	90	18	100	5		9	90	Start slab 1	Skip No.	BP1	BP2	BP3	BP4
40	18	100	31	100	17	100	9	100	*Crane walks for conc.	1	(1.30h)*	1.09	9.30	9.75
45	18	100	31	110	5		19	100	gang 8.21	2	10.80	11.90	12.75	12.95
50	18	110	18	90	17	90	4		*1 lift for joiners 3.60	3	13.90	15.25	16.15	16.40
55	18	110	13	100	10T	100	17	100	mins	4	17.25*	21.80	22.40	22.80
2.00	10T	110	14	100	6	50	18	110	* wait for concrete	5	23.80	24.90	25.60	26.10
05	18	120	31	100	17	60	19	100	4.13 mins		26.60	27.60	Lift for joiners	
10	16	100	31	70	18	100	4			6	*(slab 2)		36.00	36.10
15	18	100	31	100	18	100	31	100	Cont. slab 1	7	34.00	34.90	35.80	36.02
20	20	100	5		5		5			8	37.20	38.00	38.70	38.90
25	20	110	31	100	6	100	5			9	39.90	40.60	41.60	41.75
2.30	20	100	31	100	5		18	100	*wait for concrete 17.9 mins		43.90*	Lift for bricklayers		
35	20	90	31	100	17	100	6	100		10	60.50	61.60	62.30	62.50
40	20	100	31	100	17	100	19	50		11 12	(2 loads to slab 2)			
45	20	120	31	110	16	110	19	100	Cont. slab 1	13	9.70	10.80	11.90	12.15
50	20	110	31	110	17	110	19	90		14	13.25	14.50	15.15	15.40
55	20	100	5		17	110	18	110		15	16.60	17.00	18.30	18.70
3.00	4		31	90	17	110	18	100	*wait for concrete 5.9 mins	16	19.80* (2 loads to slab 2)			
										17				
10	20	110	31	110	17	110	19	110	cont. slab 1 *wait for	18	32.25*	34.30	34.80	35.10
15	20	110	31	100	17	80	18	110	conc. gang 0.45 mins	19	36.30	37.40	37.80	37.90
20	20	100	4		17	80	19	100		20	39.20*	43.20	43.90	44.20
25	20	110	31	110	17	110	17	100	*wait for conc. lorries	21	45.10	46.20	47.00	47.20
30	20	110	31	110	5		18	100	to change	22	48.40	48.90	49.85	50.20
35	20	110	31	90	17	100	5		over 3.30 mins	23	51.30	52.30	52.90	53.40
40	20	110	4		17	110	6	90	*wait for	24	54.25	55.35	56.40	56.80
45	19	100	18	100	17	100	6	100	concrete 13.4 mins	25	57.80*	Lift for bricklayers		
50	19	100	18	100	17	100	19	100		25	11.20	12.60	13.20	13.95
55										26	15.00	16.50	18.60	18.95
										27	19.75	21.00	21.90	22.15

Figure 2 Activity Sampling Study Sheet

Methods of measuring production times for construction work

Figure 2: Continued

Code	Additional Activities	Code	Break points	Comments
16c	Pour concrete to columns	BP1	Skip arrives at lorry area	Pour dimensions L x W x D
17c	Vibrate columns	BP2	Skip leaves lorry area	
		BP3	Skip arrives at pour area	Slab 1 2.0mx44.2m x 0.34m
		BP4	Skip leaves pour area	Slab 2 2.0m x 5.3m x 0.35m
				Columns (0.5 x 0.5m x 3.8m) x 2
				Joints (0.1 x 0.05 x 3.6) x 12
				Skip 1m³

TIME	Man 1 Act	Man 1 Rate	Man 2 Act	Man 2 Rate	Man 3 Act	Man 3 Rate	Man 4 Act	Man 4 Rate	Comments	Cum No.	BP1	BP2	BP3	BP4
3.55	20	19	31	110	13	110	18	100			PLANT: TOWER CRANE			
4.00	20	100	31	110	17	100	1		*Wait for concrete 5.1 mins		BP1	BP2	BP3	BP4
05	20	100	6	8	16	100	1			28	23.25	24.25	25.40	25.70
10	18	100	6	8	17	110	1			29	27.30*	33.40	35.00	35.40
15	18	100	4	1	17	9	5			30	36.60	37.40	38.00	38.50
20	18	100	16	100	17	100	1			31	39.60	40.10	41.20	42.90
25x	20	100	5		5		5			32	43.25	44.10	44.90	45.60
30	5		5		5		5			33	47.05	48.05	48.60	49.50
35	20	110	5		5		5		*Wait for concrete 28.7 mins	34	50.15	51.25	51.80	52.90
40	20	110	5		5		5				53.50*	Lifts for bricklayers		
45	20	110	5		5		5			35	22.20	Not recorded		
50	5		5		5		5			36	25.80	Not recorded		
55	20	110	5		17	100	18	100		37	Cols			
5.00	20	100	19	100	17c	100	16c	110	At 5.00 pm 2 men concrete columns	38	Cols			
05	20	100	19	100	17c	100	16c	100		39	Joints			
10	20	110	19	110	17c	100	16c	100						
15	20	100	4		17c	100	10	100		Note — a blank space indicates a missing observation				
20	20	100	4		10T	110	4							
25	21	100	17c	100	18	100	21	100						
30	21	100	17c	100	18	100	21	100						

TIME	Man 5 Act	Man 5 Rate	Man 6 Act	Man 6 Rate	Man 7 Act	Man 7 Rate	Man 8 Act	Man 8 Rate	Comments
5.15	21	100	21	120	21	120	21	120	At 5.15pm 4 more men help to cover concrete
20	4		21	90	21	9	21	9	
25	21	100	21	100	21	110	21	120	
30	21	100	21	110	21	100	21	9	

146

(b) Cumulative timing

Cumulative timing was more appropriate for machine dominated operations because the activities were performed in a regular pattern. Thus, the work done by the crane being of a cyclic nature, cumulative times were simply taken at the start and finish of the loading and unloading of the skip. These results were entered in the last five columns.

(c) Comments

Any events which could not be recorded in the activity sampling or cumulative timing section were noted in the comments column, eg, delays caused by the late arrival of concrete.

Data analysis (concrete gang)

The Activity Sampling Summary Sheet (Figure 3) is used to establish the standard times for the study and is completed as follows:

(1) The activity sampling results are transferred into column C1 by summating the number of occasions an activity was observed being performed at each of the specified rates (for practical reasons the rating values are recorded in steps of 10 units). It can be seen that the activity 'pour' was observed on four occasions as being carried out at standard rate (100) and once at a rate of 110. In this example the activities relating to the slabs are kept separate from those relating to the columns. From these values the average rate (C2) can be obtained for each activity.

Table 2 Concrete gang activities and codes

Code	Activity
0	Break
1	Away
2	Walk
3	Talk
4	Recover
5	Wait
6	General
7	Other work
8	Internal delay
9	Prepare work area
10	Clear away
11	Carry
12	Search
13	Instructions/drawings
14	Redo work
15	Clean shutters
16	Pour concrete
17	Vibrate
18	Shovel
19	Tamp
20	Trowel
21	Cover
22	Jack hammer

(2) The overall time (C3) is the number of occasions each activity was observed multiplied by the observation interval (0.1). The basic time (C4) is the product of the overall time (C3) and average rate (C2) divided by 100.

Methods of measuring production times for construction work

The basic time per unit (C6) is equal to C4/C5, the units in this example are slabs and columns; in the following synthetic build up of times the units of work are measured in terms of area and volume of concrete.

(3) The total relaxation allowance (C11) includes basic (C7), standing (C8), posture (C9) and load (C10) and is calculated for each activity.

(4) The standard time (C12) is expressed per unit and is equal to the basic time (C6) plus the appropriate relaxation allowance (C6 x C11/100).

(5) Site factors are calculated as shown at the bottom of the sheet, and are applied to the basic times to obtain realistic planning times.

Build up of synthetic data (concrete gang)

In practice the shape and size of the concrete pour might vary from the example shown. Therefore, for general application of the data, it would be necessary to relate the basic element times to a set of variables. For example, a slab might have a large surface area and be shallow, whereas a foundation base or column usually exhibits the opposite characteristics. Clearly, area and volume would be important controlling variables. In order to establish the relationships between these variables and basic times, over eighty concrete pours have been combined and statistically analysed. However, for simplicity, only the data given in the Study Sheet (Figure 2) will be used to demonstrate the principles of synthesis.

Example of synthesis

The following example describes the synthesis of the element times for a concrete slab. The elements relating to the volume of concrete are pour and vibrate, and those relating to area are shovel, tamp, trowel and cover. The time spent on general work is allocated in proportion to the time spent on each of these six elements. The preparation, clearing and transferring of tools depends upon the number of occasions they are performed and are treated separately as one off items.

Basic element time (concrete gang)

By substituting into C5 of Figure 3 the units of work expressed in terms of the variable to which the element relates, basic element times shown in Table 3 are obtained.

The 25 minutes spent on excess work (C13, ie, the general activity code 6) has been added to elements (d) and (i) in the ratio of the time spent on each; this gives a basic time inclusive of excess work per activity. The basic time for a concrete pour can therefore be expressed as follows:

Basic time = 45 + 4.38 Vol. + 4.61 Area .... (1a)
(first pour)

Basic time = 5 + 4.38 Vol. + 4.61 Area .... (1b)
(subsequent pours)

The basic times given by equations (1a) and (1b) include trowelling and covering which are not always required; when these situations arise the following expressions can be used:

Basic time = 45 + 4.38 Vol. + 4.62 Area
in man minutes (trowel and cover)
= 45 + 4.38 Vol. + 3.57 Area
(trowel but not cover)
= 45 + 4.38 Vol. + 3.06 Area
(cover but not trowel)
= 45 + 4.38 Vol. + 2.02 Area
(neither trowel nor cover)

To illustrate how the slab depth can effect the basic time per m^3 the area in equation 1a can be removed to obtain equation 2 which is represented graphically in Figure 4.

$$\text{basic time} = 45 + 4.38V + 4.61 \frac{V}{D} \quad \dots \dots \dots \dots \dots \dots \dots \dots \dots \dots \dots \quad (2)$$

or basic time = 45 + 50.5V when D = 0.10m
= 45 + 22.8V when D = 0.25m
= 45 + 15.9V when D = 0.40m

$$V = \text{volume in } m^3$$
$$D = \text{depth in m}$$

Data analysis (tower crane)

As the work done by the tower crane is of a cyclic nature the work elements contained within each cycle can be defined as the time between the following break points:

Break points	Code (see Figure 2)
Skip arrives at lorry area	(BP1)
Skip leaves lorry area	(BP2)
Skip arrives at pour area	(BP3)
Skip leaves pour area	(BP4)

From the cumulative timing results for 34 skip loads given in Figure 2 the following average element times were obtained for the site transportation cycle:

Elements	Break points	Average time (min)
A Pour into skip	BP2-BP1	1.02
B Lift full skip	BP3-BP2	0.81
C Pour into shutters	BP4-BP3	0.42
D Return empty skip	BP1-BP4	1.02
		3.27 x 34
Total working time		= 111.2 min

In addition to the working cycle, the crane was subjected to delays caused by waiting for concrete lorries or diverted on to other work. These delays were recorded in the comments section as shown in Figure 2 and can be summarised as follows:

ACTIVITY	CODE	\(C1\) 50	60	70	80	90	100	110	120	130	C2	C3	C4	C5	C6	C7 Const.	C8 Stand.	C9 Post.	C10 Load.	C11	C12	C12/C5
General	6	1			2	1	2				83	30	25		25	9	2	0	2	13	28.2	
Prepare tools	9					3	2	3			100	40	40		40	9	2	0	2	13	45.2	
Clear tools	10					3	1				100	5	5		5	9	2	0	2	13	5.7	5.7
Pour	16						4	1			102	25	25.5		25.5	9	2	0	2	13	28.8	
Vibrate	17		1		2	2	7	11			100	115	115		115	9	2	0	5	16	134.0	758
Shovel	18					1	17	5	1		102	120	123		123	9	2	2	7	20	147.0	
Tamp	19	1				1	8	4			99	70	69		69	9	2	3	5	19	82.1	
Trowl	20					2	11	14	1		105	140	147		147	9	2	4	5	20	176.4	
Cover	21					4	9	2	4		103	95	98		98	9	2	0	2	13	110.7	
Transfer tools	10T						3				100	15	15	3	5	9	2	0	2	13	17.0	5.7
General Col	6						1				100	5	5		5	9	2	2	2	13	5.7	
PC unit joints	31			1		2	11	11			102	125	128	12	10.5	9	2	2	7	20	153.6	
Pour columns	16C						3				100	15	15	2	7.5	9	2	0	2	13	17.0	12.8
Vibrate cols	17C						6				100	30	30	2	15	9	2	0	5	16	34.8	8.5
Total (or Aver)											101	830	841	-	-	-	-		Av =	17.2%	986.00	17.4

Column notes:
- C1: Observations taken at each rate (Observation interval (OI) = 5 mins)
- C7–C10: Relaxation allowances (C7 Const., C8 Stand., C9 Post., C10 Load.)
- C11: TOTAL C7 + C8 + C9 + C10
- C12: C4 × (100 + c11)/100 STANDARD TIME (mins)

NOTE.

Wait = 33x5 = 155 mins
Recover = 10x5 = 50 mins
Idle time = 205 mins

Away = 4x5 = 20 mins
Instructions = 1x5 = 5 mins
Ancillary work = 25 mins

$$\text{Ancillary factor} = \frac{830 + 25}{830} = 1.03$$

$$\text{Site factor} = \frac{855 + 205}{855} = 1.24$$

Figure 3 Activity Sampling Summary Sheet

Table 3 Basic element times

Activity	Variable	C4 Basic time (mins)	C13 General (mins)	C4 + C13 B.T. + G. (mins)	C5' Quantity	C6 = C4'/C5' B.T./Unit	
(a) Prepare tools	Work area	40	—	—	1	40 min	
(b) Clear tools	Work area	5	—	—	1	5 min	
(c) Transfer tools	Transfer	15	—	—	3	5 min	
(d) Pour concrete	Volume	22.5	1.1	26.6	33.6m³	0.792	4.38 min/m³
(e) Vibrate concrete	Volume	115.5	5.0	120.5	33.6m³	3.586	
(f) Shovel concrete	Area	123.0	5.3	128.3	98.9m²	1.300	4.61 min/m²
(g) Tamp concrete	Area	69	3.0	72.0	98.9m²	0.728	
(h) Trowel concrete	Area	147	6.4	153.4	98.9m²	1.550	
(i) Cover concrete	Area	98	4.2	102.2	98.9m²	1.034	

(a) External delays caused by concrete lorries

Wait for concrete lorries	$= 0 + 4.13 + 17.9 + 5.91 +$
	$3.30 + 13.40 + 5.1 + 28.7$
Total	$= 78.4$ mins

$$\text{Average delay per lorry} = \frac{78.4}{8} = 9.8 \text{ min}$$

$$\text{Average delay per skip} = 9.8 \times \frac{\text{skip size}}{\text{lorry size}} \quad \dots\dots\dots\dots\dots\dots\dots (3)$$

(b) Internal delays caused by interference

Wait for concrete gang	$= 0.45 + 8.21 =$	8.7
Diverted on to other work		3.6 min
		12.3 min

$$\text{Interference internal delay allowance} \quad \frac{12.3}{78.4 + 111.2} = 7\%$$

Build up of synthetic data for tower crane

It is necessary to combine basic element times with delays for the tower crane to obtain synthetic data suitable for other pours. For example, it was found that for a given skip size the pouring time per unit of measure remained fairly constant over a series of pours, but its travelling time depended upon the distance between pick up point to discharge rate, thus:

Pour into 1m³ skip and pour slab $= 1.02 + 0.42 = 1.44$ min

Lift and return skip $= 0.81 + 1.02 = 1.83$ min

Methods of measuring production times for construction work

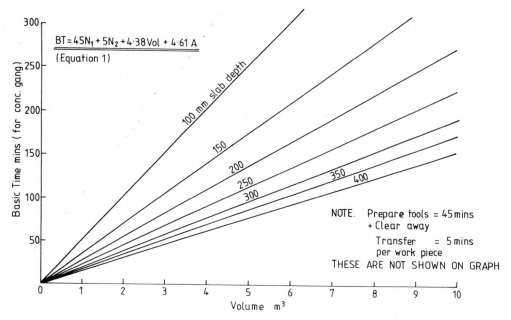

$$BT = 45N_1 + 5N_2 + 4\cdot38\,Vol + 4\cdot61\,A$$
(Equation 1)

NOTE. Prepare tools = 45 mins
+ Clear away

Transfer   = 5 mins
per work piece
THESE ARE NOT SHOWN ON GRAPH

Figure 4 Basic times (mins) for concrete gang against volume of concrete m³ for various slab depths

The average distance between the pour area and concrete lorries was 38.5m, thus the no delay cycle time can be expressed in general terms by equation 4.

$$\text{No delay cycle time} = 1.44 + 1.83 \times \frac{D}{38.5} \quad \text{when } D = \text{Distance in metres concrete is conveyed}$$

$$= 1.44 + \frac{D}{21.0} \quad\dots\dots\dots\dots\dots\dots\dots\dots\dots\dots\dots\dots\dots\dots\dots\dots\dots\dots (4)$$

The total basic crane time (TBCT) per cycle however, includes the time to place the concrete, plus delays caused by waiting for the lorries and having to give assistance to other gangs when concrete is available and is thus a restricted operation. Therefore, the basic time per cycle can be expressed as:

$$
\text{TBCT per cycle} \left(= \begin{array}{l} \text{Load and} \\ \text{unload skip} \end{array} + \begin{array}{l} \text{Travel time} \\ \text{for skip} \end{array} + \begin{array}{l} \text{Internal delay} \\ \text{due to lorries} \end{array} \times \begin{array}{l} \text{Interference} \\ \text{internal delay} \\ \text{allowance} \end{array} \right)
$$

$$
\text{TBCT per cycle} \left(= 1.44 + \frac{\text{Distance}}{21.0} + 9.8 \times \frac{\text{Skip size}}{\text{Lorry size}} \times \begin{array}{l} 1.07 \text{ mins} \\ \dots\dots\dots(5) \end{array} \right)
$$

This relationship is used to plot Figure 5 from which the crane time per cycle can be obtained for any skip size or distance hoisted, given that the time to load and unload concrete is 1.44 mins and the delay per lorry is 9.8 mins.

Calculation of gang size

As the concrete gang is working in conjunction with the crane the pouring, shovelling, vibrating, tamping and trowelling are done during the machine controlled time, and preparation and transfer of tools can be ignored. Thus, using data from Table 2 the basic time for the crane controlled work (BTCC) can therefore be expressed as:

152

$$BTCC = Vol\left\{ 4.38 + \frac{3.57}{d} \right\} \dots\dots\dots\dots\dots\dots\dots\dots (6)$$

where d is the depth of the slab in metres.

To ensure that the concrete gang is balanced with the crane the output rates of both, during the basic machine controlled time, can be equated to give the optimum gang size, thus:

$$\text{Output rate for crane} = \frac{\text{Skip size (m3) x 60}}{\text{TBCT}} \text{ (m}^3\text{/hr)} \dots\dots\dots\dots\dots\dots\dots (7)$$

$$\text{Output rate for men} = \frac{\text{Volume x gang size x 60}}{\text{BTCC}} \text{ m}^3\text{/hr)} \dots\dots\dots\dots\dots\dots (8)$$

By combining equations (7) and (8) the gang size can be determined from either Figure 6 or from the following equation:

$$\text{Gang size} = \frac{\text{Skip size}}{\text{TBCT}} \times \frac{\text{BTCC}}{\text{volume}}$$

Note: Relaxation allowances have been omitted.

TESTING THE RESULTS WITH CURRENT PRACTICE

In practice synthetic data would be taken from a comprehensive data bank of element times and delays rather than from a single study as illustrated. The results from over eighty concrete pours have been combined and analysed. As a result a series of regression equations have been developed. These equations can be used to determine basic times for different shapes and sizes of concrete pours[3].

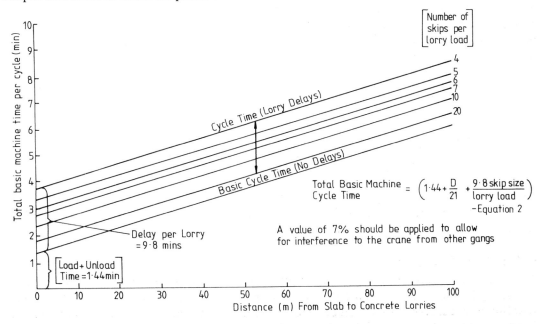

Figure 5 Cycle time (mins) against distance hoisted for various skip sizes, (with 9.8 mins delay per lorry

Methods of measuring production times for construction work

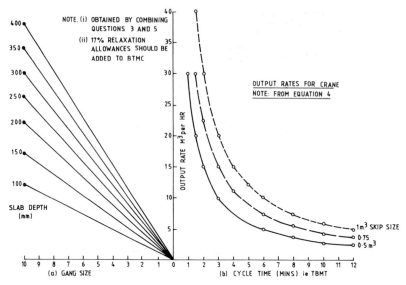

Figure 6 Output rate during machine controlled time (BTMC) against
(a) Gang size and depth of slab for concrete gang
(b) Cycle time and skip size for crane

PLANNING TIMES
Site factors (F1.F2.F3.F4)

It can be seen from equations (7) and (8) that synthesized values are expressed as basic rather than as standard times. In practice standard time is rarely achieved due to factors relating to operative motivation (ie, the relaxation period taken often exceeds that which is required for recovery). Thus, actual operation times would be more realistic for production planning and estimating purposes. In order to obtain these values, adjustment should be made to the total basic time for changes in work rate, idle time, and breaks. They may be expressed as factors and can be determined from activity sampling or time study results, as shown in Figure 7.

The relationship between the Working Day (WD) and Total Basic Time (TBT) can, therefore, be expressed as:

$$WD = TBT \times F1 \times F2 \times F3 \times F4$$

It should be noted that in practice F1 is the subjective rating value assessed by the observer.

In the example given in Figure 2 a site factor was obtained for F1 and F2 only, as F3 and F4 have to be calculated over a longer period of time. It can be seen in Figure 3 that a value of 1.26 resulted during the four hour study.

Thus, the build up of planning times using the data bank of basic element times of operation and site factors is as follows:

(a) determine the basic operation time by combining the basic element times, work contingency allowances and internal delay allowances where appropriate;

(b) apply site factors F1, F2, F3 and F4 to obtain planning times.

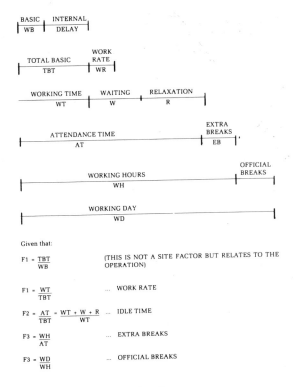

Figure 7 Site Factors

When the site factors are applied to equations (7) and (8) the output rates become:

$$\text{Output rate for crane} = \frac{\text{skip size } (m^3) \times 60}{\text{TBCT} + \text{WAITING}} \quad (m^3/hr)$$

Note: Waiting includes idle period except internal delays.

$$\text{Output rate for men} = \frac{\text{Volume} \times \text{gang size} \times 60}{\text{BTCC} \times \text{SITE FACTOR}} \quad (m^3/hr)$$

The results from the sites visited to date indicate a range of site factors between 2.50 and 1.11 can be expected for most trades (ie, sites are 40% to 90% efficient).

CONCLUSIONS

The results of the investigations indicate that work study techniques can be adopted to meet the requirements of most construction operations, sites and companies. The key to the portability lies in the application of site efficiency factors which isolate the basic times for operations and enables the application of computers for recording and handling data to be simplified. The second aspect is particularly important with the introduction of portable personalised microcomputers. In this way data can be collected on site and used to immediately update existing output values. The levels of efficiency observed on many sites and the main causes of variation have been discussed at length elsewhere[1,2,3].

As a final point it is hoped that interested parties will suggest improvements so that a pool of data might emerge for use by those responsible for planning and estimating.

Methods of measuring production times for construction work

REFERENCES

1. An evaluation of production output for construction labour and plant. SERC Research Grant GR/B/5138. Dept. of Civil Engineering, Loughborough University of Technology. 1981-84
2. An evaluation of production output for labour and plant on specialised construction work. SERC Research Grant GR/D/04366. Dept. of Civil Engineering, Loughborough University of Technology. 1985-1988
3. PRICE A D F (1987) An evaluation of production output for in situ concrete work. PhD Thesis, Loughborough University of Technology

BIBLIOGRAPHY

1. CORNWALL COUNTY COUNCIL HIGHWAYS DEPARTMENT (1982) Composite work values, highways construction and maintenance
2. LINCOLNSHIRE COUNTY COUNCIL (1978) Standard minute values.
3. SAOB. Richttijden voor bouwactwiteiten. Standard times for building operations used in Holland
4. BYGGFORBUNDET. Method och data. Standard times for building operations used in Sweden
5. Estimating and planning data for construction work. Disc file storage. Dept. of Civil Engineering, Loughborough University of Technology
6. HARRIS F C and McCAFFER R (1989) Modern construction management. BSP (3rd edition)
7. ASHWORTH A (1980) The source, nature and comparison of published cost information. *Building Technology & Management*, 18 March, pp33-37
3. MASON C (1979) The limitations of estimators' data bases, MSc project report, Dept. of Civil Engineering, Loughborough University of Technology
9. EMSLEY M W and HARRIS F C (1986) Computerised collection of work study data. *Building Technology & Management*, April
10. WALLACE W A. Management organisation and operative performance, MSc project report, Dept of Civil Engineering, Loughborough University of Technology
11. BROOMFIELD J, PRICE A and HARRIS F (1984) Production analysis applied to work improvements. *Proceedings, Institution of Civil Engineers* Part 2, 77 September
12. LAMSAC (1971) Manual work within local authorities, relaxation allowances. Report of the O&M and Work Study Panel
13. WIJESUNDERA D A and HARRIS F C (1986) Using expert systems on construction. *Construction Computing*, Autumn
14. OLOMOLAIYE P O, PRICE A D F and WAHAB K A (1987) Problems influencing craftsmen's productivity in Nigeria. *Building and Environment*. 22, (4), pp317-323
15. WIJESUNDERA D A and HARRIS F C (1987) The application of portable computers to production analysis. 5th International symposium. The organisation and management of construction, CIB, London. September
16. WIJESUNDERA D A and HARRIS F C (1989) The selection of materials handling methods in construction by simulation. *Construction Management and Economics, 7,* pp95-102
17. AL-KASS S and HARRIS F C (1988) An expert system for the selection of equipment in road construction. ASCE, *Journal of the Engineering Construction & Management,* ASCE, 114, No.3
18. HARRIS F C (1989) Modern construction equipment and methods. Longman 19. EMSLEY M W and HARRIS F C (1990) Methods and rates for structural steelwork erection. CIOB Technical Information Service Paper No.114

MEASURING SITE PRODUCTIVITY BY WORK SAMPLING

by A T Baxendale, BSc(Hons), MPhil, MCIOB

INTRODUCTION
Job management effectiveness

During the progress of a building project it is necessary to measure the effectiveness of the management on site. This can be done with a cost control system which compares the actual job costs with the project estimate. However, as estimates can be approximate it may be incorrect to concentrate on meeting the estimate, rather than investigating the efficiency of the job operations themselves. Delays in producing cost reports will mean that operations are considered only in retrospect and it could be that operations being carried out below the estimated cost are those where the greater cost reduction can still be made. Cost reports do not always highlight the cause of any problems, since they are a combination of elements such as standards of supervision, labour output, plant performance, site conditions, weather etc.

The output of a cost control system will be of use to the estimator and to the cost surveyor, but site management will be more interested in resource utilisation. As the resources available to the site are usually predetermined before the contract starts the site manager is mainly concerned with the control of these resources. Work sampling or activity sampling procedures can measure the level of activity of work being carried out on site and can be extended to measure the level of productivity.

Work sampling

Work sampling is a means of measuring the effectiveness of the site in using its resources. The results of the sampling can be available at the end of the working day, so allowing managers to make decisions on operations that are not reaching an acceptable standard. Observations can be made by a technician thereby avoiding the need to tie up valuable site management time. Work sampling involves observing and classifying a small percentage of the total activity of a project. With a large enough sample, the level of activity of the whole site or any part of it can be estimated; the accuracy of this estimate will improve as the number of samples increases.

By definition activity is a necessary part of completing work. However, more detailed analysis can reveal if the activity maintains or improves productivity.

Activity sampling (also called random observation studies or snap observation studies) is used to determine the activity levels of operatives and plant. Readings are taken at intervals of time and the percentage of readings taken for each category or element of work will give a result close to the actual percentage, provided that the sample is big enough. The reading must be made at the instant that the subject is observed. An observer can record the work of one gang or even a whole site, with readings taken at fixed or random time intervals. Readings can also be carried out continuously throughout the working day.

Statistical basis

As an illustration of the statistical theory of sampling a delivery of facing bricks has been made to site and they are required to be checked for damage. The total number of damaged

Measuring site productivity by work sampling

bricks will not be known until each one has been inspected. However, as more and more bricks are inspected it would be more certain that the total number of damaged bricks would follow the trend of the sample. All samples taken must have consistent characteristics so that:

(a) the condition of each brick is independent of any other;

(b) each brick has an equal chance of being selected;

(c) the characteristics of each brick remain constant.

An example of taking samples that do not agree with these rules is where the bricks are taken only from the top or sides of the pallets.

To express the degree of certainty between the results from sampling and the findings if all items were examined, the following concepts are used: confidence limit, limit of error and proportion of the sample having the characteristics being observed (category proportion).

The confidence limit expresses the dependability of the result, it being sufficiently accurate to use a 95% confidence limit with building work. That is to say that for 95% of the time, purely as a matter of chance, the answer can be relied upon.

Table 1 Sample size required for 95% confidence limit

Category proportion (%)	Limit of error ($\pm$%)				
	1	3	5	7	10
90	3456	384	138	71	50*
80	6144	683	246	125	61
70	8064	896	323	165	81
60	9216	1024	369	188	92
50	9600	1067	384	196	96
40	9216	1024	369	188	92
30	8064	896	323	165	81
20	6144	683	246	125	61
10	3456	384	138	71	50*

The limit of error is the accuracy of the estimated value given as a percentage either side of the result obtained by sampling. It is generally recognised that a limit of error of $\pm$5% is adequate for work sampling from construction operations. For example, if an error of 10% is obtained, the estimate of damaged bricks will fall within plus or minus 10% of the total number of bricks that are actually damaged. This result could also be depended upon 95 times out of 100 (95% confidence limit).

The category proportion will depend on the circumstances being sampled. For example, if 100 of the bricks examined are damaged out of a sample of 500, the category proportion of damaged bricks would be 20%.

Table 1 gives the sample size required, or number of observations to be made in relation to category proportion (%) and limit of error (±%) for a 95% confidence limit. If a limit of error of ±5% is required where the proportion of damaged bricks is 20%, the sample size would have to be at least 246 to meet the requirements. Rather than using the table, the number of observations required can be calculated by using a formula with a simplified constant that gives slightly more readings than are required for any given accuracy.

Number of observations required N $= \dfrac{4P(100 - P)}{L^2}$

where P is the percentage occurrence of the activity
L is the limit of accuracy required (±%).

Limit of accuracy L $= \sqrt{\dfrac{2P(100 - P)}{N}}$

Rules for taking samples

The following general rules can be made for work sampling when observing construction operations:

- a sample should have at least 384 observations. This will give the maximum number of observations required for a ±5% limit of error irrespective of category proportion;

- the observations must be categorised at the instant the operation is observed (snap observation). The observer should not rationalise what, for example, an operative has just completed or is about the start;

- every activity must have the same chance of being observed at any time;

- observations must have no sequential relationship. It follows from this that cyclic operations will have to be randomly observed;

- the basic characteristics of the work must remain the same while observations are being carried out. This is important if comparisons are to be made with different sets of observations.

ACTIVITY COUNT

A number of activity sampling methods exist which vary in their degree of sophistication. The simplest is the activity count or field count, which classifies the activity as 'working' or 'not working'. Further detail can be added by subdividing 'working' into operational elements and 'not working' as resting, walking to and from the place of work, and any other relevant activities.

Activity counts can be made with any number of operatives. When a large building project is being observed a few tours around the job will obtain the required number of observations. The operations of a single gang will have to be observed repeatedly to obtain the required number of observations. For example, if a gang was observed 400 times and 180 of these were classified as working, it can be concluded that the gang is active from between 40 and 50% of the observed time (see Table 1, with ±5% limit of error and 95% confidence). It is generally considered that less than 60% job activity is unsatisfactory and will require further investigation.

Measuring site productivity by work sampling

In setting up an activity sampling exercise it is necessary to decide on the categories of activity to be recorded, although these can be decided during the course of the study. Observations should then be made in accordance with the rules and as discreetly as possible so that the subject under observation does not react to being observed. An example of an activity count is given in Figure 1. When an adequate number of observations have been made (a minimum of 384 is recommended from Table 1) the percentage time spent 'working' and 'not working' can be calculated for each gang and any necessary action taken.

ACTIVITY COUNT			
Contract Multi-storey office block Notes: Weather cold but dry	Study No: 1 Sheet No: 1 Date : March 1985 Observer: ATB		
Operatives/Gangs: 12 on formwork, 3 on reinforcement, 5 on concreting.			
TIME	GANG	WORKING	NOT WORKING
8.10	Formwork Reinforcement Concreting		
8.26	Formwork Reinforcement Concreting		
8.45	Formwork Reinforcement Concreting		
9.05	Formwork Reinforcement Concreting		
9.25	Formwork Reinforcement Concreting		
10.04	Formwork Reinforcement Concreting		

Figure 1 An extract from an activity count study

PRODUCTIVITY STUDIES

Inefficiencies in construction operations often exist because site management is too close to the work and too busy to see them. What is required, therefore, is an approach where work can be investigated in a planned way to improve productivity. Productivity studies are a means of categorising the work in more detail than with activity counts. Using the technique three main categories of work can be identified:

● directly productive, which is an expenditure of operatives' time in making the building grow;

- indirectly productive which is an expenditure of operatives' time in preparing to make the building grow;
- non-productive.

Work is only effective when it is directly adding to the completed product. Each obsever will have to define effective work, depending on the operations being sampled. The classification of work is not rigid and may change depending on circumstances, the aim being to show where there are inefficiencies.

An example of a productivity study is given in Figure 2, together with a summary in Figure 3. The contract and operational details are entered at the top of the work sampling sheet.

WORK SAMPLING					
Contract: Low rise dwellings in timber frame, Hereford			Reference : Timber frame 1 Sheet No. : 1 Date: : December 1984 Observer : ATB		
Operatives/Plant: A & B Carpenters			Notes: Fixing units from stack at first floor level. Weather dull with occasional rain.		
TIME	OP'TIVE	ACTIVITY	TIME	OP'TIVE	ACTIVITY
8.30	A B	} Absent — not started work	9.20	A B	} Fix internal unit Idle
8.35	A B	} Fix window to External unit	9.25	A B	} Fix internal unit
8.40	A B	} Idle	9.30	A B	} Idle
8.45	A B	} Move units by hand	9.35	A B	} Fix window to external unit
8.50	A B	} Move units	9.40	A B	} Fix external unit
8.55	A B	} Move units	9.45	A B	} Fix external Idle
9.00	A B	} Fix external Unit	9.50	A B	} Idle
9.05	A B	} Idle	9.55	A B	} Absent-break
9.10	A B	} Fix external unit	10.30	A B	} Move units
9.15	A B	} Fix internal unit	10.35	A B	} Fix external unit

Figure 2 An extract from a work sampling study showing 20 observations on each operation

Measuring site productivity by work sampling

The time is entered when each round of observations are started. Operatives can be identified by name, occupation or simply by letter. Snap observations of each operative's activity are noted. Any unusual conditions can be included in the notes at the top of the page. Continuation sheets will only require reference and sheet number details. The work sampling summary sheet totals all the observations made and expresses them as a percentage. In the sample given it can be concluded from the summary of the total days observations that 10% of the time is being spent on extended breaks and that 10% of the time is being spent moving units. Site management can be alerted to these points and improvements made by supervising break times and organising the stacking of units for each dwelling - a potential improvement in productivity of at least 20%.

The motivation of the carpenters with an idle/relaxation time of 24% is also open to question. A check on the accuracy of these percentages will show that more observations may have to be made to confirm the results (the idle/relaxation time has an accuracy of ±6.4%).

Using productivity studies continuous samples can be taken with a round interval of five to ten minutes, where a group of less than ten operatives located in the same area are being observed. Random samples can be taken with the interval between rounds chosen in relation to random numbers. This type of study is best used when applied to the whole site or where there are more than ten operatives. The sample must be representative and the study is, therefore, of long duration.

Rated work sampling can also be used when studying operatives. Rating is carried out in a similar way to that with time study. However, rating is not recommended in the context of work sampling as applied here, as a considerable amount of experience and cross checking is required for an observer to maintain accuracy. All productive operations are generally

WORK SAMPLING SUMMARY				
Observation interal: 5 minutes Operatives/Plant: 2 Carpenters		Reference: Timber frame 1 Date: December 1984		
ACTIVITY	OBSERVATIONS	TOTAL	%	QTY
Fix external units	ШІ ШІ ШІ	15	8	
Fix internal units	ШІ ШІ ШІ ШІ ШІ etc.	45	28	
Fix windows to external units	ШІ IIII	9	5	
Move units by hand	ШІ ШІ ШІ ШІ ШІ etc.	34	19	
Idle/relaxation	ШІ ШІ ШІ ШІ ШІ etc.	43	24	
Remedial work	ШІ ШІ I	11	6	
Consult drawings	III	3	2	
Absent	ШІ ШІ ШІ III	18	10	
	TOTAL	178	100	

Figure 3 Work sampling summary

carried out in a British Standard Work Study rating of between 90 and 110 and only an experienced observer can rate within this band to an accuracy of 5. In addition the breaking down of work into small categories or elements enables the assumption to be made that an operative will always work at a standard rate; it is the 'not working' element that is of interest to site management.

Once the observer has become confident in gathering data a format can be used to record the information more compactly as shown in Figure 4; together with a summary in Figure 5. Operation details are entered as before with the start and finish times of the study entered together with the time interval at which each round is commenced. This will enable a check to be made between the number of observations that have actually been recorded and those that should have been recorded. Observed activites are recorded for each operative (or machine), space being allowed for up to twenty observations on one line against any operative activity. Another sheet can be used when there is no more space. To aid data collection some categories of activity can be noted or pre-selected before recording work starts. A final summary will have to be made totalling all observations against each category of activity. The total number of observations that should have been made can be calculated by dividing the total time by the round interval. This figure can then be compared with the actual number of observations recorded and the percentage error calculated. This error should not be greater than about 2% for reasonable accuracy.

WORK SAMPLING						
Contract: 3 storey office block, Bristol			Reference: Formwork 3 Sheet No: 1 Date: April 1985 Observer: ATB			
Start time 13.30		Finish time: 17.00			Round interval 3 minutes	
Operatives/Plant: A, B, C, D, E and F shutterhands				Notes: Fixing formwork to first floor. Weather fine.		
A	B	C	D	E	F	ACTIVITY
II	II	II	I	III	III	Fix metal props on ground floor slab.
II	II	II		II	II	Fix timber bearers on props.
II	III	II		III	III	Fix waffle moulds on bearers.
I		I	II	II	IIII	Strike metal props from previous pour.
I	II	II	III	II	II	Strike timber bearers from previous pour.
II	II	II	IIII			Strike waffle moulds from previous pour.
LHT LHT	LHT III	LHT IIII	LHT LHT	LHT II	LHT	Idle/relaxation.
						Absent, extended break.

Figure 4 An extract from a work sampling study showing 20 observations on each operative

Measuring site productivity by work sampling

The general conclusion to be drawn from the example is that the method of working appears to be reasonably efficient, but the idle/relaxation time is excessive and must account for what was a low rate of production and the original reason for the investigation. An agreed relaxation time will have to be agreed and maintained to say 20%. The accuracy of the study taken over five days for the idle/relaxation time is ±1.4%.

WORK SAMPLING SUMMARY			
Operatives/Plant: 6 shutterhands	Reference: Formwork 1, 2, 3 4 & 5 Date: April 1985 Round interval: 3 minutes		
Total time: 40 hours Total obs: 4788		% Error: − 0.25	
ACTIVITY	TOTAL	PERCENTAGE	QUANTITY
PRODUCTIVE (DIRECT)			
Fix metal props	507	11	
Fix timber bearers	512	11	
Fix waffle moulds	486	10	56m²
Strike metal props	186	4	
Strike timber bearers	89	2	
Strike waffle moulds	86	2	16m²
PRODUCTIVE (INDIRECT)			
Move formwork	195	4	
NON-PRODUCTIVE			
Idle/relaxation	2538	53	
Absent	179	3	
Totals	4788	100	

Figure 5   Work sampling summary

CONCLUSION

The results of productivity studies can be used in a number of ways but they are best used to point out to site management where work can be carried out more effectively. Productivity studies are simple and can be carried out by junior site management, thereby acting as the basis of a useful education in observing building operations. If productivity studies are used improperly, say, as a basis for discipline, they can create antagonisms which defeat their purpose. Productivity studies are not questioning the operatives' work but the site organisation.

BIBLIOGRAPHY

BIRD, J. O. and MAY, A. J. C. (1981) Statistics for technicians. Longman
FORSTER, G. (1981) Construction site studies - production, administration and personnel. Longman

OXLEY, R. and POSKITT, J. (1980) Management techniques applied to the construction industry. 3rd edition. Granada

PARKER, H. W. and OGLESBY, G.H. (1972) Methods improvement for construction managers. McGraw-Hill

PILCHER, R. (1976) Principles of construction management. 2nd edition. McGraw-Hill

PRODUCTIVITY, THE KEY TO CONTROL
by R M W Horner PhD MICE

INTRODUCTION
The payment of lip-service to proposals to increase productivity is a payment which any industry can ill-afford to make. In the construction industry, where gross profits are in the region of 5% of turnover[1], it is a payment which cannot be afforded at all. And yet, when the promise of increased productivity is exchanged for the promise of higher earnings, by how much does output go up? When the cost of preliminaries is reduced by cutting the number of staff, what increase (or decrease) in net profit ensues? When overtime payments for staff are introduced, how much longer do they remain at their places of work, and how much more (or less) do they achieve? In most cases, the answer to all these questions is the same, 'Pass?'.

One of the reasons for this state of ignorance is the industry's obsession with cash. Boards of Directors are responsible to shareholders for the profitability of their companies. A pre-occupation with profit is therefore understandable. It is wrong, however, to presume that because profit is a measure of the success of a company, it can also be used as an indicator of performance on site. Profit is a function of price, over which the site agent has no control. How can he be held responsible for an estimator's error? Even unit costs are subject to factors outside the agent's sway. The quality and proximity of local suppliers, the level of nationally agreed wage-rates, the attitudes of statutory undertakings, of inspectorates, of clients themselves all militate against the use of unit costs as a means of control, even though much can be learned by the discerning contracts manager from a careful, comparative analysis. It can be concluded, therefore, that the key to control lies not in the consideration of costs, but in the pursuit of productivity.

It is a happy coincidence that site management has the greatest control over the resource which arguably has the gratest effect on productivity; the resource that is, of labour. This paper described a long-term reserch programme aimed at determining the relationship between labour productivity and the factors on which it depends.

A common approach to this type of problem is to vary one parameter at a time whilst holding everything else constant. In our case, whilst the approach might have been idealogically perfect, it was also totally impractical. Nevertheless, to make any meaningful comparisons at all, it was clearly going to be necessary at least to identify those parameters which affect productivity.

FACTORS AFFECTING PRODUCTIVITY
An extensive literature review conducted by Danladi[2] identified the factors affecting productivity shown in Table 1.

The effect of a labourer's intrinsic skill on his output is clear, and although no proof exists, there is a widespread feeling that large labour forces lead to low unit outputs.

That an unbalanced labour force reduces productivity is indisputable. Balance is required both within and between trades. If the ratio of bricklayers to mates is not optimised, work rates will fall. If there are too few carpenters on site, the steelfixers will be delayed.

Table 1: Factors affecting labour productivity

1. Quality, number and balance of labour force
2. Motivation of labour force
3. Degree of mechanisation
4. Continuity of work as affected by:
    (a) supply of materials;
    (b) performance of other contractors or sub-contractors;
    (c) availability and adequacy of technical information;
    (d) variations.
5. Complexity of project
6. Required quality of finished work
7. Method of construction
8. Type of contract
9. Quality and number of managers
10. Weather

But why do people work at all? The question of motivation has taxed the minds of many and although hard data is limited, what is available does yield surprising results. Parker and Oglesby[3] reported the effects of overtime working in industrial construction and concluded that although overtime increased total output for short periods, overtime worked consistently results in a return to, or even a reduction in, the original output. (This point is illustrated in Figure 1.) Indeed, experience shows that men are loath to work any longer or harder than it takes to achieve their own, subjective target earnings. It is no coincidence that absenteeism is greatest on the Monday following a working weekend.

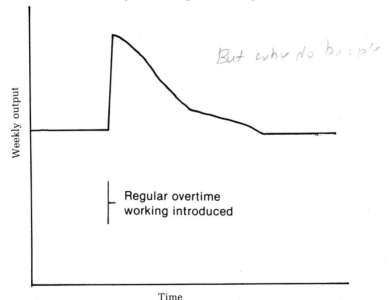

Figure 1 Relationship between hours worked and output

Mechanisation and continuity are not parameters which can be optimised. Machines would not be introduced if they did not increase productivity and discontinuities are as popular as cube failures. However, if labour productivity on different sites is to be compares, some measure of the degree of mechanisation and the severity of discontinuities must be fed into the equations.

The complexity of a project and the required quality of finished work are not factors over which site management has any control. But architects would undoubtedly benefit from a quantitative knowledge of the effect on productivity (and cost) of designing circular rather than rectangular columns; of specifying class E rather than class C surface finishes. In any case the influence of complexity and quality on output must be eliminated from the calculations.

It is obvious that the method of construction must affect productivity and there is good cause for early consultation between designer and contractor to ensure that the contractor's resources and expertise are used to their full advantage. Indeed, from this philosophy sprang the concept of the turnkey contract. It is not so obvious that other forms of contract can have an effect on ouput. However, financial security can promote an emphasis on optimisation rather than expedience. If a contractor knows that his costs are going to be reimbursed, he may well choose to hire the formwork most appropriate to his job instead of making do with the standard panels in his plant yard. Fortunately, the use of standard methods of construction and forms of contract is so widespread that by judicious choice of project, departures can be avoided.

The influence of management control on productivity was regarded as being of major significance and became the subject of a pilot study.

PILOT STUDY

Preliminaries are undoubtedly a significant element of cost and staff costs a significant element of preliminaries. Therefore, it is not surprising that in periods of high competition, the level of site management comes under the closest scrutiny at tender meetings and site agents find their site teams truncated by the razor edge of the contract manager's axe. Do such swingeing policies minimise costs? The answer may well be 'No'. A reduction in site staff can result in less checking and more mistakes. No data exist from which a mean percentage cost of error can be calculated, but the sort of mistake enshrined in safety statistics can give some idea of scale. It is not difficult to believe that the average cost of a reportable accident may be several thousand pounds. If it were just £1000, and if it is assumed that the cost of labour at £3/h represents 30% of turnover and that a profit margin of 5% prevails, it would take over one man year's work to recoup the loss. At 30000 reportable accidents in 1978[4] that would be a loss equivalent to some 7 million man days' work a year. This may be compared with the 297000 man days lost due to industrial disputes during the same period[5]. Obviously, not all accidents can be prevented but just as obviously not all mistakes result in injury.

An inadequate site management team will be unable to provide sufficient control to ensure maximum productivity and minimum cost. Too much site management will cause confusion and excessive interference. Logic dictates that somewhere between the two extremes lies the optimum solution. A hypothetical relationship between management control and productivity is shown in Figure 2.

In seeking to determine the relationship between productivity and management control, it was necessary to identify those factors which might influence the degree of control, in much the same way as those factors affecting productivity. The results are presented in Table 2.

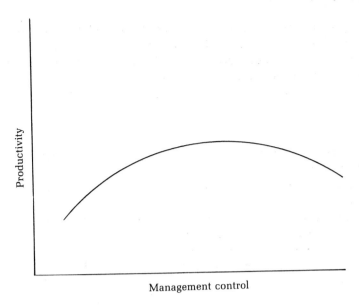

Figure 2 Hypothetical relationship between productivity and management control

Table 2 Factors affecting site management control

---

1. Quality, number and balance of labour force
2. Nature of incentive schemes
3. Quality and amount of plant
4. Procurement difficulties
5. Quality and number of sub-contractors
6. Number and nature of variations
7. Complexity of project
8. Required quality of finished work
9. Speed of construction
10. Type of contract
11. Amount of co-ordination required with:
    (a) local government agencies;
    (b) client;
    (c) consulting engineer;
    (d) other contractors.
12. Degree of monitoring and reporting
13. Amount of support from head office organisation
14. Amount of public/political interest

It was recognised at the outset that data collection would be difficult but that two options were available:
(a) to measure productivity and control on working sites;
(b) to generate data from existing site records of completed projects.

Because of the need to collate as much data as possible with a minimum expenditure of resources, the second alternative was chosen.

Productivity, the key to control

To reduce the effects of extraneous differences, the three projects selected for analysis were simple, steel-framed buildings located in the same geographical region. Similar incentive schemes operated on all sites. To reduce the incidence and effect of post-contractual variations, earthworks and finishing phases were excluded from the study.

Information on work rates was obtained from bonus return sheets. It is well known that such data are liable to gross error but after considerable manipulation, it proved possible in most cases to abstract, on a weekly basis, realistic work rates for bricklayers, carpenters and labourers.

The acquisition of management control data proved to be more difficult. It quickly became apparent that inter-site comparisons would have to await the generation of more comprehensive records. The only relevant information available at the time was the number of hours spent on site each week by the agent, who happened to be, in all cases, the sole member of the site management team. Nevertheless, by assuming that the number of hours spent supervising an activity is proportional to the number of productive hours spent by the labour executing it, some estimate of management control could be made. Numbers of hours spent supervising actvities are not, by themselves, however, an effective measure of management control. For the purpose of comparison control indexes were defined in two ways:

(a) as the ratio of management hours spent supervising an activity to the productive hours worked by the operatives in that trade;
(b) as the ratio of management hours spent supervising an activity to the total (productive plus non-productive) hours worked by the operatives in that trade.

In this way, the variation of control index with time and activity could be determined.

Graphs of management control index against work rate could now be drawn as shown in Figure 3. The large scatter of results reflected the low quality of the data. Because of this, and for the sake of simplicity, a linear regression analysis was conducted on each set of data, and the 'best fit' straight line computed. In 17 of the 18 cases, the results showed a positive relationship between amangement control and producitivity[6].

THE WAY AHEAD
Objectives
To effect control, productivity must be forecast under any given set of conditions and variances from that forecast monitored. The aim was to produce a simple accurate technique for measuring and predicting work rates, so that a single set of data could be used for estimating and for control. The pilot sutdy highlighted the paucity of existing data. It was clear that much more accurate and reliable informatin was needed to stand any chance of developing predictive models.

Data generation
The basic problem is the great multiplicity of site activities. It is plainly impractical to measure work rates for each one, nor is it necessary. Logcher and Collins[7] introduced the concept of surrogate activities in 1978. A surrogate activity is one which characterises a number of other activities. Thus, standards of performance in a surrogate activity are indicative of standards of performance in a group of activities.

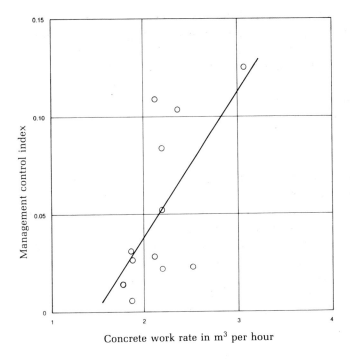

Figure 3: Relationship between management control index
and concrete work rate

It has been reported[8] that 80% of the value is contained in 20% of the items in a bill of quantities, and current analyses confirm these findings. Consequently, for the purpose of control 80% of the items of a bill can be discarded with impunity. Further, it is likely that the surrogate activities will be amongst the remaining 20% of items which are cost significant. Establishing which 20% is not simple.

One way to proceed is to arrange every item in decending order of value, and to calculate progressively the cumulative sum. The method is tedious, even with the aid of computers. However, analysis of some 25 bills of quantities describing widely differing projects suggested the existence of a consistent relationship between the number of items, the total bill value, and the value of the cost significant items[9]. Linear regression analysis led to the conclusion that the formula

$$A = \frac{V}{N}$$

could be used very successfully to predict which items were cost significant. A is the value of that item which when added to all items of higher value, would yield at least 80% of the total bill value V, and N is the number of items. Thus, for a bill of value £1m containing 2000 items, the sum of all items of value higher than £500 is expected to be greater than £800,000.

Inspection of the cost-significant items isolated using this equation usually indicated that some activities were common to a number of items. By abstracting these similar activities, the number of bill items to be monitored could often be reduced by 95%. Ina typical example, a bill of 1000 items was reduced to 39 cost significant activities.

The future

The method needs to be tested over a wide range of bills in order to define its limits of applicability. An extensive, on-site monitoring programme is planned to investigate whether or not these few cost-significant itms will serve as surrogates for the performance of the whole site. If expectations are fulfilled, sufficient reliable and accurate data can be collected for a meaningful, statistical analysis. Using multiple regression, it should then be possible to determine a mathematical relationship between productivity and the factors on which it depends. The capacity for predictive modelling would exist.

CONCLUSION

The development of predictive models is a daunting, but not impossible task. Some might also consider it thankless, since it is unlikely that any model could predict pricise performance in every set of circumstances. Provided it can be ensured that actual results are reasonably narrowly distributed about the predicted mean, this would not matter. What was lost on the roundabout would be gained on the wings.

The ignorance as to cause and effect which pervades the construction industry stems from a health fear of the cost, both in human and financial terms, of collecting adquate data. If surrogate activities can be simply identified, then the day can be anticipated when the site clerk monitors tens rather than hundreds of activities, punches data straight into a minicomputer rather tha on to an allocation sheet, and calculates both bonus and workrates 'in a twinkling of an eye'. Data will be stored on disc, fed to the company data bank and used for future tenders.

If 1½ million men are employed in the construction industry, each costing £150 a week, a 10% improvement in productivity would reprsent an annual national saving of £1000m. The pursuit of productivity may not lead us to the promised land, but it may save us from indolent mediocrity. The cake is surely worth the candle.

REFERENCES

1. (1989) The industry's leaders. *Construction News* January 31

2. *DANLADI S K (1979) Management control and construction efficiency in building and civil engineering contracts. Thesis presented to the University of Dundee in partial fulfilment of the requirements for the degree of MSc.*

3. PARKER H W and OGLESBY C H (1972) Methods improvements for construction managers. McGray-Hill

4. HEALTH AND SAFETY EXECUTIVE (1978) Manufacturing and service industries. 1977. HMSO

5. CENTRAL STATISTICAL OFFICE (1978) Annual Abstract of Statistics 1979 Edition. HMSO

6. DANLADI S K and HORNER R M W (1981) Management control and construction efficiency. J.Cons.Div. Proc. ASCE *107* CO4 pp705-718

7. LOGCHER R D and COLLINS W W (1977) Management impacts on labour productivity. *ASCE Journal of the Construction Division 104* (CO4) December pp447-461

POSTSCRIPT

COST-SIGNIFICANCE

Much progress has been made since writing this paper. The philosophy of cost-significance has led to a new method of estimating[10] and to simplified cost models for predicting and controlling project costs[11]. The new method, called iterative estimating, allows the value of a blank bill of quantities to be predicted to an accuracy of $\pm$ 5% without pricing more than 30% of the items in it. The price of a reinforced concrete bridge can be forecast to the same level of accuracy using a cost model containing only 10% of the number of items contained in a conventional bill. The systems are easily computerised, and form the basis of the software package BRIDGET[13] developed in conjunction with Babtie Shaw and Morton and issued by the Scottish Development Department to all Scottish Regional Councils.

PRODUCTIVITY

Management control and financial incentives

Productivity research has moved on apace too. An SERC sponsored site monitoring programme conducted between 1982 and 1984 failed to reveal the existence of surrogate activities, but did expose a strong, positive relationship between productivity and the management control index[14]. A dozen sites throughout the UK were each continuously observed for periods of about three weeks. The principal variables studied were management control and financial incentives. Sites were carefully selected to eliminate as many of the other variables as possible. The results are shown in Fig.4. Sites 1, 2 and 3 were steel framed buildings; sites 4 to 10 were reinforced concrete. Sites 6 and 10 were known to be overstaffed and have been ignored in calculating the correlation coefficient. No correlation was found between productivity and the level of financial incentive.

Causes of variability

A subsequent study of bricklayers' productivity carried out during 1987 and 1988 highlighted the enormous variability in productivities. Fig.5 illustrates how the productivity of one gang may vary by as much as 200% from one day to the next, and how one gang's productivity may average 50% more than that of another. Figs.6a and 6b show how hugely important is the effect of interruptions on productivity. On average, they cause a loss of 35% during the time that men remain at work. Moreover, the longer the delay, the greater the loss of productivity. Fig. 3 also illustrates how sub-contract labour paid by the piece is 40% more productive than directly employed labour working on the same activity at the same time.

Disruptions (defined as conditions sufficiently adverse to cause a reduction in productivity lasting at least half a day, but insufficient to cause an interruption to work) cause an average loss of productivity of 26%. Table 3 shows the effects of various types of disruptions.

Table 3 Effect of disruptions on productivity

Category	Productivity loss (%)
Management related	19
Site related	22
Design related	50
Weather related	9

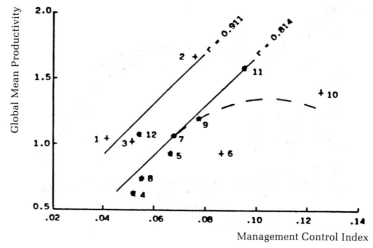

Figure 4 Productivity and management control

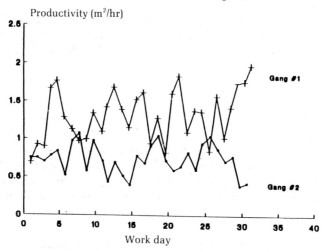

Figure 5 Productivity variability

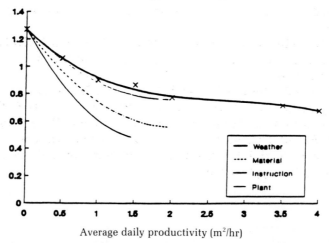

Figure 6 Effect of interruption   **Figure 6a** Directly employed labour

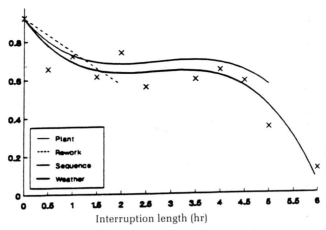

Interruption length (hr)

Figure 6b Sub-contract labour

I Mprovement

LESSONS TO BE LEARNED

If productivity is to be improved - select labour with care - minimise the number and durations of interruptions - reduce the incidence of disruptions - employ labour-only sub-contractors (provided you care not a jot for quality!) All of which boils down to improving the quality of site management!

WHITHER NEXT?

In collaboration with the Pennsylvania State University and six other research teams dispersed throughout the world, the Construction Management Research Unit at Dundee University is contributing to the establishment of a productivity database sufficiently large to isolate the effects of all the major variables. Through the principle of cost-significance, new and simple ways of measuring productivity for the purpose of site control are under development. The promised land may not yet be in sight, but we travel in hope!

REFERENCES

11. SAKET M M, McKAY K J and HORNER R M W (1986) Some applications of the principle of cost-significance. *Int J Construction Management & Technology 1*, (3), pp5-22
12. ASIF, M and HORNER R M W (1989) Economical construction design using simple cost models. Proc 3rd Yugoslav Symposium on Construction Management. Dubrovnik, April, pp691-699
13. MURRAY M, HORNER R M W and McLAUGHLIN A. BRIDGED (1990) A cost estimating suite for highway structures. *Journal of the Institution of Highways and Transportation 37(5)*, pp14-18.
14. HORNER R M W, TALHOUNI B T and WHITEHEAD R C (1987) Measurement of factors affecting labour productivity on construction sites. Proc CIB W65 5th Int Symposium on Organisation and Management of Construction, Vol.2, London, pp669-680
15. HORNER R M W and TALHOUNI B T (1990) Causes of variability in bricklayers' productivity. Proc W65 6th Int Symposium on Organisation and Management of Construction, Vol.6, Sydney, pp238-250

SITE MANAGEMENT AND CONSTRUCTION HEALTH AND SAFETY

by L Golob, BSc, PhD

The construction industry has invested much time, effort and money in trying to improve its health and safety performance. This investment has had success in the past in achieving significant improvements in the industry's record but these improvements have not been maintained in recent years and, now, construction is one of the most dangerous industries with an average of 150 deaths and 20,000 injuries reported to the Health and Safety Executive (HSE) each year.

HSE's report *Blackspot Construction*[1] analysed over 700 fatal accidents which had occurred over a five year period. Based on the investigation reports of factory inspectors, the analysis showed that 90% of these accidents were judged 'preventable' and that for 70% of them, action by management could have saved lives.

PROBLEMS OF MANAGEMENT

Site management has a crucial role to play in reducing this toll of death and injury. Sites are where the activities leading to accidents take place; site mangement is there on site to exercise control over those activities. If 'action by management' could have saved lives in a majority of fatal accidents then it would seem that, in those cases, site management failed to play its role successfully. This is not to say that the failures did not involve other levels of the management hierarchy, they almost invariably do.

However, it has to be admitted that the task of construction management in general and site management in particular has become much more difficult in recent years. In particular there has been a fragmentation of the industry due to an increase in the number of firms and the amount of sub-contracting. This fragmentation, along with growth in casual working and self-employment, has made it much harder for site management to control work on site and to ensure good health and safety standards for construction workers. For it is much more difficult for site management to exercise control over those who do not work directly for them and who might consider the exercise of such control to be unwarranted.

LEGAL RESPONSIBILITIES

Despite being faced with these difficulties, site management must succeed in exercising its control of site activities to the full. Besides the very practical effect it can have on the incidence of accidents — being on the spot as it were — site management has legal duties under the Health and Safety at Work Act. In particular, Section 4(2) of the Act places a duty on any person having control to any extent of premises made available as a place of work to others not in that person's employment. The person in control must exercise it, to the extent that he or she is reasonably able, to ensure, so far as is reasonably practicable, that the others using the premises are not exposed to risks to their health and safety.

In a recent case, a contracts manager was prosecuted and found guilty under Section 4(2) as a consequence of the continuing poor standards of scaffolding at a site where he had day-to-day control over site activities (the firm he worked for and its managing director were also prosecuted and found guilty of offences arising out of the same circumstances). Managers can also be prosecuted under Section 36 of the Health and Safety at Work Act where their act or default causes some other party, say their employer, to be guilty of an offence. The unfortunate contracts manager was also prosecuted for his company's breach of the Construction Regulations relating to scaffolding (as was the company itself). Inspectors increasingly take the view that because decisions affecting health and safety are actually taken by individuals, those individuals must also be held accountable for poor health and safety standards. So it is likely that there will be more prosecutions of individuals by HSE in the future — though there is no intention to pick out site managers for particular attention. The inspectors will look critically at the performance of managers at all levels of the management hierarchy when carrying out investigations.

MANAGING HEALTH AND SAFETY IN CONSTRUCTION

Clearly, site managers cannot achieve anything in isolation; they must receive proper support from their employer in creating the right conditions for management of site work. The means to create the right conditions are discussed in detail in publications produced by the Health and Safety Commission's Construction Industry Advisory Committee (CONIAC)[2]. Much of the advice in these publications is concerned with pre-construction planning.

PRE-CONSTRUCTION PLANNING

The main/managing contractor should identify hazards likely to be met during the construction project so as to enable the client to make proper financial provision for the arrangements that will have to be made on site, eg, provision of access equipment, materials for edge protection, welfare facilities, to ensure healthy and safe working conditions. The main/managing contractor should assess the abilities of sub-contractors who tender for work and appoint them, considering not only their competence in technical matters but also that in health and safety. The latter competence may be judged on the sub-contractor's attitude to health and safety, their health and safety policy (eg, is it available at all, does it show evidence of careful thought or does it contain only general platitudes?), and their past record.

With safety-conscious and safety-adept sub-contractors on site and an adequate health and safety budget for the project, the site manager's task is considerably eased.

There must also be a detailed discussion, with each appointed sub-contractor, of the specific arrangements the sub-contractor proposes to make to enable the work to be carried out in a healthy and safe manner. Such arrangements should cover not just risks to the sub-contractor's own employees but also to others on site and members of the public. Where complex or high risk activities, eg, erection of structural frames, demolition and roofwork, are to be undertaken by a sub-contractor, that sub-contractor should provide a written method statement which site managers can refer to during the course of work. Sub-contractors using substances hazardous to health should also be asked for their assessments, required under the Control of Substances Hazardous to Health (COSHH) Regulations,[3] which should include an assessment of the risks from each substance as it is to be used on site and the measures necessary to control the risk.

SITE ARRANGEMENTS FOR HEALTH AND SAFETY

The planning and discussions considered in the preceding paragraphs will involve site management amongst others but it is in the practical application of day-to-day arrangements for health and safety that site management makes its most important contribution.
Therefore, it is important for site management to establish clear and unambiguous site rules, divisions of responsibilities and any other arrangements necessary for the inclusion of health

and safety in site activities. Site arrangements for health and safety should not be an after-thought to site activities but as much a part as the arrangements, say, for ensuring work is carried out to an acceptable quality and at an acceptable speed. This attitude to health and safety should be part of the work culture of everyone on site, and site management is in the best position to promote it.

The arrangements for health and safety should be in force from the outset of work and be capable of development to match the progress of the work. Lines of communication should be established and arrangements made for liaison with nominated representatives of sub-contractors and the workforce; the arrangements should be reviewed with them as the work progresses.

An example of rule-making is that which the person in control of the site may do under the Construction (Head Protection) Regulations[4] to regulate the wearing of hard hats on site. Other arrangements might be those covering the shared use of welfare facilities or plant and equipment. Clearly, to leave such arrangements to be worked out piecemeal between the various sub-contractors, a group which varies as the work progresses, is impracticable and the co-ordination of site arrangements for health and safety is one of the important aspects of site management's role.

There is little point in setting out rules etc without some form of supervision to see whether the rules are being obeyed, the arrangements are working to control the health and safety risks and so on. Site management must, therefore, have procedures for monitoring all aspects of health and safety on site, both on a day-to-day basis and in the longer term. These may include environmental monitoring for hazardous dust, fume, noise, etc. Individual sub-contractors will often have a specific legal responsibility for monitoring their own activities, eg, under COSHH or Noise at Work Regulations[5], but in many cases where work activities involve or affect the workers of more than one sub-contractor, site management will have to co-ordinate monitoring and it would often be more effective and efficient for site management to carry out the monitoring on behalf of the individual site contractors.

Finally, site management must be able, and be seen, to exercise some form of sanction on companies or individuals who do not follow the rules or stick to the arrangements.

Planning, co-ordination, supervision and exercising sanctions summarise the role of site management in ensuring good health and safety standards on site. To carry out this role, each site manager will need knowledge and understanding of construction health and safety. Trained safety advisors should be able to provide site managers with a lot of assistance and back-up but often they will have to rely on their own judgement. Thus, it is essential that, if construction companies expect site management to look after health and safety on site, they make sure their site managers receive sufficient training and instruction in health and safety matters. Such training must be on a continuing basis since site managers need refresher courses and need to be acquainted with changes in legislation, standards etc.

In many respects, site managers can only be as effective as their employers allow them to be; a proper attitude to health and safety on the company's part is a prime requisite. Companies already have general duties as employers under the Health and Safety at Work Act to manage health and safety on site. However, the Health and Safety Commission believes it is necessary to strengthen the legal requirements by making the management of construction sites the subject of a set of specific regulations. The Commission set out its views on this matter in a consultative document issued in Autumn 1989 which also contained the draft Construction (Management and Miscellaneous Duties) Regulations and associated Approved Code of Practice. The proposed Regulations require that all parties to the construction process who are able to exercise control over that process should do so. Such parties include not just main/management contractors but works contractors, sub-contractors — and client groups, who in providing the finance are able to exert considerable influence, either directly or via their professional advisors, over site activities.

The legislation that finally emerges will have to take into account the Directive on construction health and safety adopted by the European Commission, which is also very much concerned with the management of health and safety on site. There is some way to go before the enactment of any such legislation, but it is likely that as a consequence of their employers' increased responsibilities, site management will have an enhanced role in the management of health and safety on site.

REFERENCES
1. HEALTH AND SAFETY EXECUTIVE (1988) Blackspot construction : a study of five years fatal accidents in the building and civil engineering industries. HMSO.
2. HEALTH AND SAFETY COMMISSION (1987) Managing health and safety in construction: Principles and application to main contractor/sub-contractor projects. HMSO.
   HEALTH AND SAFETY COMMISSION (1988) Managing health and safety in construction: management contracting. HMSO.
3. HEALTH AND SAFETY COMMISSION (1989) (CONIAC) The control of substances hazardous to health in the construction industry. HMSO.
4. HEALTH AND SAFETY EXECUTIVE (1989) Construction (Head Protection) Regulations 1989 Guidance on Regulations. HMSO.
5. HEALTH AND SAFETY EXECUTIVE (1989) Noise at work, Guide 1 Legal duties of employers to prevent damage to hearing. HMSO.
   CONSTRUCTION INDUSTRY RESEARCH AND INFORMATION ASSOCIATION (1990) Report 120 A guide to reducing the exposure of construction workers to noise.

ABSENTEEISM IN THE BUILDING INDUSTRY

by T Burch, PhD, MSc, FCIOB

INTRODUCTION

'In a country which relies for its wealth and hence the standard of living of its citizens not upon its natural resources but upon the performance of its industry, any impediment to the rhythm of that performance must adversely affect the added value of the product output. We have become accustomed to deplore and even to accept apathetically such interruptions as strikes, embargoes, work to rule and go-slows, while ignoring the continuous background of disruption - absences from work. This relegation of an important feature of modern industrial life derives partly from the diffuseness of its impact on the working economy, but partly because there is an implicit assumption that it is principally due to the incidence of ill-health, a condition which cannot be completely catered for in organising industry'.

Sir Monty Finniston FRS

This paper reviews certain important issues concerning the absence behaviour of workers in the construction industry. The concept of absenteeism is discussed in relation to the principal reasons for absence and a theoretical model is produced in an attempt to explain the fundamental process associated with the decision to be absent. The factors controlling absence and their effect are considered and aspects of involuntary and voluntary absence are examined.

CONCEPT OF ABSENTEEISM

It is one of the paradoxes of postwar western industrial development that, in spite of a gradual rise in real earnings, improved working conditions and major advances in the field of preventative medicine, there has been a steady increase in the level of industrial absence.* In common with other industries, absence from work creates problems for the building industry seriously affecting planning by reducing the effectiveness of teamwork, and output, and by causing machines to stand idle.

Since effective co-ordination of trades and operations on site is essential, absenteeism* can be a major factor in reducing the productivity of those operatives who turn up for work, as well as representing a serious loss of overall potential output. This is true particularly where the absentee is a member of a gang. For example, the productivity of a bricklaying gang is drastically reduced by a tradesman having to take on the work of an absent bricklayer's labourer.

Absence can never be completely eliminated and its level will vary from contractor and from site to site. Those factors which have been shown to have an effect on absenteeism are shown in Figure 1.

---

* The term 'absence' and 'absenteeism' are used inter-changeably throughout this paper, and the subsequent quantitative measure is termed the 'absence rate': this is computed as follows:

Absence rate = Number of man-days absent

Absence rate = Number of man-days possible x 100%

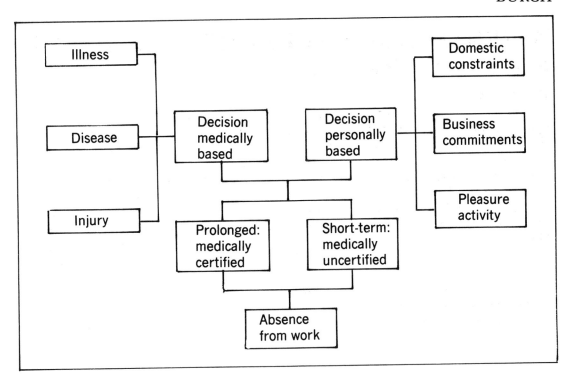

Figure 1 : Factors associated with absence behaviour

Clearly, absence will either be largely for medical reasons, often resulting in prolonged absence (exceeding three days), or for personal reasons which are frequently only of short-term duration (three days or less). Greenwood[1] pointed out the managerial problems associated with the two forms of absence:

'Casual short-term absentees cause the most inconvenience because of their unpredictability and because they are spread throughout the company. The long-term persistent absentee, on the other hand, causes less trouble because he is away for recognised causes, usually sickness, of which the duration can be estimated ... Neither category is open to much remedy; the sick cannot be hurried, and the persistent absentees are not very sensitive to coercive measures'.

Gibson[2] recognised that short-term absence represents a serious problem to industry and this fact has been endorsed by a number of studies. Negrine[3] reported that 89% of all absences were of short duration and represented half the total working days list; Taylor[4] confirmed that 60% of all absences were of less than four days' duration, and Behrend[5] found that 82% of all absence spells were of short duration. With regard to the practical effect, Porter, Lawler and Hackman[6] note that short-term absence disrupts schedules, creates the necessity of over-staffing (owing to a need for surplus labour pool to cover for absentees) and reduces productivity. Other studies[2,7,8] have all asserted that a large proportion of short absences is voluntary and is, therefore, within the control of management.

Any short absence from work is believed to be caused by one or more non-medical factors; the factors associated with voluntary absence being dependent upon the inter-relationship

Absenteeism in the building industry

of several groups of variables which are mainly psychological, sociological, domestic or economic in nature. These may be broadly categorised as:

Intra-organisational   :   factors within the employing organisation that affect the employee or groups of employees in the work situation.

Extra-organisational   :   factors outside the employing organisation that affect the labour market and employing situation,

Personal variables   :   factors within or appertaining to the individual employee.

From these basic categories of factors, a theoretical model (Figure 2) may be derived in an attempt to explain the fundamental process associated with the decision influencing absence behaviour.

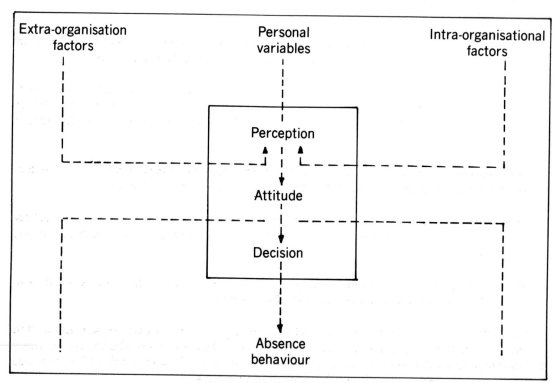

Figure 2 : Associated process of absence behaviour

Haire[9] concluded that individual attitudes are a dominant factor in shaping behaviour and if attitudes were to change then behaviour would change. This implies that in any measurement of behaviour it is necessary to isolate those attitudes that may be responsible for particular behaviour patterns. Additionally, an individual's perception of a given situation may be completely different from the perception of another. It is considered, therefore, that these two psychological characteristics of attitude and perception are mutually dependant in their formation and application and cannot be ignored in any behavioural problem. It is inferred, therefore, that individuals with differing absence behavioural patterns should display different individual characteristics and have different concepts and attitudes towards intra-organisational and extra-organisational factors and also towards their own personal backgrounds. The identification of these individual characteristics and factors associated with organisation design might thus provide a framework leading to the control of certain forms of absence behaviour.

EFFECT AND CONTROLLING FACTORS OF ABSENCE

The achievements of doctors in occupational medicine, and more recently, industrial psychologists, have clearly demonstrated the value of employing workers who are both healthy and positively motivated towards their jobs[4]. Health and motivation might thus be considered the basic criteria associated with employee attendance. However, if an employee is absent from work, for whatever reason, there are a number of consequences to the employer. If the absence is not pre-planned, as is usually the case with sickness absence, there are additional consequences. Therefore, it is in all employers' interests to do whatever they can to reduce the disruptive effect of absence. The main effects of sickness absence alone may be summarised as follows:

(a) the need to cover the duties/tasks of absent employees, either by holding a 'reserve' of labour on-site or by means of overtime work. Both lead to additional costs;

(b) the loss of production, or reduction in services offered, as a result of absence, to the extent that the employer cannot, or is not prepared to, cover planned work - resulting to delays on site leading to non-completion of contract and perhaps extension of time/liquidated damages;

(c) the direct cost falling on the employer of an occupational sick scheme, and of the cost of contributing to the state for the provision of state sickness benefit;

(d) the unproductive administrative effort involved in arrangements for covering duties/tasks, adjustment of remuneration and other documentation involved when employees are absent;

(e) the adverse effect on those remaining at work, both in relation to general morale on site and to the direct effects of absence upon teamwork.

The overall volume of absence is such that these items represent a very substantial burden falling on employers, sufficient to justify considerable effort to reduce the level of sickness absence together with other forms of absence behaviour not directly attributed to sickness or injury, a large part of which may be voluntary in nature.

It must be recognised that there will always be a level of sickness absence which is irreducible and a number of circumstances where the individual should be encouraged to stay at home rather than be a danger to himself or his colleagues. In attempting to reduce the level of absenteeism on site the contractor should consider implementing three main strategies:

- improving and maintaining the health of the operatives; this may relate particularly to the working environment on site in connection with various safety, health and welfare measures, and regular health screening checks for employees[10];

- ensuring that persons selected for employment are, at the time of acceptance, fit to undertake the work concerned;

- eliminating unnecessary absence, particularly voluntary absence which is largely short-term in nature; control procedures in this connection might include the registration and regular monitoring of all short absences. A number of previous studies including those by Borcherding and Oglesby[11,12] Waters and Roach[13], and Talacchi[14], have all confirmed strong inverse relationships between absence and job satisfaction. Therefore, factors directly associated with the satisfaction aspects of employment (pay, work itself, promotion prospects, colleagues and supervision) should be given particular attention in connection with the control of short-term/voluntary absence.

Although the strategies outlined above would certainly help in the control of absence on site, nevertheless there still exists a wide gap between theory and practice. There is often the paradoxical situation in which, despite substantial betterment of working conditions, wages, medical care and social security, there is yet a serious problem of individual productivity on the construction site, most obvious in the rising trend of absenteeism.

INVOLUNTARY/VOLUNTARY ABSENCE

Absence behaviour is often referred to as being either involuntary or voluntary[15]. Involuntary absence involves illness, disease and injury, where the absence decision is primarily medical rather than one of individual choice. Conversely, voluntary absence is non-medical, the absence decision being entirely that of the individual. In this connection, it has been found[16] that voluntary absence was in excess of 25% in two large construction organisations. While allowing for cases of malingering and unqualified or 'uncertain' short absences causes could be realistically in the region of 33%. Behrend[15] also confirmed voluntary absence to be of this order. There are also considerable absences associated with certain business commitments and domestic constraints, which are often classified into the grey area of absence behaviour. However, such cases would in reality constitute either involuntary or voluntary absence, depending upon the circumstances.

Certain cases of sickness absence, (malingering) may often extend from the initial involuntary to the voluntary, due to the employee not wishing to return to work. Malingering, which the dictionary defines as 'the pretence of illness or the production or protraction of disease in order to escape duty', frequently afflicts industry. While some individuals may fabricate the symptoms of a disease in order to obtain a medical certificate, this appears to be nowhere near as common as the exaggeration of mild symptoms which actually exist. Although this distinction may appear semantic, it is the latter form of malingering which is more difficult to expose and causes real problems to both employers and doctors alike. A doctor is trained to recognise when someone is seriously ill and in need of treatment, but it is very difficult to decide exactly when someone is well enough for work on site. Indeed, in almost every case the doctor will have to rely substantially upon the patient's own opinion - and this can be influenced by matters not related to physical fitness and the demands of the job.

This fact is ratified by Taylor[17], who says:

'Whether the employer likes it or not, it is the patient himself who decides upon his fitness for work in all but exceptional circumstances. Influence people's attitudes to work and you will influence rates of sickness absence - in either direction'.

ABSENTEEISM IN THE BUILDING INDUSTRY

BURCH

PERSONAL AND ORGANISATIONAL FACTORS ASSOCIATED WITH ABSENCE BEHAVIOUR

A limited amount of research has been carried out concerning the causes and associated factors of sickness absence but there is little related to the construction industry. However, certain valuable findings have been reported which, if translated into policy, could reduce the level and impact of industrial absence in general.

It is known that both certified and uncertified absences are non-random phenomena; some persons consistently take more time off than others and thus absence rates are to some extent predictable in accordance with certain personal and organisational factors which influence the individual's perceptions, attitudes and ultimately, absence behaviour. For example, the personal factors of age[18], length of service[19], wage level and family responsibility[20], skill level[16] and job satisfaction have all been found to have a significant bearing on industrial absence rates. Similarly, relationships have been confirmed between absence and certain inter-organisational factors such as supervision[21], group-size[22], and the physical conditions of work[23], while the more important extra-organisational factors to influence absence rates have been found to concern general employment conditions and local levels of employment[15]. Meteorological and climatological factors have also been found to have a considerable influence upon industrial absence[19], and associations have been confirmed in connection with seasonal changes[24] and daily variations[25].

Bearing all of these points in mind, considerable variations have been found in relation to different industries. A survey carried out by the Incomes Data Services Ltd[26] found that the highest absence rate of 17% was in mining and quarrying, while the lowest rate of 5% was in insurance, banking and finance. Though little data exist for the building industry, a comparative evaluation of absenteeism on selected construction sites[16] produced absence rates for a large private contractor of 10% and for a direct works department of 6%.

MEDICAL AND HEALTH FACTORS ASSOCIATED WITH ABSENCE BEHAVIOUR

Assessment of the health of workers in any industry depend upon the ability to measure and collect the necessary data. Unlike a number of other European countries, the UK construction industry has developed only rudimentary occupation health facilities for its employees[27]. There is no organised body which might make a continuous assessment of the risks to health entailed in construction site activities, and undertake the routine monitoring of employees' health standards and related absence behaviour.

Despite the lack of data concerning the health standards, prolonged illness and sickness absence of construction workers, certain pointers do exist. For example, it has been found[16] that the nature of work and the working environment of the building industry significantly related to certain medical causes of absence, namely, respiratory diseases, musculo-skeletal disorders, sprains and strains, and injury. Sorrie[27] reported:

> '... construction workers die from lung cancer and other respiratory diseases more frequently than can be explained by factors other than occupation. Certain groups of construction workers also die from a high occurrence of cancer of the stomach ... many conditions, which though painful and disabling, do not in themselves contribute directly to death. Dermatitis and other skin conditions, and muscle-joint conditions, fall into this category, though both are known to be common to construction workers'.

Magnuson[28] also reported a number of health hazards which frequently led to loss of time from work for construction workers and a number of causes of ill-health are identified with particular trades and occupations.

It seems reasonable to conclude that if an attempt is to be made to reduce sickness absence in the construction industry, companies must be prepared to give more consideration to the well-being of their workers and health standards generally. This is certainly one of the more important implications of the Health and Safety at Work Act 1974.

CONCLUSIONS

It has been demonstrated that there are a number of interacting factors which influence and affect the absence behaviour of workers in the construction industry. An understanding of these factors will assist management in achieving closer control of absence on site. Research has been carried out in an attempt to produce a framework of reference concerning the absence behaviour of workers in the construction industry upon which decisions might be based in connection with certain construction management practices, policies and control procedures[29]. High levels of absenteeism mostly ascribed to sickness and voluntary causes necessitate a thorough understanding and control. Some attribute such absence to the welfare state in which sickness absence is encouraged by substantial sickness and other benefits, which cost the country in the region of £550 million a year. Others blame the stresses and strains of life in an industrial society and the conditions under which people work. Yet other opinions blame the medical profession which is responsible for the issuing of the necessary certificates of incapacity.

Several studies made of absenteeism show that national insurance sickness benefits are now as high in relation to earnings as can safely be allowed, although it may only be in a minority of cases that they are high enough to provide a further stimulus to absenteeism. It is also clear that although the level of benefits and the ease with which medical certificates can be obtained must be regarded as factors contributing to the growth of absenteeism, the problem is considerably more complex. There are cases of outright malingering and voluntary absence which have to be counted, but there are many cases of sickness absence which occur, where medical reasons play a part in giving rise to absence, but also in which other factors are present. Questions of job satisfaction, personal attitudes, work patterns, group organisation and work motivation can have considerable effects on the extent to which a worker will succumb to the medical considerations which might be used to justify this absence.

It seems likely that contractors could do much by providing more positive motivation[29,30] and stimulating their employees' interest in their work, and thereby counteract the most critical forces giving rise to absenteeism. But before company management can make a serious attack on the problem they must first ascertain the extent and causes of absence among their employees by keeping adequate records - a procedure sadly lacking in the building industry. Full knowledge of the incidence and causes of absence would thus constitute an important part of the framework concerning absence control; subsequent reductions in absenteeism would thus be reflected in improved productivituy on site.

POSTSCRIPT

Research concerning sickness and other forms of absence is complex and the phenomenon is such that there is still only a very partial understanding of its nature. However, one fact that is apparent is that sickness absence is not synonymous with any objective measure of true morbidity. Indeed, the existence of this distinction between 'sickness absence' and 'sickness' accounts for much of the research interest devoted to the former of which a substantial amount has been medically orientated. Few attempts, however, appear to have been made in the past to identify that part of absence which is neither serious nor justified, yet causes tremendous production losses to industry.

It seems that absence from work is generally considered in broad terms as being non-medical or medical though there will, however, always by a 'grey area' in between. More recent research[29] has attempted to shed some light on the nature and extent of both genuine sickness and unjustified absence amongst construction workers, and a number of explanatory variables, attitudinal and motivational concepts have, to some extent, served to explain absence behavioural trends. A summary of this study has been published by the Chartered Institute of Building[32].

Like most research, a number of problems are to be expected which create difficulties in research design and method. The problem of absence-research field work particularly creates some difficulty: few people have the desire to be questioned on their personal and job related circumstances and on their attitudes towards their job and work situation. Still fewer people wish their absenteeism and sickness histories to be recorded and analysed, particularly when the objective of the exercise is to interrelate this information with their individual circumstances both at home and at work, despite assurances of confidentiality. A co-operative sample study-population will, therefore, always be a major issue concerning objective absence.

Another major problem in studying absenteeism is that any researcher has some difficulty in approaching the subject in an impartial and objective manner. The fact is that everyone has at some time experienced some degree of sickness and other absence both in himself and in his colleagues. From this subjective experience the researcher is bound to form some opinion as to the reasons for its occurrence, be they legitimate or otherwise.

In conclusion, to put the absence problem into perspective, it should be considered that the relatively high absence rates experienced by industrial organisations are often the result of a small number of periodic events occurring among a small proportion of the working population. It is a mistake, therefore, to evaluate absence trends without some understanding of the interwoven web of factors associated with absence behaviour. It is a major issue that management seek to understand these relationships with the sole aim of controlling absence.

It should be further considered that absence may, at some time, be beneficial both for the individual and the organisation, and at other times be problematic for both parties and symptomatic of poor psychological and social adjustment. In either case, the principal aims in analysing, understanding and controlling absence should be the enhancement of the well-being of employees both at home and at work and to procure greater effectiveness in production.

REFERENCES

1. GREENWOOD G B(1951) Is absenteeism a problem? *Journal of the Institute of Personal Management 33* (313), pp160-169
2. GIBSON J O (1966) Towards a conceptualisation of absence behaviour of personnel in organisations. *Administration Science Quarterly 11*, pp107-133
3. EGRINE N (1974) Job satisfaction and absenteeism. MPhil Thesis, Brunel University
4. AYLOR P J (1969) Return to work industry's responsibility. *Journal of Occupational Medicine 11*, pp678-682

5. EHREND H (1978) How to monitor absence from work from headcount to computer. Institute of Personnel Management 11

6. PORTER L W, LAWLER E E and HACKMAN J R (1975) Behaviour in organisations. McGraw-Hill

7. FROGGATT P (1970) Short-term absence from industry. *British Journal of Industrial Medicine* 27, pp199-224

8. BEHREND H (1951) Absence under full employment. Research Board, Faculty of Commerce, University of Birmingham

9. HAIRE M (1970) Psychology in management. McGraw-Hill

10. BAILEY A R and WRIGHT H B (1980) Joint Industry Board Health Screening Project. Press Statement No.122 April

11. BORCHERDING J D and OGLESBY C H (1974) Construction productivity and job satisfaction. *American Society of Civil Engineers. Journal of Construction Division 100,* September, pp413-431

12. BORCHERDING J D and OGLESBY C H (1975) Job dissatisfaction in construction work. *American Society of Civil Engineers. Journal of Construction Division 101,* May, pp415-434

13. WATERS L K and ROACH D (1971) Relationships between job attitudes and two forms of withdrawal from the work situation. *Journal of Applied Psychology 55,* pp92-94

14. TALACCHI S (1980) Organisational size, individual attitudes and behaviour, an empirical study. *Administrative Science Quarterly 5,* pp398-420

15. BEHREND H (1959) Voluntary absence from work. *International Labour Review 79,* pp109-140

16. BURCH T (1979) An assessment of the causes of delays encountered on construction projects. MSc Thesis, University of Salford

17. TAYLOR P J (1970) The English sickness. *Industrial Society 52,* pp8-9, 26

18. KAHNE H R and OTHERS (1957) Age and absenteeism. *Archives of Industrial Health 15,* pp135-147

19. POCOCK S J (1972) Relationship between sickness absence and meteorological factors. *British Journal of Preventative Social Medicine 26,* pp239-245

20. SHEPHERD R D and WALKER J (1958) Absence from work in relation to wage level and family responsibility. *British Journal of Industrial Medicine 15,* pp52-61 12

21. FOX J B and SCOTT J F (1943) Absenteeism : a management problem. *Harvard Business Research Studies 29*

22. ARGYLE M (1972) The social psychology of work. Penguin Press

23. SHEPHERD R D and WALKER J (1957) Absence and the physical conditions of work. *British Journal of Industrial Medicine 14,* pp266-274

24. BRITISH INSTITUTE OF MANAGEMENT (1961) Absence from work - incidence, cost and control. BIM

25. GORDON C and OTHERS (1959) Patterns of sickness absence in a railway population. *British Journal of Industrial Medicine 16,* pp230-243

26. INCOMES DATA SERVICES LTD (1975) Absence. IDS Study No.111 December

27. SORRIE G S (1980) Health hazards in the construction industry. *Proceedings of the Institution of Civil Engineers.* January, pp5-11

28. MAGNUSON H J (1961) Health hazards in the construction industry. *Journal of Occupational Medicine 3,* (7) July, pp321-325

29. BURCH T (1982) Investigations concerning the absence behaviour of workers in the construction industry. PhD Thesis, University of Salford

30. MASON A (1978) Worker motivation in building. Institute of Building Occasional Paper No.19

31. SCHRADER C R (1972) Motivation of construction craftsmen. American Society of Civil Engineers. *Journal of the Construction Division* 98, (CO2), pp257-273

32. BURCH T (1983) Absence behaviour of construction workers. CIOB Occasional Paper No.28 February

RESPONSIBILITIES OF CONTRACTORS, SUB-CONTRACTORS AND OTHERS UNDER THE HEALTH AND SAFETY AT WORK, ETC. ACT 1974

by B R Norton, Solicitor

INTRODUCTION

The construction industry continues to have an unacceptably high accident rate despite more than adequate legislative provisions. The Health and Safety at Work, etc. Act 1974 (HASAWA) and the various detailed Construction Regulations provide a very comprehensive legal framework for the industry and it is an undisputable fact that if the law were observed nearly all the reported accidents, both fatal and non-fatal, would be prevented.

There is no lack of detailed guidance available for the industry either. There are many excellent publications giving practical advice on a wide range of subjects. It would seem, however, that one area where advice has until recently been sadly lacking is in the need to co-ordinate the separate activities and contractors to be found on a construction site.

The Construction Industry Advisory Committee (CONIAC) has published two Guidance Notes — *Managing health and safety in construction : principles and application to main contractor/sub-contractor projects* (ISBN 0 11 883918 7 available from HMSO) and *Managing health and safety in construction : management contracting* (IBSN 0 11 883989 6 available from HMSO) — which offer good advice on this aspect of construction. These are being revised and updated.

This paper gives the legal framework within which contractors operate on construction sites.

RESPONSIBILITIES

A contractor undertaking a construction project takes on many responsibilities. Some are assumed voluntarily and others are imposed either by the common law or by statute. They are not all of the same kind and give rise to different obligations and sanctions.

It is the purpose of this paper to examine those responsibilities (or duties) imposed on a contractor by virtue of HASAWA. To avoid any confusion other responsibilities will be identified and differences from those under HASAWA described.

Contractual responsibilities

When a contract is entered into the parties voluntarily assume responsibilities which are usually set out in some detail in the contract documents. The contractor assumes the responsibility of constructing the works in accordance with the specification, drawings, bill of quantities, etc, and the employer is responsible for paying in accordance with the conditions of contract. These responsibilities are contractual, and if one party fails to comply with his side of the bargain the other party can claim for breach of contract. Such a claim is a CIVIL claim which would be dealt with in the civil courts.

It is important to remember that the remedy for a breach of contract is usually an action for damages, ie, an award of a sum of money to compensate the aggrieved party for whatever loss was caused by the other party's breach. It is only in very rare cases that the courts will make an order forcing the party in default to perform his contractual obligations. Such orders, known as orders for 'specific performance' are in the discretion of the court and will be granted only in cases such as the failure of a dealer to deliver a unique work of art where an award of damages is inappropriate. It is most unlikely that the courts would order specific performance of the terms of a construction contract.

Whilst it is, therefore, helpful to insert into contract documents clauses making it clear that the contractor or sub-contractor is to be responsible for providing his employees with safe places of work, appropriate safety equipment, etc, failure to comply with such conditions will only enable the other party to the contract to claim for breach of contract and an award of damages. Contract clauses cannot be used as a means of transferring or contracting out of one's responsibilities in criminal law, such as the Construction Regulations or HASAWA. A clause providing that sub-contractors using the scaffolding erected by the main contractor 'do so at their own risk' or 'gives rise to no liability on the part of the main contractor', will in no way protect the main contractor from a criminal prosecution if he has failed to erect the scaffolding in accordance with the requirements of the Construction (Working Places) Regulations 1966. There are certain special defences available to protect parties who are put in breach of the law by the actions of someone else, but so far as contract clauses are concerned it is not possible to contract out of criminal statutory duties.

What can be done, however, is to transfer the financial consequences (except the fines imposed by the court) of a breach of the criminal law by appropriately worded contract clauses. It is not uncommon to find a clause requiring one party to indemnify the other against any loss or damage caused by a breach of the law by either party. Thus, in the case of *Smith v Vange Scaffolding and Engineering Co. Ltd.*[3], whilst both Vange who employed Smith and the main contractors Constructors John Brown Ltd (CJB) were in breach of the Construction (Working Places) Regulations 1966 and were both found liable to pay damages to Smith for injuries he received as a result of those breaches of statutory duty, CJB were able to recover their portion of the damages from Vange because of an indemnity clause in their contract. This indemnity clause, however, could have no effect on any prosecutions or fines imposed on either party under the criminal provisions of those Regulations.

Tort

Tort is 'a wrong which gives rise to an action for damages'. There must be a breach of a duty owed by one person to another which causes that other damage. The two elements — breach of duty and damage — must be present before a civil action can be brought. The case of *Smith v Vange Scaffolding and Engineering Co. Ltd.* illustrates this principle. Smith was able to show that Vange (and CJB) owed him a duty to take care for his safety, were in breach of that duty, and he (Smith) suffered damage both financial and in the form of personal injuries.

Just as in the case of contractual liabilities it is possible to transfer liability for the consequences of a breach of this duty by an appropriately worded condition of contract. It is not possible, however, as a result of the Unfair Contract Terms Act 1977 contractually to exclude liability for personal injuries caused by one's own negligence. It is however, perfectly legal to obtain indemnity against any damages to be paid for such personal injuries.

Employer's liability

The most common action in tort which contractors are likely to come across is an action

Responsibility of contractors, subcontractors and others

claiming damages for personal injuries sustained by an employee at work (employer's liability).

To succeed in such a claim the employee must establish one or more of the following:

(a) that the employer was negligent (eg, failing to provide a safe place of work); or

(b) that another employee of the same employer was negligent (eg, dropping a hammer from a height on to someone's head); or

(c) that the employer was in breach of statutory duty (eg, contravened one of the requirements of the Construction Regulations). (NB. For this purpose a breach of the general duties imposed by the Health and Safety at Work, etc, Act 1974 will not found a claim.)

However, an employee who succeeds in establishing one of the above will, for the purposes of the civil claim, have his own blameworthiness taken into account.

Contributory negligence

Under Section 1(1) of the Law Reform (Contributory Negligence) Act 1945, where a person as a result partly of his own fault is injured, his damages are reduced by whatever amount is just and equitable, having regard to his share in the responsibility for the damage. In other words, if a man is injured and the court finds that the employer was 75% to blame and the man 25% to blame and damges are assessed at £2,000, the man will only be awarded £1,500.

Criminal courts are in no way concerned with this sort of apportionment of blame.

Criminal responsibilities

Quite apart from those responsibilities assumed voluntarily under the terms of the contract and those imposed by the common law (eg, tort) and statute mentioned above which give rise to claims in the Civil Courts for damages, contractors also have imposed on them many responsibiltities under the criminal law.

This distinction between civil law and criminal law lies in the purposes behind the two systems. Civil law is concerned with the recovery of money or property, or the enforcement of a right or advantage. Criminal law is concerned with the punishment of someone who has committed a crime.

The purposes are quite distinct and so are the courts which deal with them. Civil Courts award damages and in doing so look at all the facts and 'on the balance of probabilities' give judgment accordingly. In claims for damages for personal injuries this will include an apportionment of blame between the various parties.

The Criminal Courts, however, punish people who have committed a crime. They do not apportion blame and must be satisfied 'beyond reasonable doubt' that the accused is guilty.

The following is a diagram (much simplified) of the two systems of courts:

Before looking at the responsibilities of contractors, sub-contractors and others under HASAWA it is important to draw attention to two significant changes to the common law principle that the prosecution must prove the accused's guilt 'beyond all reasonable doubt', to be found in HASAWA.

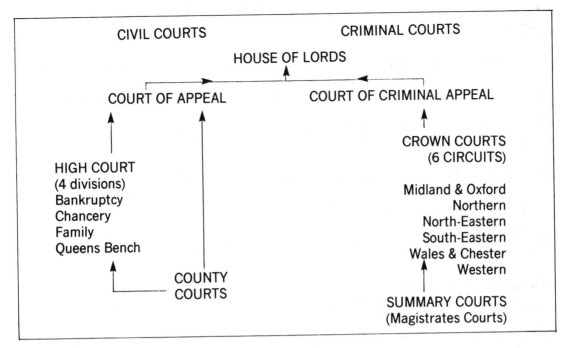

CIVIL COURTS   CRIMINAL COURTS

HOUSE OF LORDS

COURT OF APPEAL   COURT OF CRIMINAL APPEAL

HIGH COURT
(4 divisions)
Bankruptcy
Chancery
Family
Queens Bench

COUNTY
COURTS

CROWN COURTS
(6 CIRCUITS)

Midland & Oxford
Northern
North-Eastern
South-Eastern
Wales & Chester
Western

SUMMARY COURTS
(Magistrates Courts)

S.40 HASAWA provides that in a prosecution under HASAWA for a failure to do something 'so far as is reasonably practicable', it is for the accused to prove that it was not reasonably practicable to do more than was actually done. The accused does not have to prove this beyond reasonable doubt, however, only that his defence is probably right (R v Carr-Bryant;[4] R v Dunbar[5]).

S.17 HASAWA provides that a breach of an 'Approved Code of Practice' (ie, one approved under S.16 HASAWA) can be taken as proof by the Court of a breach of the requirement to which it relates, unless the accused can show that he complied with the Statutory Provision in some other way (S.17). Both these sections place the onus of proof on the accused instead of the prosecution.

RESPONSIBILITIES UNDER HASAWA AND THE RELEVANT STATUTORY PROVISIONS

Introduction

The construction industry has changed, with traditional patterns of employment no longer being followed. Rarely is everyone on a site employed (in the master and servant sense of the word) by the contractor whose name is on the contract documents or on the hoarding round the site. It is now more common to find a main contractor, who may or may not be doing a substantial part of the work himself, and a plethora of sub-contractors. Some sub-contractors are themselves sub-contracted to sub-contractors and the chain may be even longer.

Legislation to implement the EC Directive on Temporary or Mobile construction sites will recognise these changes and impose new duties on both clients and contractors to co-ordinate the activities of all persons on site.

In earlier legislation great emphasis was placed on the master and servant relationship. Thus, the Court of Appeal in Smith v George Wimpey & Co. Ltd.[6] held that the employee of a

Responsibility of contractors, subcontractors and others

sub-contractor is owed no duty by the main contractor under Regulation 3(1)(a) of the Construction (General Provision) Regulations 1961. Similarly, a sub-contractor owes no duty under this Regulation to an independent contractor engaged by him (*Clare v Whittaker & Son (London) Ltd.*[7]). This approach was changed by the Health and Safety at Work, etc, Act 1974 (HASAWA). The Act covers all who are either at work or are likely to be affected by the operations of those at work. Employees, other contractors' employees, trainees, and members of the public are all embraced by HASAWA. Even in HASAWA however, the master and servant relationship is given special consideration.

Employee — definition in HASAWA

It is singularly unfortunate that the English legal system admits of more than one definition of employee. 'Employee' is defined for a variety of purposes. It may be important to decide if this 'label' can be attached to a man for purposes of redundancy pay or unfair dismissal. The purpose may be to decide who is liable to deduct income tax, or pay statutory sick pay or pay damages for personal injury. There are numerous cases using different criteria. It would take too long and in the end be quite unhelpful in the context of this paper to review these. HASAWA defines 'employee' as 'an individual who works under a contract of employment' (but see reference to YTS trainees post) (HASAWA Sect.53), and for the purpose of the rest of this paper that is the definition that will be applied. However, it should be remembered that a Civil Court looking at an employer's liability for personal injuries is not bound by that definition and is going to look at the reality of the situation, not the 'label' the parties chose to attach to an individual.

Relevant statutory provisions

It is essential to appreciate that the statutory provisions, and the regulations made under them, which existed at the time when HASAWA was enacted in 1974 are still in force (except insofar as they have been specifically repealed or revoked). These statutes and regulations, together with Part 1 HASAWA and safety regulations made under HASAWA, are defined in the Act as 'the relevant statutory provisions'. A breach of any of these is an offence under S.33 HASAWA.

Summary conviction now carries a fine of up to £2,000. Convictions on indictment in the Crown Court carries unlimited fines and/or imprisonment. The HSE are tending to bring more prosecutions on indictment to secure higher penalties.

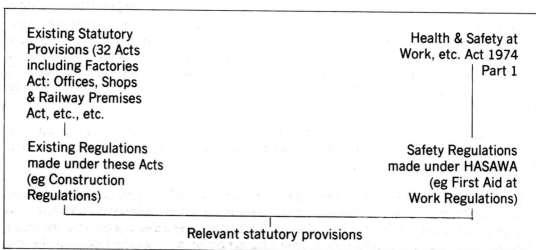

Existing Statutory Provisions (32 Acts including Factories Act: Offices, Shops & Railway Premises Act, etc., etc.)
│
Existing Regulations made under these Acts (eg Construction Regulations)

Health & Safety at Work, etc. Act 1974 │ Part 1
│
Safety Regulations made under HASAWA (eg First Aid at Work Regulations)

Relevant statutory provisions

Duties to 'employees'

The duties to employees under S.2 HASAWA are the common law duties which existed prior to 1974 but are now in statutory form. The statute (HASAWA) is a criminal statute, the purpose of which is to punish offenders and indirectly deter possible offenders. (See S.2 HASAWA in the Apppendix.)

Duties to others (S.3 HASAWA)

HASAWA broke new ground in the provisions of S.3 (and to some extent in S.4 which is discussed later), by placing a duty on employers to ensure so far as is reasonably practicable that persons not in their employment who may be affected by what they do are not exposed to risks to their health and safety. (See S.3 HASAWA in the Appendix.) Therefore, it will be seen that the question of whether a person is a sub-contractor, self-employed or a member of the public FOR THE PURPOSES OF S.3 HASAWA is really irrelevant. If he is not an employee (as defined in the Act) but is 'affected' by the undertaking then the contractor has a (criminal) responsibility for his safety. The duties under S.2 and 3 were brought home with some force in the case of R v Swan Hunter and Telemeter Installations Ltd[2].

Swan Hunter and its effect

Briefly, the facts of the *Swan Hunter* case are as follows:

During 1976 HMS Glasgow was under construction by Swan Hunter and by September was being fitted out. On 25 September 1976 an explosion occurred when a welder was lighting his torch in a small compartment on the lowest deck forward of the bridge. Of eleven men below deck only three escaped with their lives. The explosion was due to the atmosphere in the compartment being oxygen enriched. This state of affairs was caused by a hose leaking oxygen into the compartment. The hose had been left in this state by Telemeter Installations Ltd, who were sub-sub-contractors and not in any contractual relationship whatever with Swan Hunter. Many of the men who died were employees of Swan Hunter.

Telemeter pleaded guilty to charges under Sections 2 and 3 HASAWA and were fined £15,000. Swan Hunter pleaded not guilty to similar charges but were found guilty and fined £3,000. Swan Hunter appealed but the appeal failed. The most significant finding of the Court of Appeal (Criminal Division) was that in discharging their duty to their OWN EMPLOYEES under Section 2 HASAWA Swan Hunter should have given INFORMATION and INSTRUCTION to the sub-sub-contractors about the dangers, WELL KNOWN to Swan Hunter, of allowing oxygen to leak into a confined space. Swan Hunter in fact issued their own employees with a booklet giving information about this danger but the booklet was never given to Telemeter. Despite many accounts in various well known safety publications which suggest otherwise, the Court of Appeal DID NOT find that Swan Hunter owed a duty to TRAIN the sub-sub-contractor's workmen.

The mixture of workers and employers in this case can be found on most construction sites. How then should contractors regard this case as defining their duties to persons not in their employ, ie, sub-contractors, etc?

In the first case a large manufacturing company engaged a small firm of specialist demolition contractors to demolish and dismantle redundant plant which included a sulphuric acid storage tank. Before any demolition of the tank began there was a meeting between the firm's manager, who was responsible for liaison and supervision of contractors, and the site supervisor of the demolition contractors. The manager told the supervisor that it was not certain whether the tanks still contained sulphuric acid or not.

He was told to ensure that before dismantling commenced the top cover should be removed in order to look inside and see if any liquid was left. If there was any liquid left then the supervisor should report back to the manager who would make arrangements for it to be dealt with. The firm accepted that there was a need for instruction and information to be given to their contractors.

Some ten days after the meeting, dismantling of the sulphuric acid tank began but the demolition contractors elected not to use their own employees but to engage sub-contractors. Two men were to do the job and one was told by the supervisor that the tank had been washed out and was clean inside and they were left to proceed without further advice and instruction about the possible hazards and the way in which they were to be dealt with. One of the men took the top cover off the tank, looked inside and saw that it was half full of liquid but not surprisingly after what he had been told assumed that it was water remaining from the washing-out operation. He went underneath the tank to try and open the drainage valve to drain the tank but the valve was corroded and would not turn. His colleague then partially cut through the drainage valve and stood on top of the bund wall which surrounded the tank, hitting it with a piece of wood until the pipe broke. The complete contents of the tank flowed out and instead of being water it proved to be 700 tons of concentrated sulphuric acid.

The second example of a failure to govern properly the relationship between the work of a main contractor and sub-contractor involved the collapse of falsework. A construction firm carrying out a development of shops and offices engaged a firm of shuttering contractors to design the falsework and formwork which was to be used during the concreting operations at the first floor level. On the morning of the pour the Senior Engineer of the main contractor noted defects in the suppport system of the formwork which were pointed out to the shuttering contractors. Some additional ties were put in but the remedial work was not completed before the pour began and there was no check on the completion of the work by either the main or sub-contractors' site managers before the pour went ahead. The whole of the shuttering collapsed while three of the main contractor's men were on top working on the pour and all three were injured. In this and the previous case, had a written 'permit to work or to load' system been in use the accidents would not have happened. In both cases both sets of parties were prosecuted under the 1974 Act and under the Construction Regulations.

One of the recent cases to underline the fact that employers owe duties under HASAWA to anyone likely to be affected by their activities and not just their own employees is the prosecution arising out of the multiple fataltiy at Carsington Reservoir[1]. The facts are briefly:

— an earth dam had been constructed at Carsington in Derbyshire to form part of a water storage reservoir. The dam was constructed of natural earth over which were crushed rock binders on the exposed faces. Incorporated in the construction of the dam were vertical drainage blankets which comprised crushed limestone at 3 to 4.5 intervals. On the waterless side of the dam a toe drain had been constructed along the whole length to take water seepage from the drainage blankets. At intervals along the toe drain, access shafts had been constructed; these shafts were between 15 and 20 feet deep;

— the consulting engineer on the project, who had designed the drains regularly sent an engineer and a contractor's employee down these shafts to carry out inspections so that any defects could be put right before completion;

— later, when construction of the dam was almost completed but the reservoir had not been filled, it was found necessary to measure the water levels in the shafts to determine what natural water seepage was taking place;

— a labourer employed by the main contractor entered one of the shafts to carry out water measurements and collapsed in the bottom of the shaft. Another labourer entered the shaft to rescue his colleage when he too collapsed and subsequently two other persons entered the shaft to try to assist and they also collapsed. The result - four fataltities. All four men died from carbon dioxide poisoning and lack of oxygen;

— carbon dioxide can be produced by a reaction between limestone and acidic water. The construction of the dam used crushed limestone in the drainage blanket and the percolation of acidic rain through the drainage blanket could have produced carbon dioxide which propogated through the toe drain and accumulated in the access shafts;

— the consulting engineers were prosecuted under S.2 HASAWA for exposing their own employees to the risk of CO_2 poisoning during the inspections of the shafts and under S.3 for similarly exposing the contractor's employee. The Court found that the risk of CO_2 poisoning in these circumstances should have been known to the consultants and they did not take reasonable practicable steps to avoid endangering their employees and others. The contractors were prosecuted under S.2 for failing to take reasonable practicable steps to protect their own workmen from the danger. The Court heard evidence that an almost identical multiple fatality occurred a few years ago so that the risks should have been known.

It should be clear from these examples that an employer is no longer only concerned with the safety of his own employees and with certain duties specifically placed on him under the regulations. He now has a duty to anyone - and the 'label' on that person does not matter - who may be affected by his undertaking. A main contractor's 'undertaking' must be looked at in the widest sense as being the whole site of operations, not merely the small contribution he is actually making in the way of construction activity. The duty is tempered with the phrase 'so far as is reasonably practicable', however, and providing steps are taken to co-ordinate the various sub-contractors and to exercise some degree of supervision to monitor performance, and steps are taken to stop dangerous practices the duty will be discharged.

Co-ordination will involve pre-planning, regular site meetings with all contractors and safety representatives and most important of all, a regular flow of information between contractors as to what is going on and how this will affect others on site.

Stress has been laid on the main contractor but sub-contractors are carrying on an undertaking and also have duties under S.3. The duty is obviously different in the case of a sub-contractor because whilst he can control the people working under him, he has no control over other sub-contractors or the overall site programme. Nevertheless, he has a duty to ensure so far as is reasonably practicable that his operations do not injure others. He, too, must make known to the main contractor and possibly other sub-contractors any special risks attached to his work which could put them or his own workmen at risk. The flow of information referred to above is not one-way.

Consultants and clients who are involved in the construction process are 'employers' and owe duties to their employees and other people who may be affected by their undertaking.

Responsibility of contractors, subcontractors and others

It is certain that in the future the responsibilities of consulting engineers, architects and planners will be subject to greater attention.

Duties in relation to control of premises (S.4 HASAWA)

Section 4 HASAWA places a duty of care on anyone who has to any extent control of non-domestic premises which are made available to non-employees either as a place of work, or where they may use plant or substances provided for their use. It is clear that when this section of the Act was debated it was not anticipated that it would 'catch' the construction industry and its tradition of sub-contracting. The intention was to control launderettes and DIY garages and similar unattended premises where non-employees might work. HSE have, however, 'stretched' the interpretation of this section and there are many examples of cases where main contractors, in their role as controllers of the non-domestic premises (the site) where non-employees work, have been prosecuted successfully.

For example, a main contractor was prosecuted under S.4(2) for an unsafe excavation in which a sub-contractor was working. The sub-contractor's employers were, of course, also guilty of breaches of the Construction (General Provisions) Regulations 1961 and S.2 HASAWA.

In two other cases, one in a factory and one on a demolition site, the occupier was prosecuted under S.4(2) as the person having control when a contractor's employee struck live electricity cables.

It is worth noting at this stage how wide the meaning of the word 'premises' is. The word is defined in S.53 as:

'Premises includes any place and, in particular, includes:

(a) any vehicle, vessel, aircraft or hovercraft;

(b) any installation on land (including the foreshore and other land intermittently covered by water), any offshore installation, and any other installation (whether floating or resting on the seabed or the sub-soil thereof, or resting on other land covered with water or the sub-soil thereof), and

(c) any tent or movable structure.'

It should also be noted that the duty under S.4 is a limited one. The controller of premises is not responsible for the behaviour of everyone on those premises. His duty, so far as is reasonably practicable, is to ensure that the premises are safe (this includes means of access and egress) and that plant and substances provided for use on them are safe.

Duties in relation to trainees

Trainees are not 'employees' as defined in S.53 HASAWA. Whilst they would be protected under S.3 HASAWA as persons likely to be affected by the undertaking of the employer to whom they went for training, they would not receive the wider protection of S.2 which includes references to training and welfare. Additionally, the definition of 'employee' has been amended by Regulations to include all persons who are on MSC sponsored training courses.

Duties of employees (Sections 7 and 8 HASAWA)

In its attempt to embrace all aspects of work, HASAWA has not overlooked the importance individuals play in safety. Nearly all accidents are attributable to human failure and although reference is made to 'contractors', 'sub-contractors', 'controllers of premises', 'employees', these are really people with different 'labels'.

S.7 places a duty on an employee to take care of himself and others who may be affected by his acts or omissions and to co-operate with his employers and others.

S.8 is an injunction against misuse or interference with things provided for safety, health and welfare.

Enforcement

In looking to see how far contractors have discharged these duties, the HSE will look at the documentary evidence to see how far the contractors have addressed their minds to the problem. Safety policy statements, contract documents, method statements, instruction booklets, safety check lists and the like are all valuable guides. Inspectors will, of course, look at reality as well. It is no use having a wonderful statement of safety policy if it is not put into practice.

CONCLUSIONS

Safety is the province of everyone. It is right that responsibilities for safety should not be limited to the master and servant situation. In today's complex society the actions of employers affect many more people than immediate employees. These facts have been recognised by Parliament and are now enshrined in the Health and Safety at Work, etc, Act 1974. An Act of Parliament on its own, however, is incapable of saving lives or protecting health and safety. It is up to everyone - client, consultant, contractor, sub-contractor and employee - to think about safety constantly and to co-operate with each other in implementing safe systems of work within the framework and spirit set down in the Act.

REFERENCES

1. unreported - source HSE - Trevett v Shepherd Hill Ltd and v. Peter G Eccles, Derek Ormerod and John E Massey (partners of G H Hill & Sons) - Ashbourne Magistrates Court
2. MANSON K (1981) Making 'reasonable practicable provisions for subbies' safety. *Building Trades Journal 182*. October 15, pp42, 46
3. (1970) 1 A11ER 249. Smith v Vange Scaffolding and Engineering Co. Ltd
4. (1943) 1 A11ER 156. R v Carr Bryant
5. (1957) 2 A11ER 737. R v Dunbar
6. (1972) 2 A11ER 723. Smith v George Wimpey & Co. Ltd
7. (1976) ICR 1. Clare v Whittaker & Son (London) Ltd.

Responsibility of contractors, subcontractors and others

APPENDIX
SECTIONS 2, 3, 4, 7, 8 and 9 HASAWA

2. — (1) It shall be the duty of every employer to ensure, so far as is reasonably practicable, the health, safety and welfare at work of all his employees.

(2) Without prejudice to the generality of an employer's duty under the preceding subsection, the matters to which that duty extends include in particular -

(a) the provision and maintenance of plant and systems of work that are, so far as is reasonably practicable, safe and without risks to health;

(b) arrangements for ensuring, so far as is reasonably practicable, safety and absence of risks to health in connection with the use, handling, storage and transport of articles and substances;

(c) the provision of such information, instruction, training and supervision as is necessary to ensure so far as is reasonably practicable, the health and safety at work of his employees;

(d) so far as is reasonably practicable as regards any place of work under the employer's control, the maintenance of it in a condition that is safe and without risks to health and the provision and maintenance of means of access to and egress from it that are safe and without such risks;

(e) the provision and maintenance of a working environment for his employees that is, so far as is reasonably practicable, safe, without risks to health, and adequate as regards facilities and arrangements for their welfare at work.

(3) Except in such cases as may be prescribed, it shall be the duty of every employer to prepare and as often as may be appropriate revise a written statement of his general policy with respect to the health and safety at work of his employees and the organisation and arrangements for the time being in force for carrying out that policy, and to bring the statement and any revision of it to the notice of all of his employees.

(4) Regulations made by the Secretary of State may provide for the appointment in prescribed cases by recognised trade unions (within the meaning of the regulations) of safety representatives from amongst the employees, and those representatives shall represent the employees in consultations with the employers under subsection (6) below and shall have such other functions as may be prescribed.

(6) It shall be the duty of every employer to consult any such representatives with a view to the making and maintenance of arrangements which will enable him and his employees to co-operate effectively in promoting and developing measures to ensure the health and safety at work of the employees, and in checking the effectiveness of such measures.

(7) In such cases as may be prescribed it shall be the duty of every employer, if requested to do so by the safety representatives mentioned in subsection (4) above, to establish,

in accordance with regulations made by the Secretary of State, a safety committee having the function of keeping under review the measures taken to ensure the health and safety at work of his employees and such other functions as may be prescribed.

3. — (1) It shall be the duty of every employer to conduct his undertaking in such a way as to ensure, so far as is reasonably practicable, that persons not in his employment who may be affected thereby are not thereby exposed to risks to their health or safety.

(2) It shall be the duty of every self-employed person to conduct his undertaking in such a way as to ensure, so far as is reasonably practicable, that he and other persons (not being his employees) who may be affected thereby are not thereby exposed to risks to their health or safety.

(3) In such cases as may be prescribed, it shall be the duty of every employer and every self-employed person, in the prescribed circumstances and in the prescribed manner, to give to persons (not being his employees) who may be affected by the way in which he conducts his undertaking the prescribed information about such aspects of the way in which he conducts his undertaking as might affect their health or safety.

4. — (1) This section has effect for imposing on persons duties in relation to those who -

(a) are not their employees; but
(b) use non-domestic premises made available to them as a place of work or as a place where they may use plant or substances provided for their use there, and applies to premises so made available and other non-domestic premises used in connection with them.

(2) It shall be the duty of each person who has, to any extent, control of premises to which this section applies or of the means of access thereto or egress there from or of any plant or substance in such premises to take such measures as it is reasonable for a person in his position to take to ensure, so far as is reasonably practicable, that the premises, all means of access thereto or egress therefrom available for use by persons using the premises and any plant or substance in the premises or, as the case may be, provided for use there, is or are safe and without risks to health.

(3) Where a person has, by virtue of any contract or tenancy, an obligation of any extent in relation to -
(a) the maintenance or repair of any premises to which this section applies or any means of access thereto or egress therefrom; or
(b) the safety of or the absence of risks to health arising from plant or substances in any such premises; that person shall be treated, for the purposes of subsection (2) above, as being a person who has control of the matters to which his obligation extends.

(4) Any reference in this section to a person having control of any premises or matter is a reference to a person having control of the premises or matter in connection with the carrying on by him of a trade, business or other undertaking (whether for profit or not).

Responsibility of contractors, subcontractors and others

7. It shall be the duty of every employee while at work -
   (a) to take responsible care for the health and safety of himself and of other persons who may be affected by his acts or omissions at work; and
   (b) as regards any duty or requirement imposed on his employer or any other person by or under any of the relevant statutory provisions, to co-operate with him so far as is necessary to enable that duty or requirement to be performed or complied with.

8. No person shall intentionally or recklessly interfere with or misuse anything provided in the interests of health, safety or welfare in pursuance of any of the relevant statutory provisions.

9. No employer shall levy or permit to be levied on any employee of his any charge in respect of anything done or provided in pursuance of any specific requirement of the relevant statutory provisions.

WORKMANSHIP : A TEST OF SUPERVISION SKILLS — an introduction to workmanship on building sites

by David Brooks, FCIOB

A CHALLENGE TO MANAGEMENT SKILLS

To be effective as a site manager it is essential to have developed the practical skills of supervision. This requires a good understanding of each trade and the basic operations normally undertaken on site, as well as the ability to talk to those engaged on the work to correct any problems that become apparent whilst work is in progress.

To do this with confidence requires practice and there is no set way in which to deal with each operation. What is certain is that knowledge and practical common sense will provide the authority to handle the situation.

The quality of the materials to be used and that they are in good order and in the right place at the right time will ensure that the conditions are right to produce good work. But there is more in it than that. What about tolerances, fixing sequences and many other aspects?

BS8000 provides a vital and authoritative reference covering most of the basic building operations and is intended primarily for use on site. This provides for the site manager, general foremen and those under training a comprehensive Code of Practice which is in 15 parts (see Appendix).

WORKMANSHIP AND QUALITY ASSURANCE

BS8000 also puts in place one of the foundation stones to quality assurance, for it is this that is now a key issue for every client to the construction industry, demanding standards no more than he has a right to expect. It affects designers and builders alike, placing on each the discipline to manage effectively everything which will ultimately ensure buildings are constructed in a sound and proper fashion.

The administration of the detail of quality assurance (QA) does not, however, in itself ensure that high standards of workmanship are achieved. Supervision of the work while it is in progress is a vital last link in the chain of quality planning and its administration. Results cannot be judged until work starts on site.

The management skills needed to run a building site are probably much higher than those for the same level of responsibility in many other industries. Selection and training of young men and women for this increasingly demanding task, in an industry which is largely now one of specialist sub-contractors, focusses not unnaturally on planning, performance and contractual responsibilities.

Clients are increasingly demanding professionally qualified people to be responsible for the whole of the building operation. It provides reassurance that the designer and chartered builder are well trained and competent.

Insurance companies also require formal qualifications and appropriate experience before they will provide professional indemnity cover. But both designer and builder have the same

Workmanship: a test of supervision skills

Achilles' heel — workmanship. Bad work should, or has to be, rectified with all the accompanying waste in terms of time, cost and attendant disruption and hassle. An ability to supervise is essential to the effective manager. His/her technical knowledge must be blended with a practical understanding of the work and an ability to talk to those engaged in the work, in order to achieve the desired result first time. Confidence and authority are born out of this ability and are the hallmark of a successful building manager.

BRITISH STANDARDS INSTITUTION (BSI)

Standards of workmanship are open to interpretation and what the lowest acceptable standards should be is a matter for debate. The British Standards Institution provides the authority for these, covering an enormous array of products throughout every industry in the country. These standards are set by consensus of experts drawn from the industry concerned and sponsored by it.

Although BSI draws most of its funds from Government, it increasingly raises its own funding from the services it provides. That much of the advice it receives is free is perhaps both its strength and weakness. It exerts no pressure that might have commercial implications, but at the same time there is no real bite to demand results to a tight schedule, a factor of increasing importance if Britain is to meet the challenge of harmonising standards within the European Market.

BACKGROUND TO BS8000

Faced some years ago with the results of a study into the reasons for many building failures, B/-, the BSI code number for the Council responsible for all the standards relating to building and civil engineering, confirmed that workmanship was crucial to many of the problems indentified. The master codes covering all areas of traditional building activities address design, materials, workmanship and maintenance and are, of course, referred to in most building contracts. But they are rarely used on site.

Buried in each code is a great deal of valuable information on workmanship which is never seen or used. Although contract specifications refer to a range of British Standards, and to workmanship requirements, these are generally described in an unsatisfactory way, many of them being repetitive and drawn from previous specifications to provide the designer with a reference of sorts in case of problems.

TECHNICAL COMMITTEE B/146

With the need clearly established and financial support available from the DOE, a technical committee B/146 was set up, charged with the job of preparing a composite code of practice covering workmanship of most of the traditional trades and operations. The code was subsequently designated BS8000.

The committee comprised representatives drawn from each of the professional institutions, PSA, BRE, trade federations, material suppliers, together with the National House Building Council and the National Building Specification.

THE BRIEF

The committee's brief is best summarised by an abstract from the standard foreword used in each of the 15 sections of the code: 'to encourage good workmanship by providing:
- the most frequently required recommendations on workmanship for building work in a readily available and convenient form to those working on site;

- assistance in the efficient preparation and administration of contacts;
- recommendations on how designers' requirements for workmanship may be satisfactorily realised;
- definitions of good practice on building sites for supervision and for training purposes; this guidance is not intended to supplant the normal training in craft skills;
- a reference for quality of workmanship on building sites.

It is recognised that design, procurement and project information should be conducive to good workmanship on site.'

THE TASK

The task entrusted to B/146 was unique in that, unlike any other British Standard, it was to draw together recommendations given in other codes of practice and to help explain the reasons for certain recommendations; commentary was also to be provided. This was the first time that this had been done.

With the assistance of the Property Services Agency (PSA), a consultant was appointed to prepare the initial drafts. Proposals were sought from a number of appropriate firms, including architects, quantity surveyors and builders. In selecting the successful firm there were two major considerations: the style of writing to be used — imperative but not imperious, and the strength of the technical and practical back-up services that could be provided.

Laing's Research and Development Company were appointed. After detailed examination of all the current information then available (primarily from the appropriate master code) their first drafts were prepared. This at times proved to be a spring cleaning exercise. Some of the clauses were found to be out of line with current practice and inappropriate to the use of new materials. A number of the draft sections were put out for field tests on Laing's building sites with encouraging results.

Finally, 19 sections were issued as drafts for public comment to 150 organisations, conjuring up some 6000 or more comments, which ranged from those apprehensive about the code's application to those enthusiastic about its usefulness on site in the efforts needed to improve quality in building.

It was agreed that in the final editing and preparation for publication, each section would be aligned with the Common Arrangements of Work Sections (CAWS). This reduced the number of sections from 19 to 15. In the final stages a strenuous revision and editing was carried out to remove unnecessary repetition and all reference to design matters. It is interesting to note that even in some of the master codes the separation between design and workmanship was not clear, and that as a result of the work of B/146, some of the master codes have had to be revised.

PUTTING BS8000 INTO USE

1. **Read it and get to know it and keep a copy on site.**

   Each of the 15 parts is arranged in the same way for ease of reference in use. These are in three sections; first a General Introduction, the next one on Material Handling and Storage and the third sets out details of the operation covered, with details of tolerances, 'do's and 'dont's, helped with illustrations and, for the first time in a Code of Practice, commentaries are used which immediately follow important statements made in order to help understand the reasons for them.

Therefore, it provides a basic reference covering the areas of work on site where things can go wrong. It is especially useful to anyone under training who would naturally lack the range of practical experience at this stage.

For the architect — again for the first time — a reference is provided giving guidance and information on the limitations of the material or operation in question. A better understanding of these at the design stage will help to ensure that the designs are more readily realised. This can have a direct impact on the standards of workmanship.

Previously there was little formal reference for any architect under training to understand and appreciate the details and limitations of the materials he would be using in the future. For the experienced architect and builder BS8000 provides nothing new that he or she did not know or should have known already!

2. **Make sure it is specified**
For its proper use on site, it must first be specified. It is not good enough for just a vague reference to be made to the use of the latest BS. If it is not stated, it is possible that the designs were prepared without a full understanding of the operation or the limitations in the use of the materials specified — not a very good start to the job!

This situation, however, can be readily redeemed. It is a good idea to seek a meeting with the architect early in the job, to go through each work section well before operations start. It will ensure that there are no misunderstandings as to what the architect intended. One will then be able to agree how these can best be realised and by doing so make a real contribution to the success of the job!

Another very important point to remember is that if BS8000 has been properly called up in the contract specification, then the architect will have or should have amended the standard specification often in general or vague terms. This will reduce its size and also ensure any specific requirement on quality can be clearly stated from the beginning. This is important to the estimator at the time of tender submission.

3. **Ensure each trade contractor has a copy of the appropriate part**
It is also very important that each sub-contractor is fully aware of the details of the work section. Certainly the foreman or charge hand should be briefed and where appropriate, brought into the discussions with the architect.

Everyone will then be aware of the level of interest in workmanship. It will help to improve both quality and performance.

4. **Use it to prepare the quality plan check list**
With or without quality assurance, such a check list is extremely valuable. It helps to reduce the snagging operations and to get it right first time.

Check lists provide a very useful way of signing off each section of the work with each trade contractor. This ensures that the quality and workmanship are always closely monitored.

SUMMARY

For the first time both designer and builder have a British Standard code of practice devoted entirely to workmanship. BS8000 will provide a better understanding of what should be done for those who are anxious to learn and improve their ability to supervise the work in progress. It will provide the opportunity for the architect, builder and specialist sub-contractor to discuss important facets of the job together.

In future, the requirements of a project specification can be made much clearer using BS8000 as a single reference point and the reduction in the size of standard specifications will now be achieved — for example the PSA are now to reference BS8000 in all their contracts and remove all workmanship clauses. A copy of BS8000 will, of course, be required on every site.

It will also be an essential subject for every syllabus where building technology and management is taught and will provide an extremely valuable grounding for every student to the industry.

Finally, any improvement in the standard of workmanship is to be encouraged for it helps the industry to provide a better service to its clients, a reduction of the wasted time, materials and general hassle in putting things right, but most important to any company large or small, it will ensure a significant improvement in its net profitability.

Thus, every enlightened building company and sub-contractor should invest in a copy of BS8000 for each site manager, foreman or trainee in his employment. Site safety manuals are now mandatory on site and a personal copy of BS8000 is just as important.

Workmanship: a test of supervision skills

APPENDIX BS8000 WORKMANSHIP ON BUILDING SITES

Part 1 : 1989   Code of practice for excavation and filling

Part 2 : 1990   Code of practice for concrete work

Part 2.2 : 1990 Code of practice for placing, compacting and curing concrete

Part 3 : 1989   Code of practice for masonry

Part 4 : 1989   Code of practice for waterproofing

Part 5 : 1990   Code of practice for carpentry, joinery and general fixings

Part 6 : 1990   Code of practice for roof, flate, tile covering and claddidng

Part 7 : 1990   Code of practice for glazing

Part 8 : 1989   Code of practice for plasterboard partitions and dry linings

Part 9 : 1989   Code of practice for cement/sand floor screeds and concrete floor toppings

Part 10 : 1989 Code of practice for plastering and rendering

Part 11 : 1989 Code of practice for wall and floor tiling

Part 11.1   Ceramic tiles, terrazo tiles and mosaics

Part 12 : 1989 Code of practice for decorative wall coverings and painting

Part 13 : 1989 Code of practice above ground drainage and sanitary appliances

Part 14 : 1989 Code of practice below ground drainage.

Part 15 : 1990 Code of practice for hot and cold water services (domestic scale)

ACHIEVING QUALITY BRICKWORK ON SITE

by Peter Allars, FCIOB

INTRODUCTION

Brickwork can be the most satisfying part of a builder's job. Many buildings today have been designed to eliminate much of the craft skills, yet there is a growing tendency for designers to use brickwork to bring humanity, but with a feeling of durability, to the built environment. For the builder's manager on site it provides perhaps the most challenging and stimulating opportunities to use his skills.

Many of the buildings currently being built are sophisticated and incorporate such complex and subtle techniques that the site manager is dependent upon specialist trades to achieve the architect's objectives. Changing trends of employment within the individual companies also mean that he does not always have a gang of men whose abilities and loyalties are known and whom he can rely on. In many cases he has to carry out the contract with men who have no loyalty to his company whatsoever. This places great emphasis on the need for tight management control and good leadership.

The special joy of brick is that it is a material with a long history that has been known and used since before Roman times. The handling of brick and masonry is one of the natural skills that man has and enjoys. This ability can lie dormant for generations, to be called upon by talented and knowledgeable managers to produce a first class job. It is not something that will 'just happen'.

Sixty years ago the problem was identified in a text book of the time where it states: 'Brick-laying may be either a mere form of manual labour or a highly developed craft implying a knowledge of the principles of construction and appreciation of art. The one is a poor thing leading its practititioners nowhere, whilst the other is a matter of pride, opening up great possibilities to men of patience and skill. Brickwork must have an appearance and appeal to the eye which is outstanding today'.

Unfortunately, the situation today is that where a man seeks to be a bricklaying craftsman, he becomes self-employed, insular and trains too few apprentices. The whole process of building a brick structure should be a team activity.

ATTRACTING THE RIGHT MEN

For site managers to attract bricklayers who are competent and skilful and able to achieve high quality work at high productivity levels they must understand that bricklayers' prime interests are their potential earnings and the effort that they will be called upon to achieve it.

In the bricklayer's mind there is a trade-off between these two factors, and before any gang starts on site — whether to be directly employed or as a sub-contractor — they will attempt to judge the situation for themselves and use their years of experience to weigh up what is being said by the site manager or his representative against what they can see, and from this their opportunities to make money and compare this to what is being offered by other builders.

Good bricklayers have become very adept at this, and it is not surprising that the success or otherwise of a small bricklaying company may well have more to do with the actual selection of the contract to work on, than arguing about the rate for the job or even daywork and extras after the work has been completed.

One factor that should not be overlooked is that some contracts by their very nature are attractive to some men for one reason or another. In these circumstances, whether it is a local hospital or theatre, the site manager should ensure that he capitalises as best he can on these sentiments. It may be a bit late at the topping out ceremony to publicise the bricklaying requirements, whereas an inspired slot on local radio may only attract trouble makers.

The site manager must seek to inspire men to want to work for him. To do this he must be firm and fair, making it clear what is expected and how they will be able to maintain a high level of productivity, in an easy and relaxed manner, over a long and sustained period without having a constant battle with the clerk of works over quality.

ENGAGING OPERATIVES
The site manager, or his representative, has to interview the bricklayers and to decide whom to engage. This may well be easier said than done. If, after all the efforts to attract men, only a few turn up, none of whom are known, then it may well seem that the only practical solution is to offer jobs to all and sundry. This is a sure recipe for further problems.

It is important that the site manager should have clearly in his mind:

- what the bricklayer will be doing when he comes on to site;
- who should interview the applicants;
- what that person is looking for from the bricklayer;
- what in the way of a package is being offered in return.

Clearly, to do this properly the site manager has to hammer out the details with his contracts manager and the personnel department and/or the quantity surveyor.

It is important to use application forms for all people that express an interest in working on the site and to try to establish:

- if the applicant has served an apprenticeship and if so, where?;
- if he has worked on a similar project and if so, when?;
- if he is known by any colleagues, or has any other references;
- if he appears fit and strong and capable of carrying out the work;
- his address and other personal details.

Depending upon the number of applicants, it may well be sensible to ask the firm's personnel department to carry out an initial sifting process. However, the role of the personnel department must not be allowed to cross over into what is rightly the site manager's sphere of responsibility. It is not always clear where the boundary exists. Nowhere is this more evident than in the employment of the bricklayers on a sub-contract basis. If it appears to the site manager that men are being engaged without his approval then he must resolve with his contracts manager how corporate authority is to be maintained and what his role is to be.

Where the company's internal politics start to impinge on the running of the site, it is usually perceived by operatives as weak site management and is rapidly turned to their short term advantage. Consequently, it is in everyone's interest to see that the peculiarities of the individual contract are hammered out at an early stage with job responsibilities clearly defined.

Attached is a series of questions that might be helpful as a basis for a check list in dealing with sub-contractors. (See Appendix). As well as picking up the political feel of a site, the experienced operative is skilled at assessing the site in his own terms. This will include:

- where he parks his car, and the state of the access to it;
- where he is going to change into his working clothes;
- where the drying room is located;
- what canteen and rest room facilities exist.

If all are one and the same, ie, his car, the operative may well consider that the site is less attractive than others. The operative will also be concerned to note:

- where the materials are stacked in relation to the workplace;
- whether there is a good and dry access from the materials to the place of work;
- the state of the scaffolding.

Undoubtedly, he will be looking at the work with a measured eye to decide which part of the contract he wants to work on and how long it might suit him to work there. It is pointless to believe that a man who feels he is good and is competent at laying facings at a fair rate will wish to lay heavy concrete blocks which he feels would better suit a younger, stronger but less experienced man. Even the younger man may not be disposed to lifting these blocks high above his head on a sustained basis and will be looking to see what lifting gear is available.

As much as anything else a tradesman will judge the site by the manager's ability to attract other men whom he knows. In this regard a foreman bricklayer with a good local reputation will have a significant effect on recruitment and the site manager's ability to secure such a man may well be the key to a successful job.

As far as brickwork is concerned it is the role of the site manager to create conditions which will motivate the bricklayers. Whilst they are clearly coming to earn money, if it is the job they want to do and can see that it is well organised and well laid out, and believe that it is reasonably paid and that they will be treated fairly, then the site will get off to a good start.

MOTIVATION AND PRODUCTIVITY

Productivity levels are notoriously emotive subjects as is the level of rate of pay that goes with it. The site manager should concern himself with minimum cost. If this means he should recommend more money than allowed for, he should argue his case logically, demonstrating how doing this would involve less overall cost than what is currently proposed. He should expect his contracts manager to be difficult to convince.

Bricks laid per hour or day is a fair measure of productivity. The fact that very high levels have been reached in competitions yet are unobtainable as a regular feature of normal site practice must be a matter of concern. That more bricks were regularly laid in a working day fifty years ago than is common today, despite the efforts of the brick manufacturer and researchers, demands examination.

To obtain a sensible idea about productivity levels one has to understand the concept of a 'job'. No matter how large the contract, it is a series of many interlocking 'jobs'. In this context 'job' is used to describe the task that a man or a group of men working together can do which has a reasonable sense of unity relative to the contract as a whole. Laying a single brick in a wall containing 10,000 bricks would not normally be a 'job'; however, if it was a key brick to be laid on its own in special circumstances it might well qualify.

Using this concept, it can be seen that a lift of brickwork to a single elevation might well be a 'job'. If the next 'job' is not close by, or if members of the gang are to be broken up to do other 'jobs, or the present 'job' does not fill a complete working period, then productivity levels will drop.

The responsibility for this state of affairs lies with the designer of the building, who will consider these matters secondary to other design issues. No matter how well trained the architect, he or she will have little experience of how men actually work. There is also a reluctance, once a drawing is issued, to withdraw it and alter what is often perceived to be a minor difficulty. Even when this is not so, the time taken to actually get such a change authorised can take much effort from the people concerned.

Where the client has employed a project manager or a package deal contractor, these problems can be avoided, as there is a single technical co-ordinator. Some architects have tried to resolve this problem by producing, before the contract starts, working drawings for each elevation. This can lead to even further confusion, as it takes a very particular skill, not usually available in an architect's office, to do this in abstract.

The foreman should make it his business to study the drawings and specification for his part of the work so that he knows as much about it as anyone else. He should be encouraged to look for alterations that would make the work easier to do, or better in a constructive way, AND MAKE RECOMMENDATIONS TO HIS SUPERIOR. IN NO CIRCUMSTANCES SHOULD HE IMPLEMENT THESE WITHOUT RECEIVING WRITTEN AUTHORITY TO DO SO.

In most circumstances the contractor has to work out, once he has secured the contract, the detail of how the structure is to be assembled as a whole, involving how the structural elements as well as the services are to be brought together. It is against this background that each individual trade has to work, including the brickwork. In earlier days this was not the case. Then, the bricklayer built the structure and others fitted around him; now walls have to be left down to facilitate other trades and the bricklayer is expected to return later to finish.

The consequences of this is not always evident, but the reality of the scaffold is that these problems have to be resolved and if they have not been sorted out before the work starts then they certainly will affect productivity.

Depending upon the number of bricks to be laid and in what period and its relative importance to the contractor, all these problems can be minimised by resources being allocated to resolve them before the men reach the workstation.

One solution is to pay the men enhanced rates, but this cannot be a long term solution. The real solution lies in recognising that large complex contracts, containing significant amounts of brickwork, require a skilled foreman, together with a team of planners finding ways and means for men to work as efficiently as possible.

It would be expected that the site manager will sit down with the bricklaying foreman and agree realistic targets for the following six week period and review progress weekly at a formal meeting attended by their own planners.

The foreman and his planner, having studied the drawings and specification, must break the work to be done in the next six weeks into groups of 'jobs'. These should be related to the brick being used, the scaffolding, access and the labour available, ensuring that materials and other resources are available as required and in a form that can easily be translated into instructions. The foreman should see that these are in line with how he wants the work done, not just the planners' ideas.

If the foreman has chargehands working between himself and the operatives then they must be involved.

One target the site manager ought to set is that the men should have recovered the basic daily rate by lunchtime, leaving the afternoon open to earn bonus without endangering themselves or others. There must be a proper inspection of what has been done by someone in authority, who signs that it has been passed. This authority should trigger payment in respect of the work done. Targets are best issued before the work is started and sensibly set so that there is sufficient incentive for the men who have to finish the last few courses.

There will always be significant differences in productivity levels between one site and other, none of which may be due to the men on site or their foreman. What can always be done is to generate productivity levels greater than they would otherwise have been.

QUALITY OF FINISH

Quality control should not be left to the clerk of works. Each company will have its own quality control procedures but for each site it is the standard set by the site manager from day one that matters. It is too late to decide, once the contract is under way, that it is unsatisfactory.

Where brickwork is concerned the site manager should ensure that he is present at key moments and set standards, such as when the first brick is being laid. On his regular daily walk of the site, he should monitor the quality of the work and if dissatisfied, call the trade foreman across to the offending work and publicly discuss the situation. Discreet discussions in the site hut can lead to the view that the management is weak on quality.

Brickwork is a large scale external cladding material utilising small units laid over a relatively long time period. Clearly, the conditions under which the bricks have been laid will vary considerably, from dry windy weather to heavy rain, all of which will affect both the bricks and those laying them.

Achieving quality brickwork on site

If the architect's ambitions as far as the brick cladding is concerned are to be realised, then the impact of weather and other matters have to be sympathetically understood by those trying to build it. For instance, a patterning can occur by one half of a wall being built by a left-handed bricklayer and the other by a right, caused by the fact that each would strike off their perpends differently.

THE BRICK ITSELF

Site management must understand the brick to be used. The best advice is to visit the brick manufacturer, failing which as much information should be obtained about the brick and arrangements made for a sample batch to be delivered to site. These should be used to build a sample panel of brickwork for discussion with the architect and clerk of works.

It is a great temptation to make a special effort with this panel: this is a mistake. It should be built properly so that it fairly represents the general appearance of the brickwork.

One of the biggest problems is that bricks vary in colour and texture. To overcome this it is not uncommon for bricks from one part of the kiln to be mixed with those from another, or even mixed between one firing and another. Clearly, if there are a considerable number of bricks requiring many firings spread over a period of months, then it is very possible for bricks that are separated in their manufacture to such an extent that the clay used is different, to appear in a wall adjacent to each other.

The practical problems involved with mixing bricks to avoid the worst effects have to be resolved between the clerk of works, the contractor and the manufacturer at the very outset.

The problem becomes more acute if there are any significant amounts of special bricks, because these are usually produced by a different process to the ordinary brick and will stand out when laid. Before a brick is laid on site everyone in the chain from manufacturer to bricklayer on site must know what is required.

BRICK HANDLING AND DISTRIBUTION

It is pointless to take trouble with the bricks if they are chipped on being unloaded or whilst being moved about the site, or splashed with mud from passing traffic. Careful consideration must be given to material storage, handling and protection.

This is dictated, to a large extent, by factors beyond the control of any single person. The constraints of the site, and other work concurrently under way, all impose restraints on where and how these issues are dealt with.

There are two choices for the storage of bricks. One is a single or a series of compounds and the other is to unload the bricks close to the point where they are to be laid. Both have different requirements.

For compounds, it is worth considering putting down blinding concrete for bricks to stand on at the start of the contract and break it up on completion. However, this is a simple matter in comparison to the problem of access roads to and from the compound and careful consideration must be given to the method of distribution. Someone must be in charge of each compound and provided with a materials schedule setting out what is coming in and where it has to be distributed. The hidden costs of compounds lie in the number of times materials are handled, the materials remaining once the contract is completed, and pilfering. To be effective, compounds need a good storeman with enough men under his control to make it unnecessary for anybody else to go into the compound. Most pilfering is done by lorry during working hours.

Storing bricks about the site is easier in many ways assuming that it is possible to get the right number of the right bricks to the right place. Usually, this is a problem because a lorry load of bricks is not the correct quantity for an individual location. It is unusual to be able to stack the bricks on a hard standing and, consequently, the bottom layers sink into the ground and become unusable. It is also more difficult to protect the stack from splashing or chipping by passing traffic.

There is no easy solution to the storage of bricklaying materials. It is extremely expensive to start work without all the necessary materials readily available on site and the site manager must see that materials deliveries are linked to the site's needs. Material schedules, capable of being understood and used by the ganger, are essential.

In most cases the requisitioning of materials is not carried out by site management. There are two problems that the foreman should address. The first is to confirm that everything has been ordered. It is as well to have a check list of the items that require ordering and to tick them off as and when the order is raised. Whoever orders the bricks will almost certainly not be doing so.

The other likely problem is incorrect quantities. This could result from the architect varying the works, from a mistake or an inadequate waste provision. Again, it is easier to deal with this if the foreman is already involved with whoever raises the orders. Having a copy of the taking-off can also be useful for site planning and facilitates the occasional spot check.

Distribution

The major methods of distributing materials about the site will almost certainly be established by the contracts manager.

Whichever method is employed it is a waste of effort to transport more materials to the workstation than is necessary. This requires from the bricklaying foreman a clear idea of what is needed.

Stock control and distribution on a large site is an important subject, playing the key role in achieving quality brickwork on site. As such the foreman must become involved in the operation of this part of the site organisation and ensure that he is being properly serviced.

Occasionally, it will be necessary to stack materials on the structure itself. The foreman must see that the structure is properly propped to withstand the loads being imposed by the dead load of the materials themselves, plus the live loads imposed by traffic servicing that area.

WORKING PRACTICES

The location in which the wall is to be built should be recognised as a working place and be clean, free from obstructions, with the work at a reasonable level. Bricklayers should not be expected to reach below the level of their feet to get at bricks or to stand on tiptoe to lay the last few. These will not be laid as well as others laid within more normal operating levels and the difference will show.

Achieving quality brickwork on site

The brick bond and joint size has to be worked out physically on site. This should be done on top of the foundation brickwork, just below the finish ground level when the building is complete. The position of all the door and window openings should be marked out accurately and then the first two courses of bricks laid dry. The position of cut bricks and broken bond can then be determined and agreed with the clerk of works and recorded in some positive way that everyone on the site understands (such as marking the position of the perpends on the top two courses of the foundation brickwork with a marker pen).

As bricks will not all be the same length it will be impossible to maintain the perpends exactly one above another unless the perpend is of sufficient width to take up the variations in the brick size. This may well spoil the look of the wall, and it would be better in those circumstances to allow some perpends to wander but insist that every fourth one should be plumb.

Gauge rods can be a most useful tool to work out and record the position of vertical courses. They consist of a length of 35mm square prepared softwood that has each face allocated to a particular part of the brickwork, with a saw cut across the face of the rod at each bed joint. In this way a control can be established before the work starts on bed joints, as well as recording where special items such as airbricks, cill and lintel levels occur. Occasionally, there will be split courses; using a rod enables a decision on where these are to be located. If at all possible this should be hidden, preferably below ground.

When trying to determine the width of the bed joint, it is important to relate it to the brick being laid; some bricks need a narrow joint and others a thick joint. The normal way is to take eight bricks at random from the sample load and stack them one above the other in a dry condition. That figure is subtracted from the total gauge for eight bricks and divided by eight to arrive at an average thickness.

On reveals and moulded returns it is impossible to get all arrises exactly plumb. So it is sensible to decide beforehand which are to be the plumb joints and see that these are followed. Similarly with bed joints, the nature of the brick will preclude every bed joint being absolutely level, but this is no reason for not selecting which bed joints should be horizontal and straight and to allow those between to wander (within reason). The bed joints to be selected to be straight should be those that will be seen at eye level from the finished building. This requires some imagination, as they will not be on view whilst building is proceeding.

The face of the wall can become twisted unless effective control is maintained. A common cause of problems is a careless approach to the use of the bricklayer's line, and the men's reluctance to alter work once it has been built. It is important that the leading hand should watch the line and see that it is used properly. These are strung from either quoins or profiles but it is easier with quoins to check that they are correct before work starts on the length of wall, but should be allowed to set before lines are strung off them. With profiles, the problem is that they can move and in doing so affect a whole length of wall. They need checking every hour or so.

With modern building regulations has come a more sophisticated approach to cavities within the wall itself. These are commonly filled with insulation and in doing so dictate how the wall is to be built. Great care must be taken to see that the inside of the wall is kept clear of mortar droppings. This will almost certainly involve a separate and distinct operation from the laying of the bricks.

At the end of the shift it is important that the brickwork is covered to protect it from either frost or overnight rain, and that the scaffold board nearest the wall is turned back, so that in the event of rain there will be no splashing from the board on to the face of the wall.

When as becomes necessary from time to time to clean down the brickwork, whether because of careless workmanship, water running off of other buildings, or simple efflorescence, reference should be made to the Brick Development Association booklet *Building Notes — cleaning of brickwork.*

MORTARS
The correct mortar is a significant factor in achieving good brickwork. The sequence of operations with on site mixing should be as follows:

(a) ensure that the machine is level and firm;
(b) set the machine to poisiton '1', which means that the drum is at an angle of 27°;
(c) load most of the water;
(d) load almost all the sand or sand/lime;
(e) load all the cement;
(f) load remainder of the sand or sand/lime;
(g) change down to position '2' (18 degrees from the horizontal left or right);
(h) add any further water to obtain the right consistency;
(i) mix for three minutes.

The specification of mortar tends to be made with little regard to its physical appearance and yet it can affect the colour of the finished wall every bit as much as the brick itself. Therefore, if the colour is important then it is necessary to see that the source of supply of all the ingredients, particularly the sand and the cement, does not change and that there is sufficient of both to carry out the works. Some very nasty changes of colour half way up high brick buildings are evidence enough that this has not always been achieved.

On other occasions the strength requirements of the wall are more important than the visual effect. This needs to be checked with both the architect and the structural engineer. Basic rules to follow are:

(a) check that the sand is of the correct grade and specification at regular intervals;
(b) check that the water and the cement is always fresh;
(c) check that the cement is used in the order it is delivered;
(d) do not allow labourers to break the tabs casually on the bags of cement;
(e) only use cement from broken bags if the cement is still fresh;
(f) check that the sand is stored in bays specifically for that purpose and is not mixed with soil or other contaminant;
(g) the mortar is properly cared for and transported after mixing.

No matter how carefully the sand and cement is treated, unless batching procedures are enforced with regard to the amount of water and the proportions of the constituents, the mortar will vary dramatically both as to colour and strength. It is important that tests are carried out on mortar regularly. When air entrainers are specified extra care is needed, as they can, if misused, weaken the mortar and undermine the strength of the whole wall.

Care must be taken to see that old mortar is not used; two hours after mixing is about the

Achieving quality brickwork on site

limit of workability. With structural brickwork there should be no re-tempering, although this may be permissible in non-critical areas.

Occasionally, the situation will arise where retarded mortars are in use. Reference should be made to the structural engineer or architect to establish under what circumstances it may or may not be used on structural brickwork. Assuming that there are some areas where it should not be used, then it is sensible to arrange for that work to have a separate mixer set up.

There can be a significant loss of water during the setting times of mortar under certain circumstances. When that occurs it is a great temptation to adjust the absorbent rate of the brick by soaking them in water before they are laid. This can lead to a significant reduction in the strength of the finished work. (See SP.56, 1975 Model Specification for Load Bearing Clay Brickwork).

The weather can also affect the rate at which mortar will set and this will in turn determine the number of courses that can be laid. This can dramatically affect the carefully laid plans of the foreman and he should, therefore, keep a careful eye on the weather and have contingency plans for the men to see that this has a minimum impact on progress.

JOINTS

The strength of a wall will depend upon all bed joints and perpends being solid. An exercise in the US found that a team of unsupervised bricklayers created a furrow with their trowels in the bed joint and as a result these were not solid and in tests on the finished work showed a reduction in strength of the wall of 23 — 37%!

Similarly, if the perpends are not solid then dampness will penetrate between the bricks.

Expansion joints

The architect will expect the site manager and his staff to understand how one part of the building will move in relation to the rest and appreciate the important of movement and expansion joints designed to take up differential movement.

This being said, architects themselves do not like the impact these have on their finished work and try to hide them. This leads to all sorts of problems. Such as not taking them continuously through the structure, or placing them too near a return or corner producing sympathetic cracking in the adjoining brickwork when they start to perform their designed function.

It can be most frustrating if, after discussion as to where and how these joints should appear, to find the effect has been reduced, leading in some cases to cracking of adjoining brickwork, because somebody has carelessly left mortar droppings within the thickness of the expansion joint.

STRUCTURAL IMPLICATIONS

Wherever a brick building involves new techniques it is sensible to ask either the architect or engineer to talk to the men who are going to carry out the work. In many cases, when faced with something completely new, bricklayers will fall back on what they know and understand and in the process negate what is trying to be achieved. In most cases the designer will welcome the opportunity to speak to the men.

Brick walls only attain their full strength when the building is completed. It is important to recognise and understand the design implications of the structure by discussing it with the designer. Brickwork will require temporary support or propping in certain circumstances until other parts of the building are able to take the loads.

The most usual example is centering for arches, but there are other less obvious situations, such as where the wall is particularly high, long or thin, and only attains its ultimate strength when the whole structure is tied to it, but in the meantime could be vulnerable to high winds. Such propping has to be identified when programming the works and normally requires designing so as not to interfere too much with following trades.

A related problem is where the walls are subjected to a form of loading during construction different from that for which they have been designed. This can fall into two types. The first is incidental, when for instance, a following trade props their work off of brickwork where the mortar has not set properly, and in the process breaks a bed joint that goes undiscovered. The other is where the techniques being employed by the contractor demand that the particular brickwork requires strengthening. This is normally well understood and the wall is redesigned to take the extra loading at the contractor's expense.

It is important to recognise that the same structural disciplines apply as for handling concrete. Loads are transferred from one part of the structure to another in a designed framework and it is essential that each wall is set out accurately at each level and plumbed between these points. Usually, one wall sits squarely on top of the wall below it. As there is normally a concrete slab between these, there must be careful setting out so that there is no eccentric loading.

DAMP-PROOFING AND FIXINGS
As much care should be taken with damp-proof courses, lintels, wall ties, anchors, and fixings as with the brickwork for a building to perform as designed. The location of all these need to be determined well in advance and form part of the normal instructions for each length of wall.

Where the cavity being used is a gutter within the walls, it should be inspected to check that the water that could be expected can be discharged without overflowing into the inside of the building. Both in these situations and where the dpc is being used as a barrier against water penetration it is important that the dpc has not been cut short and that the laps are of the correct length. These are the points where water will penetrate, particularly round doors and windows.

CONCLUSION
To achieve quality brickwork on site it is necessary to develop skills and abilities at all levels within the site organisation, but it has to be led from the top. Sites are not a one man show and the co-operation and integration of all those working on it is essential. It is no good the bricklayers performing to a high standard if the carpenters are going to sweep their shavings into the cavities, or attempt to create an artificial world of near perfection in which real people cannot exist or work economically.

It is important that the team develop a 'natural authority' which commands respect, so that it is able to discuss calmly real problems with the architect and 'design in' solutions in an amicable manner.

Achieving quality brickwork on site

Quality brickwork is not perfect brickwork. Brickwork by its nature has too many variables for that to happen; the architect must understand that by specifying brick he is bringing a variegated finish which develops its own character, which can only be done by men of art and craft working together.

BIBLIOGRAPHY

1. CURTIN W S et al (1983) Structural masonry designer's manual. Granada
2. CURTIN W S et al (1982) Design of brick diaphragm walls. Revised edition. Brick Development Association
3. CURTIN W S et al (1980) Design of brick fin walls in tall single storey buildings. Brick Development Association
4. HASELTINE B A. Bricks and their properties
5. HASELTINE B A. The design of calculated loadbearing brickwork. Brick Development Assocation
6. HANDISYDE C C and HASELTINE B A. Bricks and brickwork. Brick Development Association
7. NASH N G (1969) Brickwork. Hutchinson (3 volumes)
8. BRITISH STANDARDS INSTITUTION BS 5628, Part 1, 1978. The structural use of masonry

   CP 111, part 2, 1970. Structural recommendations for loadbearing walls

   CP 121, part 1, 1973. Walling — brick and block masonry

   BS 3921, 1974. Specification for clay bricks and blocks

   BS 187, part 2, 1970. Specification for calcium silicate (sand lime and flint lime bricks)

   BS 4729, 1971. Shapes and dimensions of special bricks

   BS 3798, 1964. Coping units of clayware, unreinforced cast concrete, unreinforced cast stone, natural stone and slate

9. BRITISH CERAMIC RESEARCH ASSOCATION

   SP 56, 1975. Model specification for loadbearing clay brickwork

   SP 9, 1977. Design guide for reinforced and prestressed clay brickwork.

APPENDIX: INTERVIEW CHECK LIST FOR BRICKLAYING SUB-CONTRACTORS, AFTER RECEIPT OF QUOTATION AND BEFORE PLACING ORDER.

1. Name, address and telephone number of sub-contractor.
2. Tax exemption certificate, check type and whether still current and take a copy.
3. Insurances: employers and public liability
    — who with
    — policy number
    — renewal date.
4. Current number of employees.
5. How long has he been trading?
6. Name and telephone number of referees.
7. Details of similar work done.
8. Name of foreman empowered to accept instructions.
9. Will the foreman be able to instruct the sub-contractor's men?
10. Who pays for taking down and replacing defective work?
11. Explain how the work is to be inspected, and by whom, are passed for payment.
12. Explain work to be done.
13. Site safety policy.
14. Sub-contractor safety policy.
15. Extent of cleaning up in the sub-contract.
16. Where the rubbish is to be put?
17. What materials are the main contractor providing?
18. How is excessive waste or damage to be dealt with?
19. What specifications are drawings are to be provided?
20. Sample panel.
21. Datum — setting out
    — bonds   } who checks?
    — rods
22. Productivity levels expected, can they be achieved and what are their implications.
23. What is the programme of work and method of operations?
24. Can the sub-contractor comply?
25. Can he get the number of men needed?
26. Where will they be coming from?
27. How will he be recruiting them, is this likely to be satisfactory?
28. How will they be employed?
29. How will they travel to this job?
30. Establish disciplinary procedures.
31. How the welfare facilities will work?
32. Site's housekeeping rules regarding toilets, starting times, etc.
33. Who attends the site manager's weekly meetings?
34. Establish what facilities the sub-contractor requires.
35. Arrangements for testing materials etc.
36. What plant will the main contractor provide and the rules for its use?
37. Who provides the scaffolding?
38. What notice does the sub-contractor require to start on site?
39. Establish who signs the daywork sheets and when (normally weekly).

QUALITY CONTROL AND TOLERANCES FOR INTERNAL FINISHES IN BUILDINGS

by I Rankin BSc MPhil MCIOB

INTRODUCTION

The quality and expertise of carrying out building work in general has, over many years, been the subject of much debate and argument. Definitive specifications have proved to be the most practical means of both defining responsibility and settling disputes.

Unfortunately, the quality of internal finishes has been traditionally less clearly defined, being based on personal opinion and vague subjective specifications.

The object of this paper is to:

- outline research carried out on tolerances[1];
- define more objectively tolerance limits for internal finishes;
- establish guidelines for site management to assess and control the standard of building work in general.

Although concentrating initially on internal finishes in new private sector housing, the procedures and systems discussed should be of interest to those involved in other sectors, as well as those having responsibilities for quality control.

THE IMPORTANCE OF QUALITY CONTROL/FINISHES

It is important for all those working in the industry to appreciate that the finished product - the building - is for a client and that the client's views and expectations must not be ignored or neglected.

In making an initial assessment of a building the client or purchaser will consider its aesthetic impact and the quality of the internal finishes. To assess the significance of this the following factors may be considered as important:

Purchaser expectation

In today's consumer orientated society there is less willingness to accept inadequate standards and a greater likelihood of disapproval being voiced about defects and quality of work.

Efficient and prompt remedying of defects, and the associated modification to any standard, is good practice but failure to give time to a complaint in the first place can create unnecessary friction and ill will. This may lead to an increase in the number of complaints and involve both senior management and maintenance operatives in frequent visits and discussions, as well as additional work.

A National Opinion Poll (NOP) survey[2] found that good internal finishes were fourth in a list of house purchasers top priorities, behind NHBC protection, loft insulation and a well designed kitchen. As part of the author's research mentioned earlier a questionnaire was given to visitors to a showhouse on a large estate. The questionnaire listed ten areas of common complaint associated with the standard of internal finishes to new houses. From the 110 replies received it was apparent that four faults caused most annoyance:

— excessive gaps between door and door frame
— wavy reveals at openings
— uneven floor levels
— junction of wall/ceiling forming a wavy line.

Reputation of builder/sales

As the quality of internal finishes is very often the only facet about which a house purchaser can make a judgment, a high quality will enhance the reputation of the builder and create a demand for his product.

Contacts who have previously purchased from the builder can influence a potential sale, depending on the opinion they express.

It is difficult to assess the quantitative effect of unfavourable comments or good referrals. However, the NOP survey did find that 46% of new purchasers gave unfavourable comment about the quality of internal finishes and 59% about the after sales service.

Financial

Any quality control system must balance the cost of the non-productive process of inspection/checking, against the risk that something might prove defective or unacceptable at a later date, requiring costly repairs. The total cost, not only of the initial contract, but also of the follow-on costs of maintenance and correcting defective work must, therefore, be taken into account.

This raises the question of whether it is cheaper to control the standard of work initially rather than pay additional costs subsequently.

At the same time it must be established if the builder and purchaser are getting value for money. This will not be the case, for example, if substandard materials are being accepted or shoddy workmanship tolerated. The sub-letting of sub-contracts can also present problems to management in exercising effective control.

Management must be aware of these points and appreciate that production cannot be completely divorced from quality control. A practical and well defined quality control system can provide substantial financial benefits.

RESEARCH ON INTERNAL FINISHES

From a review of existing specifications at the start of the research it was clear that internal finishes tended to be defined in vague, ambiguous or subjective terms.

Therefore, the prime objective was to establish more precise and clear specifications on tolerance limits, which could be used as a practical means of quality control and for settling disputes.

(a) Establishing current standards

Before assessing and defining tolerances which would be fair and capable of being achieved at site level it was considered essential to appraise the standards currently being employed.

Observation through normal duties and a pilot survey of over 100 private development sites throughout the country, allowed the conclusion that the range of standards would not vary

Quality control and tolerences for internal finishes in buildings

to any significant degree, and that some form of sampling would give a general indication of the quality of work being produced.

Consequently, detailed measurements were taken on 125 houses spread over 20 medium to large housing estates. The sites were chosen at random and covered an area from Essex to Glamorgan.

Measurements were restricted to internal finishes, deviations from level, square or plumb, blemishes, marks etc and in all about 50 items were measured or assessed in each house just prior to occupation.

(b) Establishing thresholds of acceptable work on tolerances

When results for each item measured were plotted on a frequency distribution diagram, it was found that they exhibited a marked skewness, typical of the shape shown in Figure 1 below, with a long tail on the high deviation side.

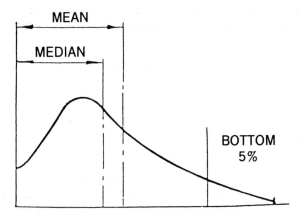

Figure 1 Frequency distribution diagram

The results indicated that the majority of houses had similar, relatively low tolerances/deviations and that a few exhibited very poor results.

The effect of the few large deviations was to worsen the MEAN (average) value achieved. It was considered, therefore, that a more realistic work tolerance or target would be found by using the MEDIAN value, ie, the tolerance achieved by half of the houses measured.

Tolerance levels based on this median value, with adjustment in some cases derived from common sense, and detailed consideration of what could practically be achieved on site are recommended. It is difficult, however, to draw a precise line dividing the unacceptable from the acceptable.

On careful examination of the results it was decided to set the level of unacceptable work at the worst 5% achieved on each item.

This means that tolerance limits are defined for each item under two distinct categories:
— Builder Guideline or Target Tolerance (median value)
— Unacceptable work tolerance (worst 5%).

RESULTS AND TOLERANCE LIMITS

As the measurements were taken on newly completed houses, the defects and dimensional deviations found related solely to site workmanship and practices. The effects of shrinkage and drying out of the structure on occupation could add to the warping, twisting and shrinking of components.

The results are set out below under the two main headings which cover several trades. The first deals with measurements of line, level, and plumb and the second covers aspects of surface finishes and blemishes.

(a) Line, level and plumb

Line, level and plumb were perhaps the easiest to define and assess. They involved taking simple measurements with normal construction equipment, such as a 1m level, 2m/3m straight edges, 500mm right angled square (steel roofing square) and measuring tape.

It should be noted that a distinction has been made between the flatness or bumpiness of a surface and the overall level or plumb. Where a tolerance of $\pm$ (x)mm from a straightedge with protruding feet is given, this can be measured by adding a block (x)mm thick to each end of the straightedge.

Additional tolerance limits are also given for three items which were not actually measured in the survey. The level of ceiling and the flatness of the floor and plastered wall finish. For these items comparison with the results obtained in other areas and the recommendations stated in British Standard, Codes of Practice[3] enabled a fair tolerance estimate to be made.

The following tables give a summary of the results defined as both target levels and tolerances for unacceptable work.

Table 1 Tolerances for floor finishes

	Builder guideline tolerance (median)	Unacceptable work tolerance
A. Joint between floor and skirting	2mm	6mm
B. Level of floor — general	3mm in 2.0m	10mm in 2.0m
C. Level of floor at doorway	3mm in 1.0m	8mm in 1.0m
D. Flatness of floor	$\pm$ 3mm under 3.0m straightedge with protruding feet. (ie 3mm block under each end) (from CP 204)	

Table 2 Tolerances for wall finishes

	Builder guideline tolerance (median)	Unacceptable work tolerance
A. Flatness of wall	$\pm$1.5mm from 2.0m straight edge with protruding feet	$\pm$ from a 2.0 straight edge with protruding feet
B. Flatness of wall (alternative to A above)	3mm from 2.0m straightedge (dervied from A. above)	
C. Plumb of wall	5mm in 1.0m (Or 10mm in floor ceiling height)	12mm in 1.0m

Quality control and tolerences for internal finishes in buildings

Table 3 Tolerances for openings and junctions in walls

	Builder guideline tolerance (median)	Unacceptable work tolerance
A. Width of reveal	3mm in full height	8mm in full height
B. Level of reveal soffit	1mm in 1.0m	5mm in 1.0m
C. Plumb of reveal	2mm in 1.0m	7mm in 1.0m
D. Reveal-soffit angle	2mm from side of 500mm sq	7mm from side of 500mm sq.
E. Vertical junction of	5mm from side of 500mm sq	10mm from side of 500mm sq

Table 4 Tolerances for ceiling finishes

	Builder guideline tolerance (median)	Unacceptable work tolerance
A. Flatness of ceilings at wall/ceiling line	$\pm$ 3mm over 2.0m straightedge with protruding feet	$\pm$ 6mm over 2.0m straightedge with protruding feet
B. Flatness of ceiling at wall/ceiling line (alternative to A.)	6mm from 2.0m straightedge provided there is no deviation of more than 5mm in 300mm (derived from A. above)	
C. Level of ceiling	10mm from a 3.0m straightedge (from PD 6440)	

Table 5 Tolerances for joinery

	Builder guideline tolerance (median)	Unacceptable work tolerance
A. Square of door frame	1mm from 500mm square edge	3mm from 500mm sq. edge
B. Door twist or bow	5mm	10mm
C. Fit of door	Shutting edge 3mm Hinged edge 2mm	7mm 5mm
D. Joint in architrave	Flush gap of 0.5mm max	Flush gap of 2mm max
E. Level of window sill	1mm in 1.0m	6mm in 1.0m
F. Plumb of window frame	2mm in 1.0m	6mm in 1.0m

(b) Quality of surface finishes and painting

The items in this section relate not to tolerance measures but to unacceptable defects which according to specifications and good building practice, should not be acceptable.

In general the tolerance for these items is that they should not be visible when viewed in normal daylight from a distance of 1.0 - 1.5m. (see Table 6).

(c) Discussion of results/important areas

(i) Line level and plumb

The results indicated that a reasonably high standard was capable of being achieved. Joinery items such as forming joints in internal trim and the level and plumb of frames and sills showed consistently good results. Some excessive deviations were, however, found in the gaps between doors and frames and the waviness of some

Table 6 Quality of surface finishes on joinery and walls

Joinery	Building guideline tolerance (median)	Unacceptable work tolerance
A Unfilled nail heads — architrave	1	4
other elements	0	0
B Splits, cracks — all elements	0	1
C Knots — all elements	0	1
D Chips, marks — frame edge/stop edge	1	
architrave face/stop face	2	
other elements	0	
E Rough or coarse patches — frames	% of dwellings with defect	
up to 1.0m in length — other elements	46	
	32	
F Grain showing through — frames	85	
stops	62	
architrave	54	
G Brush marks — frames	61	
stops	46	
architraves	52	
H Paint runs — all elements	25	
I Bare or starved patches — frames	58	
other elements	15	
Walls		
J Damaged area to main wall	3	
K Damaged area to reveal	1	7
L Rough patches	% of dwellings with defect 59	
M Paint runs	13	
N Brush marks	41	
O Bare or starved patches	20	

ceiling finishes. This indicated a little more care, dexterity, time and effort may be needed in these areas to achieve a more consistent standard. It was concluded that more care should be given to the fixing and alignment of hinges and ironmongery to doors and to the levelling and support of plasterboard to ceilings, especially at perimeters.

Similarly, greater care or checks should be encouraged in both bricklaying standards and plastering in respect of opening reveals (variation in width) and the squareness of internal corners or junctions.

(ii) Painting and surface finish
Indications from the results and comparison with standard or traditional specifications, was that much of the preparatory work in connection with achieving

a good quality paint finish was being omitted, especially where painting of wood-work was concerned. Economics will probably dictate the standard for the future. Improved site supervision and control should, however, limit the number of unfilled nail heads, chips, marks, damages and missed areas. Excessive roughness in timber could be improved by insisting that the painter rubs down with sandpaper and cleans prior to painting. Raised grain could be reduced by ensuring external joinery or components to be fixed to damp surfaces are adequately primed and that all joinery is stored under cover and protected. Consideration could be given to the use of templates to form openings, rather than building in components at an early stage, which would be subjected to the elements, and possible physical damage.

(iii) Important areas
From the results of the survey and consideration of purchaser expectation, the following four areas were highlighted as being of particular concern:
- fixing of doors to frames or linings, and the resulting gaps, twist or slope of the door leaf and the fixing of ironmongery;
- waviness of the ceiling, at the wall/ceiling line junction;
- quality of woodwork finish and painting, particularly with regard to damage marks, rough patches and the thickness and application of paint;
- quality of plastering finish, particularly in respect of the waviness of opening reveals and the squareness of internal corners or junctions.

(d) Variability of items and overall standards
The results were analysed by computer, and the general trends which proved significant and are of particular interest were:

(i) standards varied more between different sites, than within each individual site;
(ii) there was a high correlation between the standards achieved by individual builders on different sites (whether high or low).

Both these factors indicated that there is a strong possibility that the policy or actions of a particular site or building company will have a bearing on the standard achieved.

Finally, to provide an overall picture, each of the items measured was allocated a de-merit rating of 5 points or 2 points according to its seriousness. De-merit points were allocated to a house if the item did not reach the guidline tolerance level (median).

Figure 2 shows the de-merit scores achieved on the houses on four of the sites. The relative quality levels are clearly apparent.

ESTABLISHING GUIDELINES FOR SITE MANAGEMENT
The classical theory of management defines the process as:
(i) forecasting/policy making;
(ii) planning;
(iii) organising;
(iv) motivating/directing;
(v) controlling;
(vi) co-ordinating;

with communication being essential at and between each actvity. Site management should never lose sight of these basic concepts. In dealing with quality control, all the above processes must apply. Forecasting, planning and organising the business and firm's structure would normally be the responsibility of top management. The other three, motivating or directing, controlling and co-ordinating, will have more direct relevance to the site manager.

Individual items

It would seem reasonable that where the tolerance achieved by an individual item falls below the guideline tolerance level, as indicated in the tables on pages 6-17, remedial action should be taken.

Whole house quality

Each of the 40 investigated items of internal finish was allocated a de-merit rating of 5 points or 2 points according to the author's opinion of its seriousness. De-merit points were allocated to a house if the item did not reach the guideline tolerance level (median). The figure below shows the de-merit scores achieved on the houses on 4 sites. The relative quality levels of the sites are clearly apparent. The achievement of high scores would indicate to the builder that improvement was needed on future houses.

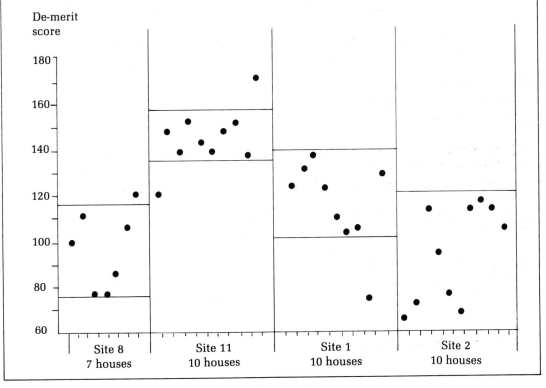

Figure 2 Application of tolerances

However, it may be that the responsibility for forecasting, planning and organising the quality control function is left entirely to site management. In practical terms, no house will be perfect. The fact that tolerance limits are under consideration presupposes this point. What then can be done to optimise both production and quality?

Quality control and tolerences for internal finishes in buildings

(a) Company policy

The overall effectivenss of production and responsibility for the finished building product and financial return must rest with senior management.

Consequently, a firm commitment to quality control, or getting the thing right first time should stem from this level. The basic approach of the firm through stated policies and procedures, clearly communicated to all levels, is essential to ensure that everyone is aware of what is expected.

Typical policies which have a bearing on quality control include:

(i) drawings and specifications to be available and checked for compliance with all regulations prior to work starting on site. All details to be communicated to site personnel;

(ii) separate or independent quality check or report to be submitted at various stages to managing director;

(iii) all major sub-contractors to be vetted and approved;

(iv) all materials to be checked against specifications upon delivery to site. Any defective/incorrect materials to be returned, or set to one side and a report submitted to head office;

(v) each property to be inspected each day and defects notified to sub-contractors or bonus payments to be authorised only after check that work is satisfactory;

(vi) completion dates to be realistic and any anticipated delays communicated to purchasers as soon as possible;

(vii) properties to be completed and checked before handover. Site supervisor to carry out walk round inspection with each purchaser and deal with any complaints within seven days.

Site management should be aware and strive to fulfill these policies. It will not always be easy with the practical day-to-day problems which arise on site. Nevertheless, a firm target or goal is to be aimed for; if no policies exist site management should set their own.

(b) Staffing

Adequate numbers of staff and an appropriate organisational structure should exist to deal with both production targets and quality control. Delegation of responsibility is inevitable. Even in the small company where the sole proprietor is in daily contact with the site, it will be necessary to delegate areas of responsibility, due to the demands of the building process.

A single site agent dealing with five subordinates does not only have direct lines of communication, but must deal with the inter-relationships between all five. Communication is a two way process and it should be borne in mind that interpretation and meanings may be misunderstood.

Site management must, therfore, be aware of the informal relationships between subordinates and try to co-ordinate their activities. Delegation of responsibililty to sub-contractors or trades foremen is necessary to limit the number of direct relationships, generally accepted as 5-6 per supervisor. It does, however, create the risk that communication will be poor, or lead to misinterpretation.

General managemement must, in addition to choosing responsible staff (and sub-contractors), run a constant check on their activities and become involved by showing an interest and looking at the work produced. Time must be set aside for this. Whereas responsibility can be delegated accountability cannot.

Education and enlightenment of staff and sub-contractors should be a prime objective of the site manager. Firm but fair leadership will establish respect and lead overall to a better quality performance, with a consequent reduction in crises.

Again, when dealing with general construction work it is better to separate the quality control function from that of production. A separate quality control manager or finishing foreman will free the site manager from part of the burden and ensure that poor workmanship or materials are spotted at an early stage and corrected.

(c) Knowledge of standards expected

It should be the duty of every site manager to know what standards and specifications are expected and this will require at times an individual effort to improve knowledge on any weak or unfamiliar areas. It is not sufficient to rely or hide behind the drawings and specifications provided, as these may be incorrect, or lack detailed information. The man in charge of construction on site has a clear responsibility to ensure things are right from the start.

Drawings should be checked and doubtful or missing information found or clarified. Full specifications should be available, together with any manufacturers' literature on the materials to be used, the method of storage, and any limitations or precautions in their use.

Where sub-contractors are employed the site manager should be aware of the legal conditions of contract, any specifications on quality contained in these and the procedure for dealing with unacceptable work.

Once the site manager knows in his own mind what standards are to be achieved he can communicate these to his workforce, thereby making sure everyone is clearly in the picture before work proceeds.

(d) Use of prescribed forms/standard procedure

The day to day activity of co-ordinating and controlling building work can be greatly simplified by establishing standard procedures and by using standard forms. The extent and nature of these procedures should stem from the firm's management policy.

The following are examples of areas which will have a direct influence on quality control:

(i) Purchasing and ordering of materials

Large companies with a separate buying department should ensure that site management is fully briefed on the specification for materials and conditions of contract. Should it become necessary to order direct from site, a prescribed order form with standard conditions of contract printed on the back will be of great assistance. The conditions of contract should include a warranty that the materials to be supplied will be as specified and fit for purpose intended. Delivery instructions should be included and a statement made relating to the procedure for payment and dealing with defective materials.

It is important when ordering that a full and adequate decription of the materials is given. If there are any doubts, the buying department or manufacturer should be contacted to confirm or seek advice on the description.

(ii) Checking and storing of materials

The well run site will have planned deliveries of materials, which will be checked on arrival and then stored and protected as necessary. If the standard conditions of contract outline procedures for payment and for dealing with defective materials little argument will result. Site management should be aware of manufacturers' instructions on storage and handling to limit the amount of damage and to restrict for example, the amount of excessive shrinkage. Much valuable practical information is given in the CIOB publication *Try reducing building waste*[4].

(iii) Selecting labour

Whether using sub-contractors for each trade or employing a direct labour force, site management will be faced at some stage with hiring labour. Hiring procedures, whether utilising standard forms or not, should include obtaining references and/or investigating the quality of previous workmanship.

Making enquiries regarding the dependability and credit rating of sub-contractors is also a sensible step.

Once having selected the workforce, it is essential that the standard of work expected is clearly laid down at the start. This can be achieved by issuing standard conditions of contract and specifications.

Sub-contractors may also be asked to certify that their work has been properly carried out and, in some cases, retention monies held for certain periods. In addition tradesmen should be responsible for reporting any items of sub-standard work, which could affect the quality of their own work. They should also be responsible for damage caused by their activities to the work of others.

(iv) Inspection control

Good supervision at the correct time is the key to achieving good standards throughout the building process. In addition to general observation of the work as it proceeds, site management should systematically check each day all work which has been carried out. Check lists can be prepared of the important areas to monitor at various stages. The detailed nature of those lists will obviously depend on the particular house type. An example of such a checklist is given at Appendix 1.

Inspection should be methodical and a system organised to deal with the correction of any defective work which is found. The *Registered house builders site manual*[5] outlines such a system on how to approach snagging a house and recording defects with standard prescribed forms. Extracts from this publication are given at Appendix 2.

(v) Completion dates handover

One of the most common complaints against builders is their failure to quote and achieve realistic completion dates. The practical problems faced on site may not always be understood or considered acceptable to a purchaser who has his expectations, and usually financial arrangements, set for a specific date. Equally, the problems and ill feeling, which often result from allowing occupation before the building is finally complete, is a frustrating experience which everyone could do without.

Site management should, therefore, pay particular attention to the building programme, redrawing this as necessary to keep an up to date picture of realistic completion dates. If weather conditions or other events hold up foundations or early superstructure work, it should be obvious that this will extend the building programme, and the sooner revised dates are given to the sales team and intended purchasers the better.

Many builders also find that a walk round inspection with the purchaser, just prior to handover, overcomes many potential problems. This can be done with the site agent, finishing foreman or a member of the sales team. Again, a checklist or prescribed form can be used and indeed signed by the purchaser, especially in relation to chips or damage in sanitary ware, and glass. A policy of correcting any agreed faults immediately, or within say, seven days, will greatly enhance the firm's reputation and improve customer relations.

CONCLUSIONS

The quality of work produced can have a marked effect on the profitability of the contract and on a firm's reputation. An organised system and commitment to quality control throughout the building process can save money on costly remedial work and lead to a more agreeable relationship and enhanced reputation between a builder and his client.

The quality of internal finishes is an aspect which is often ignored or thought to be of minor importance. Often the excuse is made that it is an area too difficult to define or control and that a client will either put up with poor finishes or accept that these can be improved at a later date. In today's more consumer orientated society such an attitude is outdated and misguided.

It has been shown that it is possible to define tolerances for internal finishes objectively. Coupled with an organised system of quality control, (which should relate to all stages of construction) it should be possible to produce better and more consistent standards, at little or no extra cost.

Site management bears the brunt of the responsibility for achieving good standards. The use of prescribed forms and an organised system of checking and inspecting can assist in the day to day control and smooth running of a site. Three steps are involved for site agents; they should:
(a) establish and know the standards they are going to accept, and what is unacceptable;
(b) communicate this information to their assistants, workforce and sub-contractors;
(c) organise a system to check that the standards expected are being achieved and establish a procedure for dealing with unacceptable materials or work.

Certain aspects of the above process may be delegated to assistants, trades foremen etc (eg, checking or taking physical measurements). The site manager must however accept overall accountability, and set aside time to check and motivate his personnel into achieving the results expected.

REFERENCES

1. HOUSING RESEARCH FOUNDATION (1978) An investigation into tolerances for the internal finish of new houses
2. NATIONAL OPINION POLL. The new houses people buy : what they think of them. Housing Research Foundation
3. BRITISH STANDARDS INSTITUTION. In situ floor finishes. BS 5606: 1978 Code of practice for accuracy in building; CP204 : In situ floor finishes; CP221 : Internal plastering
4. CHARTERED INSTITUTE OF BUILDING (1980) Try reducing building waste
5. NATIONAL HOUSE BUILDING COUNCIL (1974) Registered house-builders site manual. A guide for site management on how to improve quality and prevent defects.

ACKNOWLEDGEMENT

The opinions and recommendations expressed in this paper are those of the author and not necessarily those of the National House Building Council or the Housing Research Foundation. He would like to express his thanks to these bodies for their kind permission to reproduce the extracts on inspecting and recording defects at Appendix 2 and the tables/diagrams on tolerances.

APPENDIX 1: INSPECTION CONTROL

Trench digging and foundations
(a) Check for hazardous conditions such as shrinkable clay, trees, sloping ground, ponds etc.
(b) Check for adequate bearing/soft spots/hard spots, and changes in level.
(c) Check trenches for width, depth, position and squareness.
(d) Check for detailed drawings and reinforcement for special designs.
(e) Check excavation is dry and quality of concrete.
(f) Check soil within overside area and assess where suspended slabs will be required.

Underbuilding up to DPC level
(a) Check location of brickwork or blockwork on foundations.
(b) Check brickwork or blockwork and mortar for quality, line and plumb.
(c) Check wall thickness (retaining fill).
(d) Check cavity fill.
(e) Check support over any service entries.
(f) Check all vegetation is removed from oversite area.
(g) Check quality of fill to trenches and oversite area.
(h) Check method of consolidating fill (in 300mm layers).
(i) Check damp proof membrane (imperforate and lapped).
(j) Check vertical tanking where applicable.
(k) Check thickness of floor slab.
(l) Check reinforcement or size of joists and spans where suspended floors apply.
(m) Check damp proofing of encasement ducts and any structural effect changes may have on oversite concrete.
(n) Check spacing and position of air bricks, where applicable.

Oversite concrete laid
(a) Check for damaged brickwork or cracks.
(b) Check oversite for quality and any signs of movement or excessive cracks.
(c) Check damp proof membrane, imperforate and adequate length to link with DPC.
(d) Check other forms of suspended ground floors for structural adequacy and damp proofing.
(e) Check storage and treatment of joinery.

1st Lift brickwork blockwork
(a) Check for line, level and plumb/bonding.
(b) Check mortar for quality and thickness of bed, (also gaps, especially with party walls).
(c) Check cavity widths, and for any bridging.
(d) Check wall ties for cleanliness and slope.
(e) Check location of all DPC's and ducts and trays through cavities.
(f) Check external window and door frames for treatment and adequate fixing.
(g) Check lintels adequate for span and load (especially over large spans, including garage).
(h) Check lintels for damage and bearing.
(i) Check chimneys and flue construction.
(j) Check cavity trays provided at abutments.

Quality control and tolerences for internal finishes in buildings

Superstructure and floor joists

(a) Check location of indents with joists.
(b) Check joist size for span.
(c) Check joist size under first floor walls.
(d) Check point loads on lintels and trimmers.
(e) Check joist trimmer size.
(f) Check end bearings.
(g) Check all joints and connections.
(h) Check quality to timber and any damaged members.
(i) Check flat roof timbers for fall and where necessary treatment.

Roofing

(a) Check roof framing.
  (i) Non standard designs
  (ii) Sizes of struts, hips, valleys, purlins etc.
  (iii) Fixing and jointing, including ladders and tops of walls.
  (iv) Span and centre spacing.
  (v) Wind bracing.
  (vi) Damaged members.
  (vii) Plumb of trusses.
(b) Check fixings to soffits/facias/barge boards.
(c) Check provision of rainwater goods.
(d) Check lap of roof felt.
(e) Check for damaged felt.
(f) Check felt at fascia for overhang and sagging.
(g) Check size and spacing of battens.
(h) Check for stagger of joints on battens.
(i) Check flashing and soakers at abutments and perforations.
(j) Check pitch and lap suitable for tile.
(k) Check fixing of tiles (nail or clips).
(l) Check general conditions of roof.
(m) Check flat roof finish and construction where appropriate.

First fixing

(a) Check door linings and windows boards, (quality of timber, line, level and plumb).
(b) Check storey heights and staircase fixing.
(c) Check notching and drilling.
(d) Check joints, and support of service pipes.
(e) Check socket outlets.
(f) Check beam filling.
(g) Check strutting or bridging.
(h) Check fixing or floor-decking (size and support).
(i) Check framing and support to cold storage tank (plus any damaged or cut trusses).
(j) Check construction of partitions.
(k) Check thickness and fixings of plasterboard.
(l) Check fire stops (especially integral garages).
(m) Check all external cladding.

Plastering

(a) Check plaster materials are suitable and compatible.
(b) Check liquid dpm if required.
(c) Check services in floor screed/thickness/cover and falls.
(d) Check finish to walls is smooth and flat (line level and plumb).
(e) Check floor screed for thickness, quality and level.
(f) Check joint taping, caulking and ceiling board finish.

2nd Fixing

(a) Check damaged joinery and general standard.
(b) Check fit of doors and hinges.
(c) Check door locks to bathroom/wc.
(d) Check skirting and ceiling lines.
(e) Check kitchen layout and storage.
(f) Check internal glazing.
(g) Re-check sanitary fittings/services/heating and electrical.
(h) Check provision of stop valves and draining down.
(i) Check heating system provided and radiator points.
(j) Check fall length of waste pipes.

Painting

(a) Check preparation of surfaces prior to painting.
(b) Check standards of painting number of coats applied and surface finish.
(c) Check ceiling finish if not already seen.

Final stages

(a) Check access to lofts and gangway boarding.
(b) Check thermal insulation to roof/pipe lagging and lid to cistern.
(c) Check steps to external doors.
(d) Check drives and paths.
(e) Check site drainage/possible waterlogging.
(f) Check clearance of garden area.
(g) Check garages/outbuildings/retaining walls.
(h) Final check outside/inside.

APPENDIX 2: SNAGGING CHECK

8. Snag every house-b

Snag as follows, and carefully look at all parts of the property:

1. On each external elevation look at:

A Chimney	H Door frame and step
B Ridge	I Sills
C Flashings	J Rendering/brickwork
D Tiling	K Dpc
E Eaves	
F Windows	
G Flat roof	

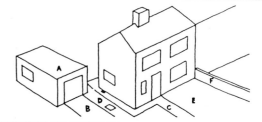

2. Around the dwelling look at:
A Garage
B Drives
C Paths
D Drains
E Garden areas
F Boundary walls

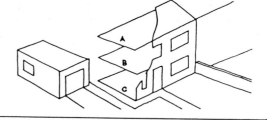

3. Inside the dwelling look at:
A Roof space
B First floor
C Ground floor

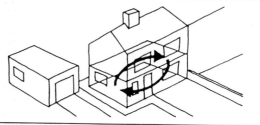

4. On every floor look at:
All rooms in a constant clockwise or anti-clockwise order.

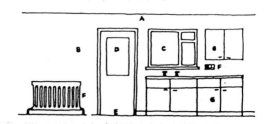

5. In every room look at:
A Ceiling
B Walls
C Windows
D Doors
E Floor
F Services
G Cupboards and fittings

APPENDIX 2: SNAGGING CHECK — FORM

8. snag every house-c

It is essential that, as you do your snagging, you write down the defects you find.

Many firms have their own snagging forms. Those wo do not can if they wish purchase such forms from the NHBC. As you make your inspection of the dwelling, note the defects you find on a form like this:

Snagging Inspection	Property	NHBC HB 72

Inspected by .... Date ....

Inspect the property systematically.

As you inspect, note all defects and outstanding work on this form.

After completing this form :—

1. Note the defects/outstanding work you have found on Trade Defects Sheets and hand top copies to trade foremen or sub-contractors.
2. Chase up work not done on time.
3. Check that work has been done satisfactorily.

Location	Defects to be rectified/outstanding work to be completed
Front elevation	
Left elevation	
Back elevation	
Right elevation	
Garage	
Drives and paths	
Garden area	

for interior see overleaf ⟶

Room	Defects to be rectified/outstanding work to be completed
Roof space	
Bed 1	
Bed 2	
Bed 3	
Bathroom	
Landing/stairs/hall	
Living	
Dining	
Kitchen	

STORAGE ON SITE

by W W Abbott, PhD, MPhil, FCIOB, FRICS

INTRODUCTION

Storage may be defined simply as the planned occupation of space by materials or goods, but it is clear that storage is a link in the chain of processes involved in building and any criteria of effectiveness must recognise this inter-dependence. For example, the use of detailed method study to reduce marginally the allowance in preliminaries for the area and cost of accommodation, temporary roads, hardstandings and perimeter compound fencing, may easily result in a loss, due to the inability to store materials properly which arrive early as a result of changed production programmes.

THE NEED

The principal factors related to the need for storage are:
(a) economic buying of materials;
(b) changes in production programmes;
(c) late deliveries by suppliers;
(d) economic production outputs;
(e) shortages of particular materials;
(f) limited period of availability of some materials;
(g) variations.

These factors are subject to variables often outside the control of the contractor, although some allowance can be made based on experience. Therefore, the allocation of space and other resources to storage should incorporate as much flexibility as possible for alternative utilization and adaptability.

This means, for example, reserving an area of the stores or compound for the awkward load or unexpected delivery; allowing for two means of access to the storage area instead of one; arranging storage areas to avoid complete dependence on one item of mechanical plant; using easily moved or mobile temporary covers for protection of materials in stacks; using sectional interlocking metal racking rather than traditional steel scaffolding.

COSTS

The total cost of storage, if considered in isolation, may appear excessive, as it could be in the order of 20 to 30 per cent of the value of the stock held per year. However, against this may be set the estimated savings which arise by having provided for the circumstances (a) to (g) listed above.

Elements of total cost include the following:

(a) cost of enclosed storage compound, protective coverings, depreciation, maintenance, transport, erection, dismantling, rates, taxes, insurance, heating, lighting;

(b) cost of salaries and wages for personnel directly and indirectly concerned;

(c) cost of mechanical and other equipment;

(d) costs arising from deterioration, wastage, theft, vandalism or their prevention.

To avoid unnecessarily high costs of storage, there must be close co-operation and co-ordination between planning, purchasing and production staff.

For instance, when deciding the economic quantity, size and weight of loads delivered to site, the purchasing department may wish to place an order based on the largest possible quantity per load, in order to obtain larger discounts. This proposed saving must be considered against the possible additional costs arising from allocating extra space to that material; protection; disruption of traffic flow; damage to road surfaces; longer duration of storage on site with consequent increased risk of deterioration, damage, pilferage and wastage.

Also, when choosing the most economical alternative method of increasing the speed of construction, an increase in the number of men on critical operations could result in lower productivity if the supply of a material was unable to meet the new demand. It may be impossible to find additional storage space for some materials, whilst others could be accommodated without incurring extra cost.

Without careful consideration of all aspects of storage, additional labour or plant costs expended on acceleration of the work may be wasted.

ORGANISATION

Storage, particularly for high rise construction, should receive detailed consideration during pre-tender planning.

The materials content of the building project should be analysed in terms of:

(a) quantities of different materials;
(b) variety of materials;
(c) unit and total value of different materials;
(d) timing and frequency of use according to the construction programme;
(e) liability of different materials to deterioration, damage, wastage and theft.

The above factors can then be related to the needs of a particular site, including geographical location, terrain, proximity to other building work, construction plant and space available for storage.

From these considerations, which may be assisted by the application of the Pareto principle, should emerge an order of priorities on which to base the materials storage arrangements. The Pareto principle in essence, states that important items in a given group tend to be concentrated into a small part of the group. For example, in many cases, about 80 per cent of the total value of materials is contained in 20 per cent of the number of items in a bill of quantities. It is also likely that about 80 per cent of the total cost of wastage and loss of a particular material is caused by about 20 per cent of the number of possible causes, which could include quantity and quality inspection at delivery, unloading, stacking, distribution, deterioration, contamination, breakage, theft and vandalism.

Storage on site

Example of inadequate site control

By recording and collating experience, a firm should be capable of distinguishing the vital aspects on which to concentrate for a particular site. Materials of high unit cost and large total quantity should, if possible, arrive packaged, or banded or cut to size.

The timing of deliveries should be very carefully programmed and progressed so that storage time on site is reduced to a minimum. Materials of lower unit cost and smaller total quantities could receive relatively less attention but they must all be available and locatable when required.

By working to a planned order of priorities staff time will be allocated most effectively.

LAYOUT

The layout plan for the enclosed stores must include provision for three distinct areas of activity:

  (i) receiving materials;
 (ii) storage;
(iii) issue or collection of materials.

The storage area should, where possible, avoid crossflows of traffic and minimise movement. Main gangways should be central and run lengthways to shorten the more difficult access ways (see Figure 1A).

Items of stock should be grouped generally into those frequently in demand, and those less frequently in demand, placing the former nearer to the point of issue. In open storage areas, where space allows, the flow of traffic should be one-way to avoid delays (see Figure 1B).

Any particularly valuable, perishable or hazardous materials should receive special attention and be located where they can be more secure and frequently supervised by the storekeeper.

The use of chain link fencing for divisions, compartments or the perimeter fence assists security, increases visibility, helps to avoid accidents and encourages a tidy site.

Whilst materials should be quickly locatable, reserving a fixed location in the storage area for each item tends to under-utilize space. The locations should be planned by reference to the construction programme, allowing alternative use of a space as soon as possible but making sure that another delivery of the previous materials is not imminent.

On restricted sites where storage space is very limited, the use of the building under construction may be facilitated by obtaining permission to strengthen, where necessary, certain floors or adjacent external paving to take the proposed additional loads of materials.

Very often full utilisation of available height in storage spaces is not achieved, and consideration should be given to proper stacking, and the use of shelves, racks and pallets appropriate to the material. Flat or mono-pitch roofed storage huts will be advantageous in this respect. Stacking can be relatively light work with the assistance of small mobile stackers, operated hydraulically by hand pumping and lightweight telescopic mobile hoists powered by compressed air. *costs?*

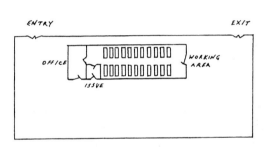

Figure 1A Storage area layout

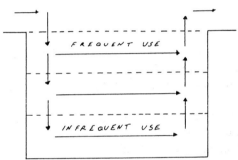

Figure 1B Storgage area layout

MAINTENANCE OF STORAGE AREAS NOT UNDER COVER

This is an area often neglected and attention to the following can have a marked effect on improving performance:

(a) cleaning is necessary to remove nails, broken glass, battens etc. Dirty and untidy conditions can cause accidents and lower morale;

(b) re-surfacing with additional hardcore or concrete should be a priority task where deep ruts or potholes develop, as these can cause accidents and delays. The maximum weight of lorries should be taken into account in the design of temporary roads and hardstandings;

(c) damaged signs or notices should be repaired quickly and painted, preferably in a bright colour;

(d) the surface areas should be laid to falls or drained so that access is maintained and damage to stock does not occur during inclement weather.

Storage on site

EQUIPMENT

A recent study of the activities of site operatives in the building services field indicated that 20% of working time was spent on walking, carrying and sorting small materials. The use of the following equipment to store materials can reduce considerably the time spent on checking, handling, locating, stock taking and will allow more effective use of the available space:

(a) bins for small items such as ironmongery;

(b) racking for materials such as timber, steel bars, plywood sheets, roof trusses. The use of racks which can be lifted by crane or fork lift truck for storage of timber roof trusses etc can also be an advantage. The A-frame type of rack with projecting brackets allows for bars etc to be handled both from the end and from the side of the racking (see Figure 2);

(c) pallets for use with cranes and fork lift trucks;

(d) tubular steel stands for storing oil drums etc.

It is important to use supports at sufficiently close intervals to avoid sagging of materials which may be deformed permanently.

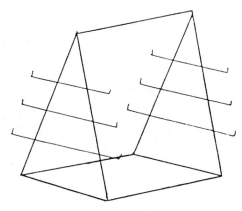

Figure 2 A-frame racking

SECURITY

The aim of store security should be to remove temptation and to make matters difficult for a potential thief.

The following check list is a guide to the action required:

- access should be limited to one or two gates;
- frequent checks must be seen to be made on materials, windows, doors as the potential thief will be deterred if he knows a theft will be quickly discovered and investigated;
- the internal layout of enclosed stores should be arranged to provide an issue counter segregated from the main storage areas;
- responsibility for the custody of keys should be clearly defined. There should be a minimum number of duplicate keys and if one is lost the lock should be changed;

- security patrols should be employed in particularly difficult areas where there is a high probability of loss;
- materials or equipment which are attractive for domestic use and easily transportable should be marked, if possible, with the company name or colours;
- in some areas stealing from open building sites is regarded by the local residents as normal practice, rather than as a crime. Some new tenants on partially completed housing estates will remove paving slabs, bricks, cement, timber, rainwater goods etc during their 'do-it-yourself' weekend activities to complete gardens, sheds and extensions. One or two tours by a security guard should have some effect and a single prosecution in the courts, well publicised, is an excellent deterrent to other potential pilferers who live on the estate.

SAFETY

Many accidents occur due to falling materials. Care and attention at the delivery stage to ensure that materials are stacked safely can relieve management of some of the continual worry about this aspect of their work, which has become so onerous due to recent legislation.

Poor stacking practice which creates safety problems

Prevention must be the watchword rather than consideration of what must be done as a minimnum to avoid blame under existing safety legislation. The presence of the company safety officer at pre-production meetings will be of considerable value in this respect. Some important points to remember are:

- avoid the need for the use of ladders wherever possible;
- stacks are potential killers if not properly bound together by using timber slats or bonding. Their protective covering should be brightly coloured to prevent collision by delivery lorries;

- protective clothing including gloves and boots with steel toe caps should be available;
- metal to metal stacking should be avoided as slippage can easily occur. This may be avoided by using timber slats between the metal objects;
- when materials are stacked, shelved or racked above arms length reach, safety helmets should be worn;

245

Storage on site

- fire extinguishers should be provided at strategic points such as junctions of gangways or centrally in a working area. First aid kits should be available;
- the local factory inspector should be consulted if in doubt about safety arrangements, particularly in relation to hazardous materials.

MOVEMENT

The costs of handling a material can be as high as 50 per cent of the total cost of labour expended in connection with the material. For example, where an item such as a precast concrete lintel or large window frame is handled without the assistance of a fork lift truck or a crane, the cost of receiving delivery, unloading, stacking, loading from stack on to a trailer or barrow, transporting to hoist, hoisting, further transport at required floor level, and unloading at fixing position could exceed the cost of the actual fixing or building in.

The need for careful planning and management is clear, and application of the following points could result in savings:

(a) manual handling should be avoided where possible. This often arises naturally where the site organisation has been designed around a tower crane but in other cases the use of fork lift trucks can prove to be very economical;

(b) any material should be handled in its largest convenient unit load. Advantage should be taken of unitisation and packaging offered by some suppliers, for example in the delivery of bricks, blocks and timber, and special arrangements with suppliers can be made for other material to be similarly treated at relatively little extra cost. These measures can help to prevent wastage, damage and deterioration which, according to recent studies, is likely otherwise to be in the order of 10-20 per cent of the cost of material. Additional handling costs associated with loose materials on site are also saved;

(c) the receiving area should be suitable for the type of vehicle and size of load expected, so that turn-round time is minimised. It is often a false economy to reduce expenditure on preliminaries by not providing proper temporary roads.

Where the site contours are suitable, consideration should be given to the use of gantries or additional excavation to allow direct horizontal movement of materials from vehicles to storage position.

It is essential to have an informed responsible person on site to receive and direct a lorry as it arrives, as considerable costs due to double handling and wastage can ensue from inefficiency at this critical time;

(d) in many cases where mechanical plant such as cranes, fork lift trucks or dumpers can be used, it will be more beneficial to have one centralised storage location, rather than several storage areas nearer the construction operations, particularly as this arrangement has the advantage of providing opportunities for improving security, stock control and wastage prevention.

Changes of location which may be required at different stages of construction may be facilitated by the use of pallets, containers, packages and mobile storage huts.

STOCK CONTROL

The system of recording deliveries to site and any transfers to other sites must be effective and simple. To achieve this satisfactorily, senior management must strictly enforce and support the rule that all drivers of delivery lorries must first report to the checker or store-keeper. If the checker or storekeeper is not available, a responsible foreman or operative must be designated and suppliers should be informed that no payment will be made if an appropriate signature is not obtained.

The contractor should also insist on deliveries being punctual to avoid disruption of planned storage.

Records of previous experience of a supplier's performance can provide a guide to the time which should be spent on checking the quantity and quality of deliveries. Expensive items and those critical to the completion date and the finished quality of the building, should receive special attention.

Immediate identification of delivered items, by labelling, colouring or numbering is important, so that time is not wasted later in searching, and also to avoid the use of a material as a substitute for another similar one which cannot be readily located when required.

Monthly measurement of unused materials on site will allow comparisons with net quantities required for the building work during the period, in order to detect excessive wastage, or inaccurate ordering of materials. This monthly stocktaking should be selective, bearing in mind the value and importance of the various materials and should be accurate to within ± 2 per cent for those selected. It should also be timed so that the quantity surveyor can use the information in the preparation of valuations.

A weekly reconciliation of the quantities of: materials received; materials transferred to other sites; materials used in building; and stocks held, will provide a system of early detection and the opportunity to avoid financial loss. It is particularly important to check more frequently the rate of usage and the allocation of materials to floor levels in multi-storey buildings, where the supply of further quantities to a floor level, the transfer of surplus material, or the removal of wasted materials can be costly and difficult. The storage and use of facing bricks should primarily accord with the designer's specification, but in the absence of full instructions facing bricks should be checked carefully for consistency of appearance when received and loads should be used in chronological order. The control of wastage is very important as it may not be possible to obtain further supplies of matching bricks. Once a quantity of plaster has been delivered to a floor level it is unlikely that any will be retrievable or transferable due to damage, setting or deterioration. Again, frequent checking is essential in order to avoid shortages.

Frequent checking of the rate of usage should also bring to light excessive allowances for unavoidable wastage or that the buying department has miscalculated and ordered larger quantities than necessary. It may, therefore, be possible to avoid surplus materials on site at the end of the contract. ↳ left over

In deciding whether to retain, sell or tip surplus materials the following factors must be considered:

Storage on site

- storage space available;
- distance from site to storage depot;
- cost of collecting, loading and unloading;
- deterioration;
- probable future use;
- scarcity;
- market price of material
- time remaining before site must be vacated.

CONTRACTUAL OBLIGATIONS

Manufacturers' advice should always be observed when storing and moving materials, particularly those liable to deterioration, breakage or damage.

The standard form of tender used by nominated suppliers specifically excludes replacement of defective goods where defects are caused by improper storage. The respective responsibilities of main contractors and sub-contractors for storage and movement of materials should be established clearly at an early stage to avoid disputes arising later.

QUALITY

The quality of finished building work is often unacceptable due to the use of damaged or contaminated materials. To avoid costly remedial work it is essential that materials are adequately protected at all stages. Where materials are unloaded and stored on site rather than in the enclosed stores, a major cause of contamination and loss is mud. The contractor should always have a supply of plastic sheeting readily available to place on the ground or paved area when materials are delivered. After stacking part of the load on the sheet the edges can be turned up the sides and secured so that the vulnerable lower area of the stack is protected. Immediate attention to deliveries is critical and the availability of a few used plywood sheets and scaffolding boards at this time often means that materials can be transported across site obstructions. Without these the materials may have to be left in vulnerable positions, often disrupting traffic and suffering damage.

The vigilance applied during storage and handling can be nullified if protection of materials is neglected between the time of incorporation into the building and handover.

Mortar and mixer setup problems

248

Damage by operatives and the effects of the weather should be prevented by the use of battens or hardboard to cover edges of stairs and corners of partitions; using tapes or lagging on sanitary goods and fittings; ensuring that fixed joinery has been primed; protecting floors with sawdust or chippings; ensuring that incomplete external plumbing does not allow water into the building; avoiding contamination and stains from soil and other materials by regularly clearing away surplus and waste; having available clear plastic sheeting to cover window openings etc in wet weather.

Although the majority of weather problems are due to water, it is well to remember that excessive heat causes damage by shrinking and cracking whilst freezing conditions can ruin the work of wet trades and make some materials brittle, increasing the risk of fracture.

PERSONNEL

With about 50 per cent of the cost of building represented by materials, it is difficult to understand why many contractors regard personnel concerned with storage as relatively unimportant.

There is some evidence that storekeepers spend a considerable amount of time on progressing, filling in forms and issuing materials, and relatively little on checking deliveries and organisation of storage. At the same time, it seems that foremen spend relatively little time on organisation of materials storage. It may be concluded that insufficient time and attention is being paid to what is an important activity and if line management is unable to give sufficient attention to the problem due to pre-occupation with other activities, then it is necessary to increase the effectiveness of staff engaged in storekeeping.

Some firms have used monetary incentive schemes for site staff, based on discovery of quantity of quality discrepancies on deliveries and also on wastage being held within a particular percentage over the period of the contract. The latter can be successfully used on repetitive work where reliable norms can be set for particular materials. However, this approach to improvement is of doubtful value and cannot be a substitute for proper selection and training of staff.

The training of storekeepers and anyone engaged in handling and storage of materials should include principles of materials handling and a basic knowledge of the characteristics of materials, including physical, chemical, mechanical and thermal properties, so that failure by excessive deformation, fracture and chemical deterioration may be minimised. If the storekeeper is to assess properly the quality of materials at delivery for acceptance on site, he must have a knowledge of common defects, blemishes and other imperfections which may occur, and the extent to which these are acceptable or unacceptable by reference to the specification and normal practice.

He must also know the appropriate method of handling, storage and protection.

For example, to deal satisfactorily with taking delivery, unloading, storage and distribution of flush doors, the person concerned would require knowledge of the following:

- moisture content related to shrinkage, swelling and twisting. Doors should arrive covered and protected from the weather and be immediately unloaded and stored in weather protected conditions. The doors should not be distributed and fixed until the building is substantially weathertight;

Storage on site

- strength and structure related to deformation and splitting. Doors should be stacked flat and surfaces kept clean to avoid scratching the veneer. Point loading and impact damage should be avoided on surfaces and edges;

- absorbtion related to vulnerability to staining. Doors should be protected from splashes of all liquids such as paint, dirty water, etc;

- fungal and beetle attack related to season of year. This is particularly important in alteration and repairs to existing properties where existing buildings may be used for storage. It is preferable to use new temporary storage. Existing property should be carefully inspected and treated if necessary before use for storage.

The pace of innovation in design and manufacture of materials and components makes frequent updating of knowledge very important. In order to avoid repetition of costly mistakes due to ignorance, a firm should circulate instructions to those concerned, showing clearly the correct methods of handling and storage of new materials which are particularly prone to damage or deterioration.

MAINTAINING AND IMPROVING PERFORMANCE

Much of the hard earned experience and knowledge gained on individual sites can be wasted if no records are kept of difficulties and successful ways of dealing with them.

Staff may move to other jobs and take their expertise, but the firm should aim to have its own written standard procedures based on collated reports from sites.

A policy for storage should be developed and standard procedures for staff, accommodation and equipment laid down, so that staff may become more efficient and effective and other resources are better utilised.

Within this general framework, each site will have particular problems which must be recognised at planning stage and resources allocated on a basis of priorities.

BIBLIOGRAPHY

1. ABBOTT, W. W. (1970) An evaluation of the criteria used in the assessment of allowances for wastage of materials in building. MPhil Thesis (CNAA).
2. ABBOTT, W. W. (1976) Quality control in buildings with special reference to workmanship and supervision. PhD Thesis (Nottingham University).
3. BAGGOTT, P. (1971) An investigation into materials management in the construction industry. BSc Building Final Year project (Coventry Polytechnic).
4. DAND, R. and FARMER, D. (1970) Purchasing in the construction industry. Gower Press.
5. (1969) Nottinghamshire builds : Project 'Research into site management'. *Architects Journal*. December 11.
6. SANSON, R. C. (1959) Organisation of building sites. HMSO
7. LONDON BUILDING PRODUCTIVITY COMMITTEE. A code of procedure for monitoring quotations for building materials.

Further useful reading: JOHNSTON J E (1981) Site control of materials. Butterworth
SKOYLES E R and J R (1987) Waste prevention on sites. Mitchell.

AN APPROACH TO REDUCING MATERIALS WASTE ON SITE

by John R Skoyles BSc(Econ), PhD

INTRODUCTION

There has been much research highlighting the incidence and problems caused by the waste of materials on building sites. Such research is pointless unless the lessons are applied in practice under everyday working conditions.

This paper gives practical guidance to site managers (in the form of a 'second pair of eyes') and emphasises for the benefit of management, that waste is a problem more serious than generally appreciated.

Inevitably, materials will be wasted in the process of building and it was with the aim of establishing the degree of inevitability that the Building Research Establishment (BRE) undertook its initial studies[1]. Based on studies of 230 sites it soon became clear that methods of accounting had to be developed especially to measure and quantify the problem[2]. But BRE only applied its findings in terms the materials lost other than in terms of the monetary losses to the contractor. Subsequently, the author undertook the studies in this field described in this paper and explained in greater detail in a comprehensive book which was based on studies on 282 sites of varying size and complexity[3].

Other approaches to the problem have been taken by Abbott[4], Wyatt[5] and Hussey[6]. The Chartered Institute of Building also considered the results of these studies and produced a Plan for action[7] to deal with the problems identified.

At current (1989) values the estimated annual value of materials waste is nearly £4 billion excluding repairs and maintenance but including labour costs of moving materials to work places. Much waste would be uneconomic in site resources to justify its saving, but this is not true of all incidences of waste. Sites which would be expected to face similar problems in avoiding waste were found to have widely differing levels of waste.

A fact of major importance emanating from these studies was that waste allowances for estimating were much too low by about 200%.

Investigations of why waste occurred under particular circumstances revealed that responsibility had to be shared among all those in the industry and materials and plant manufacturers.

Materials wastage has an adverse effect on the profitability of a contract and is an obvious area for detailed attention by building management, in particular site managers.

Waste occurs on site at all stages of materials handling; when they arrive, when they are stacked, when they are moved, and when they are fixed or placed.

The key to the control of waste lies in persuading site management of its importance. Since it is a problem unlikely to create major difficulties it tends to receive low priority or is ignored altogether.

An approach to reducing materials waste on site

ENCOURAGING SITE MANAGERS TO BE WASTE CONSCIOUS

Site managers are often unaware of the opportunities of cutting waste. Many believe that it cannot be effectively controlled and that it is more efficient to allow losses to occur than to involve the use of extra resources in reducing or preventing waste.

Table 1. Comparison of results with those previously published

MATERIAL	Norm %	BRE RESULTS			SPECIAL STUDIES			
		No of sites	% Range	Overall Waste	No of sites	% Range	Overall Waste	Mode
TRADITIONAL CONSTRUCTION								
Concrete: ready mixed	2½	110	3-18	5	12	3-10	4½	5
Bricks: commons	4	144	1-20	8[4]	12	1-20	9	9[4]
facings	5	138	1-22[1]	12[3]	12	1-21	12	11
engineering	2½	10	9-11[2]	10	12	7-11	9	9
Blocks: lightweight	5	65	1-22	10[5]	12	1-21	10	11[5]
dense	5	25	1-18	9	12	1-16	9	10
STEEL CASED BEAM CONSTRUCTION								
Concrete: ready mixed	2½	4	1-11	9.5	4	1-12	9	8
Bricks: commons	4	4	5-13[6]	7[7]	4	5-12	8[6]	8
facings	5	4	2-9[6]	5	4	1-10	6[6]	6
engineering	2½	4	3-9	6[8]	4	3-8	6	6
Blocks: lightweight	5	4	7½-8	8[9]	4	6-9	8	8½[9]
REINFORCED CONCRETE CONSTRUCTION								
Concrete: ready mixed	2½	8	4-13	6[10]	8	2-15	7	8
site mixed	2½	3	3-8	5	1	4-12	6[14]	6
Bricks: commons	4	11	5-13[11]	7[12]	9	4-12	8	8
facing	5	11	5-11	9	9	4-12	9	9
Blocks: lightweight	5	11	1-11	9[13]	9	1-13	9	9

(1) One site had a result of over 27%. It was considered to be exceptional and has been omitted from the calculations.

(2) These results are based on contracts using large quantities. On many sites the quantities of engineering bricks used were small and not accounted for.

(3) Indirect waste was 2% for facing bricks (as an average) and 5% for bricks (both commons and facing bricks in blockwork as substitution).

(4) Losses at stack averaged 4%.

(5) Losses at stack averaged 4% and at work place averaged 4% (excluding design waste).

(6) Samples showed no incidence of design waste.

(7) Waste at stack was between 0% - 2%.

(8) Waste at stack of all results was less than 1%.

(9) Losses were high at stack but varied due to whether packs were banded or banded and on pallets.

(10) Special studies indicated the spillage at columns to be about 1.5%.

(11) Samples had no incidence of design waste.

(12) Losses at stacks and at work place averaged 3%.

It is important to emphasise that preventing waste can be cost effective. This can be achieved by stimulating an awareness of the waste problem rather than by identifying the particular means of achieving lower waste levels. This is not to deny the importance of such instruction. Posters produced by BRE and by the then National Federation of Building Trades Employers (now Building Employers Confederation) had an impact when first issued but their effectiveness declined rapidly. What is needed is a constant reminder to site management throughout the contract.

Towards this end BRE urged that accounting for materials usage on site be an essential feature of site management. However, such accounting will cost something to implement and will provide results only after the loss has occurred.

What is needed is a system that warns the site manager before wastage is likely to occur. One solution is the judgement of the site's waste problems, or potential problems, by another pair of eyes. This would allow site management an opportunity to relate problems on waste control to someone independent of the running of the site but with an awareness of its problems in the same way that the services of safety officers are used. If not given priority by special means, waste, like safety, will tend to be ignored as a result of the everyday pressures experienced in running the site.

→ costs money

The 'second pair of eyes' would draw attention to those features of waste which would lack immediate consequences if ignored.

How far could observations of this type made on regular and short site visits reflect the general performance of a site? Was the performance on waste control so variable that its improvement or deterioration could be masked by other factors? It was important, therefore, to find how far waste levels were consistent over the duration of a project and to establish if qualitative observations could reflect a quantitive level of waste.

DESCRIPTION OF STUDIES AND RESULTS

To establish these points studies were carried out around 1984 with traditional construction and reinforced concrete structures. Data were obtained from direct observations of the work, materials handling and fixing on seven sites in the London area over the duration of each project. On a further 18 sites observations were restricted to specific materials and self-contained parts of the work. Frequent observations were made, including measurement of waste and a detailed breakdown of waste for a number of different causes. On every visit the observer toured the site to note examples of waste and any influencing factors. These observations ultimately demonstrated that the same events occurred regularly on the same sites.

Waste levels of the principal materials examined are given in Tables 1 and 2. Adjustments have been made, where applicable, for substitution.

An approach to reducing materials waste on site

Table 1 gives the results of the study and compares these with BRE work. They confirm the results of the earlier work and show that waste levels have not been reduced, thereby bringing into the question the 'application' of earlier research.

Table 2 Results of BRE research compared to CIOB sponsored studies

MATERIAL	Mode	Lower quartile			No.	Upper quartile		No.	Middle	%
		No.	Range	O/A		Range	O/A		Range	O/A
TRADITIONAL CONSTRUCTION										
Concrete	5	3	3-4	3	5	11-18	14	4	6-10	8
Bricks: commons	9	2	1-3	2½	3	16-20	17	7	8-11	8
facings	11	2	1-3	2½ (Mode)	4	16-22	18 (Mode)	6	9-12	10 (Mode)
engineering	9	2	-	9	3	-	11	7	-	10
Blocks: lightweight	11	1	-	1	5	14-22	-	6	6-9	8
dense	10	1	-	1	5	14-18	-	6	6-10	8
STEEL CASED BEAM CONSTRUCTION										
Concrete	8	1	-	1	2	10-11	10½	1	-	7
Bricks: commons	8	1	-	5	2	12-13	12½	1	-	8
facings	6	1	-	2	2	8-9	8½	1	-	6
engineering	6	1	-	3	2	8-9	8½ (Mode)	1	-	6
Blocks	8½	1	-	7½	3	-	8	-	-	-
REINFORCED CONCRETE CONSTRUCTION										
Concrete: ready mixed	8	2	4-4½	4	2	10-13	11½	4	5-9	7½
silo mixed	6	-	-	-	1	-	6	-	-	-
Bricks: commons	8	1	-	4	2	11-12	11½	4	7-9	8
facings	9	3	4-6	5	4	11-12	11½	2	8-9	8½
Blocks: lightweight	9	2	4	4	3	12-13	12½	4	7-10	8½

Notes: Results demonstrate the isolation of the lower and upper quantities as being broadly common for all materials.

In Table 1 the overall waste for each material is expressed as a percentage of aggregated deliveries. The results confirm, for the major materials, that waste is about twice that assumed.

Table 2 shows the correlation between the upper and lower levels of the results, clearly indicating the 'grouping' of waste levels.

Table 3 presents the results from four sites and shows that changes in waste levels only occurred when personnel changed.

Observations were made on the frequency of a number of activities, (eg, spillage of concrete when unloading, breakage of packs when moving bricks etc) to establish the variability of performance on site. A strong correlation was found between the waste levels on site and each event, providing that no changes occurred either in personnel, line management, operatives or plant. Hence, it can be assumed that an occasional visit to site will be representative and thereby provide a forecast of likely waste performance. For example, bricks will always be dropped while unloading or moving stacks.

Any change in performance is only likely to occur when personnel and/or the methods of packaging, delivery or handling are changed.

MATERIALS		OBSERVATIONS 1	2	3	4	5	6	7	8	9	10	11	12
CONCRETE													
site mixed	A												
	B	2	1½	2	2	2	1½	2	2	2	2	1½	2
ready mix	A	1½	1½	2	1½	2	2	1½	1	1	1½	1	1
	B	1½	1½	2	-	2	2	2	1½	1	1	1	½
Bricks													
Commons	A	3	3	2	3	3	2	3	2	2	3	3	2
	B	4	4	4	4	8	8	8	8	7	8	7	8
Facings 1	A	2	2	2	2	2	3	3	3	3	1½	2	2
	B	4	4	4	4	9	9	9	8	8	9	8	9
Facings 2	A	3	3	3	3	3	2[5]						
	B	4	4	4	4	4	5						
Facings 3	A						[4]4	4	3	4	3	4	
	B						3	5	5	6	7	5	-
Facings 4	A												
	B												
Engineering	A	2	3	-	-	4							
	B	3	2	3	2	3							
BLOCKS													
Lightweight 1	A	2	2	1½	2	1½	2	2	2	2	3	-	-
	B	4	4	4	4	-	8	7	-	-	8	9	-
Lightweight 2	A	4	4	5	5	4	4	5	4	4[5]			
	B	4	4	4	4	-	4	5	-	5	6		
Dense 3	A	4	5	5	5	6							
	B	4	4	4	3	4	4	5	4[2]				
TIMBER													
Carcassing	A												
	B												

MATERIALS		OBSERVATIONS 1	2	3	4	5	6	7	8	9	10	11	12
CONCRETE													
site mixed	A												
	B												
ready mix	A												
	B	1½	2	2	1	2							
Bricks													
Commons	A	5	6	7	5	5	5	6	7	5[5]	-	-	-
	B	4	-	4	-	5	-	-	6	8	-	-	-
Facings 1	A	5	6	6	4	5	6	-	-	-	-	-	-
	B	4	-	4	-	4	-	-	4	-	-	-	-
Facings 2	A												
	B												
Facings 3	A												
	B												
Facings 4	A												
	B												
Engineering	A												
	B												
BLOCKS													
Lightweight 1	A	3	4	4	4	4	4	4	4	3	4	-	-
	B	3	4	3	-	-	-	-	4	-	-	-	-
Lightweight 2	A	4	5	4	4	5	4	-	3	-	-	-	-
	B	3	-	3	-	4	-	-	4	-	-	-	-
Dense 3	A												
	B												
TIMBER													
Carcassing	A												
	B												

MATERIALS		OBSERVATIONS 1	2	3	4	5	6	7	8	9	10	11	12
CONCRETE													
site mixed	A												
	B												
ready mix	A	1	1	1	0	0	1	0	1	0	1	0	1
	B	1	1	1	1	1	1	1	1	1	1	1	1
Bricks													
Commons	A	2	2	2	2	2	2	2	2	2[5]			
	B	3	3	2	3	3	2	2	3	2			
Facings 1	A	5	4	4	4	4	3	4	5[5]				
	B	3	2	3	3	3	3	3	3				
Facings 2	A												
	B												
Facings 3	A												
	B												
Facings 4	A	0	0	0	0	0	0	0	0	0[5]			
	B	3	2	3	2	3	3	2	2	2			
Engineering	A	2	2	2	2	2	2	2	2	2[5]			
	B	3	3	4	4	4	4	4	4	4			
BLOCKS													
Lightweight 1	A	3	3	3	4	2	3	2[5]					
	B	4	4	4	3	3	3	-					
Lightweight 2	A	3	2	2	2	2	2	2[5]					
	B	4	4	4	4	5	4	4					
Dense 3	A												
	B												
TIMBER													
Carcassing	A	2	2	3	2	2	3	2	2[5]				
	B	4	5	5	6	7	7	7	6				

MATERIALS		OBSERVATIONS 1	2	3	4	5	6	7	8	9	10	11	12
CONCRETE													
site mixed	A												
	B												
ready mix	A	4	5	5	6	4	4[1]	2	2	3	2		
	B												
Bricks													
Commons	A	4	5	6	6	6	5[1]	2	2	3	2		
	B	5	5	4	5	5	5	5	5	5	5		
Facings 1	A	4	4	5	5	5	5[1]	2	2	2	3		
	B	5	4	6	6	6	5	5	5	5	5		
Facings 2	A	3	3	3	3	3	3[1]	1	1	1	2		
	B	5	5	5	6	5	6	5	6	6	6		
Facings 3	A					3	4[1]	2	1	2	2		
	B					6	6	6[3]	5	3	3		
Facings 4	A	1	1	1	1	1	1[2]	0	0	0	0		
	B						6	6	6	5	6		
Engineering	A												
	B												
BLOCKS													
Lightweight 1	A	4	5	5	5	4	4[1]	1	1	2	1		
	B	5	5	6	6	5	5	5[3]	2	3	3		
Lightweight 2	A												
	B												
Dense 3	A												
	B												
TIMBER													
Carcassing	A												
	B												

Table 3 Results from four sites which demonstrate strong similarities until the incidence of specific events.

Notes:

1. New site manager appointed and briefed to improve waste performance. The site was cleaned up - by mainly bulldozing into the ground wasted bricks and blocks - but standards of stacking overall improved. However, no change was made in the losses at the work face due to the same gangs of men being employed.
2. Due to the above reason low levels of waste of shrinkwrapped bricks was improved by better stacking.
3. Improvement of waste levels occurred due to change in composition of one of the laying gangs.
4. Operation commenced at this stage.
5. Operation ended at this stage.

CONCRETE

A. Waste at delivery (for ready mix concrete), or at the batching plant, usually spillage.
B. Waste placing concrete in foundation.

BRICKS AND BLOCKS

A. Waste at unloading and stacking position.
B. Waste at work place - usually cutting, dropping and rejection.

FORM OF DELIVERY

1. Banded in blades with top and bottom edges protected with arris strips.
2. Banded with blades but only top arris protected with arris strips.
3. Bricks delivered loose.
4. Shrink wrapped polythene packs.

1. Blocks banded laterally on pallets.
2. Blocks banded not on pallets.
3. Blocks delivered loose.

An approach to reducing materials waste on site

Means to reduce losses include:
- providing adequate storage positions;
- providing a clear site layout plan showing materials locations;
- unpacking materials carefully to prevent collapse when the packaging is removed or the bands are cut;
- collecting pallets and returning them to the supplier when the goods are off-loaded;
- effecting general tidiness and unstacking at the areas concerned.

These are points which largely go unnoticed.

A further example of waste is the dropping of materials at the workplace through their being spoilt or unsuitable. Frequently such materials are not reported and hence the credit which could be obtained from the supplier is lost; or a variation order is not claimed because of the difference of the bill specification and the materials actually delivered.

The presence of surplus mortar and plaster in excess of the day's requirement, tools left lying about and torn bags of cement and plaster left in the open, all reflect poor site practice, as do surplus materials not being passed on to the next workplace and idling plant.

The effectiveness of materials control will be related to the interest of the site manager. Ideally, he should delegate responsibility as appropriate to ensure that there is effective control throughout all stages of materials handling and storage.

DISCUSSION

It is chastening to find that the levels of waste found in this study were of the same order as that reported elsewhere. Yet, on half the sites the site managers had attended courses on waste control. On three sites a Materials Controller was measuring waste regularly and on five others ad hoc waste control procedures were employed.

The role of the 'waste observer' or 'second pair eyes' proposed earlier is important and should be fulfilled by a member of the contractor's staff who has to visit sites regularly, for example

the contracts manager. Alternatively, it could be delegated to a special materials controller or the quantity surveyor.

An on-going independent evaluation of site performance in relation to waste should be supplementary to the site accounting of materials and should be a regular feature of the firm's materials management policy. The 'waste observer' who focuses the site manager's attention on waste, will provide the opportunity for discussion and allow any formal tuition on waste to be related to actual site situations and problems. It could be an advantage for courses on waste control to be linked to site visits so that the lessons in the classroom can be related and revived later on site. This approach could prevent the tendency for the effects of knowledge gained to be dissipated with time.

CONCLUSION

The control of waste on site involves several people apart from visiting senior management, from the site manager to those handling the materials, but the keypoint for success is changing the attitude of the industry. This change will not come about quickly, for the industry is conservative in nature and pressures for output often override the care and handling of materials. Moreover, sites are usually complex in their layout and it is often physically impossible for the site manager to have all parts of the project under constant view. However, the site manager is expected to run a profitable job and any waste is a factor on the adverse side of the equation. Management should emphasise where improvements are possible and give site management its support.

The potential has been shown of using a 'waste observer' to review performance. Readers are also directed to the CIOB publication *Try reducing building waste*[9] which illustrates the main areas where costs are incurred, how they occur and how they might be avoided. It also provides a useful check list on storage and handling methods.

An approach to reducing materials waste on site

REFERENCES

1. SKOYLES E R (1976) Materials waste - a misuse of resources. *Building Research and Practice* 4, July/August, pp232-243, BRE Current Paper CP67/76

2. SKOYLES E R (1978) Site accounting for waste of materials, BRE Current Paper CP5/78, pp26

3. SKOYLES E R and SKOYLES J R (1987) Waste prevention on site. Mitchell Professional Library.

4. ABBOTT W W (1970) Reducing materials waste on site. *Building Technology and Management, November*

5. WYATT D P (1978) The control of materials on housing sites; CIOB Site Management Information Service Paper No.75
   WYATT D P. Materials management. Parts 1 and 2. CIOB Occasional Papers No.18 and 23

6. HUSSEY H J (1973) An examination of waste of bricks and blocks with particular reference to interchangeability, September (Institute of Building Thesis)

7. INSTITUTE OF BUILDING (1980) Materials control and waste in building - A plan for action prepared by the Site Management Practice Committee of the Chartered Institute of Building. October

8. SKOYLES J R (1983) Site managers can increase profits by reducing waste. *Building Trades Journal*, August 11

9. CHARTERED INSTITUTE OF BUILDING (1990) Try reducing waste. Site Management Practice Committee

STEELWORK TODAY

by P H Allen, CEng, MICE, MWeldI

INTRODUCTION

The aim of this paper is to describe the development of steels and steel products used in the construction steelwork industry for structural and civil engineering purposes. It covers the development of steel from the early days, sets out its current qualities, products and application, reviews the stages in the construction of a building and lists the advantages that come from its use.

It continues by considering modern developments in corrosion prevention, fire protection, fire engineering, and construction methods and concludes by looking at future activities and giving sources where further advice can be obtained.

HISTORICAL BACKGROUND

Iron has been known since very early days. It was probably first discovered by chance by heating iron ore in a charcoal fire. It was found that the fire burned more efficiently when the wind was blowing. This led to the use of a forced draft by means of bellows to increase the air supply and produce the iron more rapidly. Such primitive furnaces were the forerunners of the modern blast furnace, the charcoal being replaced by coke, a product of coal and not wood.

Iron produced by early primitive methods did not actually become molten but could be forged and shaped by hammering. On the other hand the molten iron produced by the blast furnace was hard and brittle since it absorbed three to five per cent of carbon from the firing medium, ie, charcoal or coke, both being almost pure carbon. These facts which are part of history, and are given in outline. The main purpose is to introduce the three iron products used in building, cast iron, wrought iron and steel. The basic chemical difference of the three is the amount of carbon and other elements included with the iron, but the mechanical properties are appreciably different.

The use of cast iron as a building material probably dates back to about the year 1800, with wrought iron being introduced a few years later. Cast iron columns were still being made for limited use in the early 1930s, though they ceased to be used in any quantity after the beginning of the century when steel took over as the main structural material.

Wrought iron, probably because it was costly to produce, began to be replaced by steel about 1850 and very little wrought iron was used after 1890, though there is evidence of some use of wrought iron sections as late as 1910, and also mixtures of wrought iron and steel in identical sizes in the same structure.

Steel structural sections were available in very limited sizes from 1850 onwards but in 1880 they had started to increase quickly in both size and quantity. By 1887 Dorman Long and Company offered a range of 99 beam sizes as well as a vast range of channel and angle shapes.

Structural steel quickly replaced wrought iron and by the year 1900 few beams or other structural shapes were rolled in wrought iron.

Developments in the manufacture and in the control of quality and strength of steel are reflected in the British Standard Specifications which have been published since 1906.

Steelwork today

QUALITY OF STEELS AND THEIR USES

The requirements for structural steels are given in BS EN 10025 and BS 4360 being the latest and most comprehensive specifications covering steels previously given in BS 15, BS 2762 and BS 3706 and appreciably increase the range of steels now available.

These specifications between them not only include the mechanical (strength) properties of the steels but also the tolerances on plates, and sets out the acceptable quantities of the various elements that can be mixed with the pure iron.

The end product is a material with ultimate tensile strength in the three basic grades of steel of 43 (mild steel) 430/580 N/mm² and 50 (high tensile steel) or 490/680 N/mm², with the option of an even higher grade tensile steel (55) and weathering grades of steel (WR) which do not need any corrosion protection.

The majority of steel produced and used in general building construction is Grade 43 but with the development and increasing availability of Grade 50 steels these are being used with considerable economy for multi-storey buildings. Here their 30% strength/weight advantage leads to smaller connections (less fabrication), lighter structures (less foundations), smaller sections (less building height with greater plan area). It is these similar qualities that lead to the use of Grades 50 and WR for bridgework and roll on/roll off ramps.

The Grade 55 steels are mainly used for wind sensitive structures where their ultra-high strength and structural form gives a high strength/surface area ratio for mast and tower construction.

PRODUCT RANGES AND APPLICATIONS

The product range of section shapes that are available and their uses are shown in Figure 1 and are available in Grades 43A and 50B for sections and 43C and 50C for hollow sections. These sections rolled are to BS 4 and BS 4848 and full details of these and the safe load bearing capacities etc can be found in the BCSA/BSC Sections book.

CONSTRUCTING THE BUILDING

Design

The requirements of the Building Regulations 1985 are deemed to be satisfied provided the structural steelwork complies with the requirements of BS 5950 *The structural use of steel in building*, the standard based on the limit state approach to steelwork design.

At the time of writing (October 1990) the following parts have been published:

Part 1 - Code of Practice for design in simple and continuous construction - hot rolled sections.

Part 2 - Specifications for materials, fabrication and erections - hot rolled sections.

Part 3.1 - Code of Practice for design in composite construction - beams.

Part 4 - Code of Practice for design of floors with profiled steel sheeting.

Part 5 - Code of Practice for design in cold formed sections.

Section	Representation	Uses	Section	Representation	Uses
Joist		Stanchions beams, supports bracings, portal frames	Structural Hollow Section (S.H.S.) Square		Roof trusses, stanchions, walkways, balustrading bracings, stiffeners, lattice work, light framed buildings.
Universal Beam (U.B.)		Supports, beams bracings, purlins, portal frames	Structural Hollow Section (S.H.S.) Rectangular		
Universal Column (U.C.)		Columns supports, stanchions, heavy fabrications.			
Channel		Bracings, purlins, walkways, roof trusses.	Structural Hollow Section (S.H.S.) Circular		
Angles Equal and Unequal		Roof trusses, purlins, bracings, stiffeners, light framed buildings lattice work.	Plates		
Tee's Short Stalk, and cut from U.C. or U.B.		Roof trusses, Roof trusses bracings, stiffeners, lattice work, hip rafters.	Universal Flates		Plate Girders Box columns, & Box girders.

Figure 1 Product range of steel

Part 8 - Code of Practice for fire resistant design.

and the following parts are scheduled to be published within the next year:

Part 6 - Code of Practice for design in light gauge sheeting, decking and cladding.

Part 7 - Specification for materials and workmanship - cold formed sections.

Provision has been made to extend the acceptability of BS 449: Part 2: 1969 - in common with other allowable stress approach standards, as an alternative method of design etc. This standard was amended for the last time so that it has generally the same 'factors of safety' as that of BS 5950.

Structural steelwork in building having five or more stories must be designed to meet the requirements of the Building Regulations.

The design of other types of steel structures are covered by the following British Standards:

BS 5400 - Steel, concrete and composite bridges.

BS 5502 - Design of buildings and structures for agriculture.

Steelwork today

BS 8100 - Lattice towers and masts (also see DD 133 for strength assessment of towers).

With the advent of the single market in Europe set for 1992, the European Commission is harmonising the whole of the construction industry. For steelwork this will involve two new design codes:

Eurocode 3 - Common unified rules for steel structures.

Eurocode 4 - Common unified rules for composite steel and concrete structures.

These will be supported by new standards for materials and workmanship being prepared by the European Committee for Standardisation (CEN). These standards will gradually take over from all countries national standards, including our own British Standards.

Fabrication

Steelwork fabrication is an efficient process enhanced by extensive use of modern technology such as numerically controlled machinery and multiple operation lines. These advanced techniques ensure that the structural steelwork is of required quality and dimensional accuracy, leading to fast, simple and positive assembly on-site. Steel is not delayed by weather conditions and progress is sure and reliable.

Details of the various fabrication processes can be found in the BCSA publication *Structural steelwork - fabrication*.

Erection

A steel frame is quicker to erect than any other form of multi-storey skeletal structure. The reasons lie in the very nature of the modern fabrication and erection processes, the essential feature of which is that fabrication is done entirely in the workshops whilst foundation work is being carried out and that subsequent on-site resource requirements are minimised.

A steel erection team of small gangs of specialist erectors needing far less equipment than that for on-site work and are able to operate effectively in much smaller working areas. This is vitally important with a multi-storey structure in a built-up city centre where site access and working areas are very restricted.

Guidance on the safe erection of structural framework can be found in BS 5531 and the Health & Safety Executive's Guidance Note GS 28 Parts 1 to 4. The BCSA has published *Structural steelwork - erection*, which gives more details how to go about erecting specific types of structures. This will be followed shortly by the *Erectors Manual* which will give advice to personnel engaged on erection tasks.

BENEFITS FROM THE USE OF STEEL

To sum up the current situation with respect to steel buildings, there are four distinct benefits that arise from the use of a steel framework:

Changes in building use and structure

A structural steel frame provides maximum flexibility for changes in use of the whole or part of a building. Where changes of use, finishes, location and partitioning and extensions necessitates structural alterations, these can be accommodated with relative ease. Where additional members are required, connections can be made to the existing frame with minimum disturbance and cost.

It is of tremendous value to both architects, whose imaginative refurbishment programmes are so often limited by the inflexibility of existing office layouts, and to property developers. In negotiating the sale or renting of office space, they are able to offer layouts to suit the needs of their various clients and to command corresponding premiums.

Reduction of foundation costs
The lightness of a structural steel framed building as compared to one with a concrete frame can result in lower foundation costs. Particularly where ground conditions are poor, this weight advantage can mean the difference between using spread foundations and short bored piles at considerably higher cost.

Maximum floor space
The high rentals characteristic of any major city in the world today mean that the maximisation of floor space is essential to profitability. Structural steelwork offers greater value for money on a nett floor area basis with its wide spans and moderate member sizes. The minimal column dimensions and long span beams permit maximum clear floor area and effective floor utilisation and, where long spans are necessary, the cost savings between steel and other forms of construction increase considerably.

Long term strength and value
Structural steel retains its original properties and strength. It has high fatigue strength and its great ductility is a vital attribute to a building's structural integrity under normal working stresses.

Competitive basic cost, swifter construction giving lower total project costs, technical advantages in fire protection all mean that structural steel has a bright new image.

CORROSION PREVENTION
Surface preparation
The performance of all paint systems is improved by good surface preparation. Grey mill scale on the surface will accelerate corrosion by galvanic action and hydrated red rust may contain soluble sulphate salts which will promote premature failure of paint coatings.

For mild conditions, when the steelwork is to be regularly over-painted for decorative reasons, it may be necessary only to remove heavy rust deposits and loose scale by scraping and wire brushing. For external exposure and aggressive conditions all mill scale and rust should be removed by blasting.

Paint coatings
Paint coatings normally have three major components. A pigment, the fine solid particles which give the paint film colour and durability; a binder, which fixes the pigment particles together and gives adhesion to the steel surface; and a solvent or thinner which enables the paint film to be applied evenly, then evaporate.

It is the binder that gives a paint film its predominant charactersistics, and it is for this reason that paints are normally classified by the binder type, eg, alkyd, chlorinated rubber, epoxy resin etc.

Metal coatings
Steel can be protected from water by coating it with another metal. Galvanising - dipping steelwork into a bath of molten zinc - is the cheapest and most common method. A zinc/iron

alloy is formed at the steel surface which gives perfect adhesion and total impermeability to moisture. Galvanising is particularly cost effective for light steelwork, eg, lattice girders.

Metal spraying, using a specially designed spray gun to throw molten metal particles on to a clean and roughened steel surface, is used for both zinc and aluminium coatings. The metal droplets freeze into the surface and form a lightly porous layer of overlapping plates which should be impregnated with a sealing treatment to give maximum corrosion resistance.

Building interiors

The easiest way of keeping water from steel and thus preventing corrosion is to put it inside a dry building. Even if the building is unheated the period of wetness from condensation is likely to be so short that uncoated steel will remain structurally sound for the lifetime of the structure.

In dry heated buildings such as offices, hospitals and schools it is unnecessary to coat the structural steelwork at all, unless for decorative reasons. Where the steel is hidden inside a dry building by fire protection materials or above suspended ceilings, it can be left bare.

If there is a danger of leakage through the external skin of hidden perimeter steelwork, a black tar based coating can be used. This is very effective and cheap, and where the steelwork is hidden, the aesthetics are unimportant.

There are, of course, cases where steelwork will become wet in service, in outside structures exposed to the weather or buildings that contain water or where wet processes are carried out - like swimming pools or breweries. In these cases some form of coating is necessary to keep water away from the steel. Guidance on the protection of steelwork from corrosion is given in BS 5493 and there are Corrosion Protection Guides for interior environments, perimeter walls as well as for refurbishment and exterior environments each of which was prepared by BCSA in association with British Steel - General Steels, Paintmakers Association/Paint Research Association, and the Zinc Development Association.

Fiat HQ, Slough   British Steel General Steels

FIRE PROTECTION
Fire Protection methods and materials
In the United Kingdom a wide variety of products are available to protect structural steel-work. They are divided into five main categories:
— sprays;
— boards;
— pre-formed casings;
— intumescent coatings;
— concrete and brick or block protection. (These methods of protection are well established and will not be discussed further.)

In total there are over 40 proprietary products available in the UK along with the generic products. Their performance and cost vary significantly, but many are expensive and have proved reliable in service. The book *Fire and steel construction - protection of structural steelwork*, available from the Steel Construction Institute, lists the products available and gives general guidance on their suitability for use in various situations.

Some of the main features of these materials are summarised below.

- Spray systems
    — range of costs from low to medium;
    — applied wet;
    — control required on thickness;
    — susceptible to over-spray thus needs masking;
    — often used for beams in conjunction with suspended ceilings.

- Boarded systems - initial costs medium to high;
    — form box encasements;
    — clean fixing;
    — often provide a finish;
    — well tested and proven;
    — factory manufactured.

- Preformed systems
    — initial cost high;
    — fast fixing;
    — resistant to damage;
    — dry and clean;
    — provides a finish;
    — choice of finishes.

- Intumescent coatings
    — range of costs from low to high;
    — applied by spray, brush or trowel when the building is weatherproof;
    — very fast application;
    — control required on thickness;
    — decorative finish.

Steelwork today

FIRE ENGINEERING

Fire engineering to BS 5950 Part 8 permits the determination of what, if any, fire protection is required by considering the significance and severity of the real fire in the building and, therefore, offers a cost effective alternative to Building Regulations. Economies can be realised by specifying fire resistance requirements more precisely and in many structures the quantity of materials available to burn) (ie, the fire load) is low and therefore the potential fire severity will not be high. Consequently, fire protection may be avoided when there is sufficient inherent fire resistance in the unprotected steel members.

The following highlights the main features of the subject:

How hot does the compartment become?
fire load, ventilation, building materials

How hot does the steel get?
heat transfer - radiation, convection, section size and shape, position, insulation

How stable is the structure?
Rate of heating, expansion, design load, elevated temperature, strength, temperature, strength, restraint, continuity and other engineering factors.

Trinity Square, Hounslow   British Steels General Steels

Fire load

The fire load of a building is a measure of the amount of material available to burn (whether as contents or as part of the structure) and is calculated by summing individual calorific values.

The fire load in a compartment is established by listing the weight of contents and the materials used in their construction. The equivalent fire loads of the various materials have been established and conversion factors are used to relate their calorific value to wood. The floor area is measured and a fire load in terms of kg of wood/m established.

In the office shown in Figure 2 the fire load would be the sum of the calorific values of the furniture, fittings, floor coverings and contents, such as paper distributed in cupboards, drawers, cabinets and on the desk.

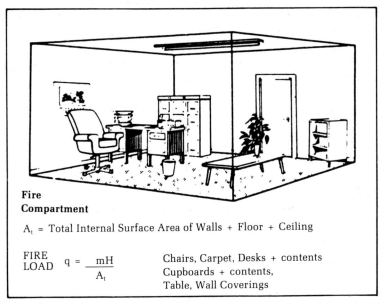

Fire
Compartment

A_t = Total Internal Surface Area of Walls + Floor + Ceiling

FIRE
LOAD $\quad q = \dfrac{mH}{A_t}$ $\qquad$ Chairs, Carpet, Desks + contents
Cupboards + contents,
Table, Wall Coverings

Figure 2 Fire compartment

For further advice on the other aspects of fire engineering reference can be made to the BSC publication *Fire protection*.

MODERN DEVELOPMENTS

Three recent developments have enhanced the inherent qualities and advantages of structural steel. Firstly, there has been a very clear shift in relative cost levels to the advantage of steel. Detailed cost summaries and construction programmes for varying types of construction for multi-storey buildings can be found in the publication *The economics of construction in the UK*, available from the Steel Construction Institute (SCI).

Secondly, developments in fire protection systems have reduced costs significantly.

The third development is the increasing use of a composite steel floor deck. This form of construction is widely used in North America and the UK and permits substantial savings in erection times.

The profiled sheets can be lifted and stacked in multiples with significant savings in crane time compared with pre-cast floors. The decking can be laid up to the highest levels to act as a safety screen and secured against the wind. Below, welding of studs, installation of reinforcement, and pouring of concrete can proceed unhindered.

Fire protection is achieved by placing normal extra reinforcement in the floor slab. This approach, which is accepted by all authorities, shows substantial savings compared with spraying the soffit.

In the provision of an economic floor design for composite slabs the following points should be considered:

Steelwork today

— the use of lightweight concrete topping;
— either power floating the floor to a level or the use of raised (computer) floor on a tamped slab finish;
— unpropped construction throughout;
— multi-span lengths of decking.

Further details on this approach are available in the SCI publication *Design recommendations for composite floors and beams using steel decks.*

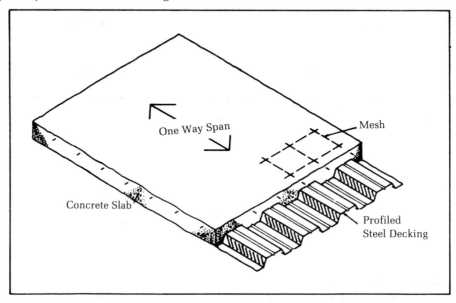

Figure 3 Typical composite floor slab considered by BS 5950 : Part 4 : 1982

THE FUTURE AND OBTAINING FURTHER ADVICE
Further research

BCSA, British Steel General Steels and SCI are all continuing to strive to make steelwork even more economic. Currently, work is going on to develop the preparation of design rules for the use of backing plates rather than stiffeners in bolted connections and fire tests on steel beams in their protected and partially protected state to further reduce construction costs.

The use of computers in design, drawing, estimating, quality control, fabrication etc of steelwork is already within the industry and their use is steadily growing as new systems are researched and developed. Advances in the application of robots and lasers to welding techniques is beginning to lead to their installation in the larger fabrication workshops.

Quality assurance schemes have been devised for both BCSA and British Steel in alignment with the central government and European emphasis on companies being able to guarantee the quality of the processes and products.

Advisory services

The BCSA and SCI both provide the design professions and the construction industry generally with information and advice on the effective use of steel in construction. In addition both the BCSA and SCI have a wide range of publications and lists are available.

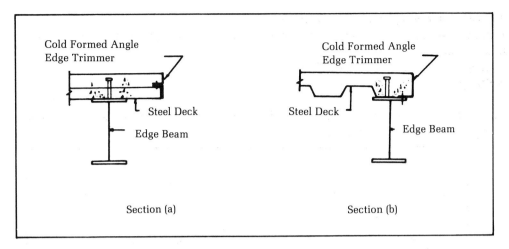

Figure 4:

The British Steel General Steels Sections at Steel House, Redcar (0642 474242), Tubes at Corby (0536 404120) and Plates at Motherwell (0698 66233) will give further advice on the effective use of their products and, if required, arrange a visit by one of their regional advisory engineers.

BCSA members and the BSC

BCSA is the officially recognised trade association of firms engaged in the design, fabrication and erection of constructional steelwork. It has the support of the companies producing 80% of the fabricated steelwork in the UK and thus many clients approach BCSA for names for their tender list knowing that it is a mark of a company's standing and as an assurance that it is up to date on all the latest developments.

INFORMATION ON SITE MANAGEMENT

CONTENTS

Site Management (1-43)   273
Training (44-48)   276
Site Organisation (49-112)   277
Communication (113-119)   281
Site Surveying (120-124)   282
Foundations and Ground Conditions   282
  (a) site investigation (125-134)   282
  (b) contaminated sites (135-140)   283
  (c) ground engineering (141-147)   283
  (d) foundations (148-165)   284
Materials Management (168-183)   285
Materials Handling   287
  (a) concrete (184-186)   287
  (b) concrete pumping (187-191)   287
Quality Control (192-204)   287
Conditions of Working (205-211)   288
Labour Relations (212-227)   289
Productivity (228-247)   291
Environmental Factors Affecting Site
Operations (248-252)   293
Site Accommodation (253-258)   293
Plant   294
  (a) management (259-272)   294
  (b) general (273-284)   295
  (c) hiring and purchasing (285-295)   296
  (d) access plant (296-302)   297
  (e) breakers and cutters (303-305)   297
  (f) compaction plant (306-309)   297
  (g) conveyors (310)   297

  (h) cranes (311-322)   298
  (i) dryers (323-325)   299
  (j) excavators and earth moving
    plant (326-340)   299
  (k) fork lift trucks (341-343)   300
  (l) dumpers (344-346)   300
  (m) hand tools (348-350)   300
  (n) lighting (351)   301
  (o) lorries (352-355)   301
  (p) loaders (356-363)   301
  (q) power supplies (364-370)   301
  (r) pumps (372-374)   302
  (s) sweepers (375)   302
  (t) washers (376)   302
Temporary Works (377-395)   302
Robotics (396-436)   303
Buildability (437-457)   307
Safety   308
  general (458-496)   308
  accident economics (470-471)   309
  accident statistics (472-478)   310
  regulations and legislation (479-490)   310
  building operations (491-524)   311
  plant (525-529)   313
  toxic materials (530-554)   314
Security (555-561)   315
Defects (562-627)   316
Pollution (628-637)   321
Demolition (638-651)   321

INFORMATION ON SITE MANAGEMENT

SITE MANAGEMENT

1
M Coomber
CLEAR SITED SWEDES
Building 1989 254 December 1, pp61-65
The approach of Skanska to site operations is claimed to lead to a cleaner, more project oriented job. As client it is emotionally and physically part of the construction management team. Its management contract with Taylor Woodrow Construction is evaluated to establish where the benefits lie.

2
V Sutherland and M Davidson
MANAGING STRESS
New Civil Engineer 1989 October 5, p40
Report of a survey into stress among site managers.

3
G Forster
CONSTRUCTION SITE STUDIES : PRODUCTION, ADMINISTRATION AND PERSONNEL
1989. 2nd edition, Longman. pp264
Covers the site supervisor's role and responsibilities in regard to communication, personnel/industrial relations, site organisation, health and safety, setting out, programme of the works, documentation and cost control.

4
SITE MANAGERS WIN, ESTIMATORS LOSE IN SALARY SWEEPSTAKES
Building 1988 253 July 29, p13
The findings of a recent survey show that, in the 12 months to the end of June, salaries for site managers increased by 17.5%, contractors surveyors by 10%, and estimators and buyers by 3%.

5
E R Fisk
CONSTRUCTION PROJECT ADMINISTRATION
1988. John Wiley. 3rd edition. pp562
Essential information is provided on how to conduct effective on-site project administration. Related to US practice.

6
D W Halligan et al
MANAGING UNFORESEEN SITE CONDITIONS
ASCE Journal of Construction Engineering & Management 1987 113 (2) June, pp273-287

The legal merits and limitations of alternate contract clauses and contract forms have received considerable attention in the literature. Here, using the data from several recent studies, it is shown that the actual contract language used is largely irrelevant to the actual costs borne by owners and contractors. More importantly, the management techniques used determine the ultimate cost to all participants. Two management techniques particularly relevant to unforeseen site conditions are the use of interpretive reports and the early resolution of claims. These techniques, their effectiveness and the overall limitations of legalistic approaches to the management of unforeseen site conditions are discussed in detail.

7
M Brensnen et al
EFFECTIVENESS OF SITE MANAGEMENT
CIOB Technical Information Service Paper No.85. 1987. pp6
The question is investigated of effective construction site management, with particular attention to the impact of variation in the leadership exercised by site managers in the performance of their role. Specific emphasis is placed upon identifying both the range of experience, background and training of site managers, as well as variation in the conditions under which site managers operate. The inter-relationship between these factors and their impact upon project performance are analysed.

8
V Powell-Smith
KEEPING COSTS DOWN BY CLEVER USE OF A KEY MAN
Contract Journal 1987 February 19, p7
The role of the site manager in keeping records to substantiate claims is discussed.

9
G Forster
BUILDING ORGANISATION AND PROCEDURES
1986. 2nd edition. Longman. pp294
Directed to BTEC level 2 syllabus in Organisation and Procedures. Covers structure of the industry: organisations; the building team; design; Regulations; safety and welfare; measurement; contractor's pre-tender operations; contractor's pre-contract work; and contract opererations.

10

V B Torrance and J P Cathcart
MOTIVATION OF CONSTRUCTION MANAGERS
Building Technology & Management 1986 24 December,
pp38-41

11

M J Bresnen et al
**LEADER ORIENTATION OF CONSTRUCTION
SITE MANAGERS**
ASCE Journal of Construction Engineering & Management 1986 112 (3) September, pp370-386
The ways are reviewed in which leadership is conceptualised and taken into consideration in the context of
construction projects. Data are presented that derive
from a study of 43 site managers in England and Wales
and that take Fiedler's Contingency Model as the point
of departure. The results indicate that variations in
leader orientation have an impact on project effectiveness and that this relationship is contingent upon a
number of factors including project length, value, and
the extent of reliance on sub-contractor labour. In particular, the results indicate that an emphasis on
relationships in site managers' leader orientations is
more likely to enhance project performance than an
emphasis on tasks. The implications of these findings
for the study of construction management are examined.

12

J Johnston
GOOD SITE PRACTICE AND CONTROL
Building Technology & Management 1986 24 June,
pp26-29
Elements of good practice on site are discussed, reference being made to supervision, checking and testing,
storage, communication and feedback.

13

R Taylor
FREEING MANAGERS TO MANAGE
Construction Computing 1986 (12) April, pp16-18
Application of on site computers to the Lloyds building
is outlined, with particular reference to planning and to
drawing control.

14

D Judge
CANADIAN APPROACH TO EFFECTIVE MANAGEMENT
Building Technology & Management 1986 24 January,
p25
The role of the Construction Management Institute,
particularly in regard to site management, is outlined.

15

P Bill
CUT ABOVE THE REST
Building 1985 269 December 6, p24
The importance of site management as seen by Higgs &
Hill and Lovell is discussed.

16

D D Turner
**SUPERVISION OF CIVIL ENGINEERING
CONTRACTS**
Civil Engineering Technology 1985 9 December, pp2-6
Aspects discussed include the role of the engineer's
representative, pre-contract meeting procedures, setting out procedures and bench marks, records and
safety.

17

Chartered Institute of Building
**PRACTICE OF SITE MANAGEMEMENT, VOLUME
3**
1985. CIOB. pp216
A collection of 18 individual contributions covering the
broad spectrum of site management from archeological
finds to performance reviews. Includes a classified
information section of over 400 items.

18

D Howard
MORE COST-EFFECTIVE BUILDING
Management Accounting 1985 64 February, pp36-38
Details are given of the construction of a computer
training centre using an architectural practice as the
design and project management consultant. The contract used was basically the AIA form and the tender list
of six was based on interviews. Particular attention is
paid to site supervision and contract administration.

19

R Holmes
**INTRODUCTION TO CIVIL ENGINEERING
CONSTRUCTION**
1983. 2nd edition. College of Estate Management. pp374
Covers site investigation; site organisation and temporary services; temporary works; plant; earthworks; piling, diaphragm and retaining wall systems; tunnelling
and underpinning; and civil engineering works.

20

P A Rougier
SITE ENGINEERING PRACTICE
1984. Longman. pp149
Covers equipment; setting up and preparing the site;
structural calculations; setting out; mass excavation;
tests; excavation plant and labour.

21

G Hedley and C Garrett
PRACTICAL SITE MANAGEMENT - ILLUSTRATED GUIDE
1983. 2nd edition. Longman. pp168

22

W A Ridgwell
**WHY EMPLOY A CLERK OF WORKS? 1.
JUSTIFYING THE COST**
Clerk of Works 1988 106 July, pp10-12

23
D Chappell
THERE MUST BE A BETTER ROLE FOR THE CLERK OF WORKS
Building Today 1988 *195* March 24, pp20-21
Consideration is given to how the best use be made of the clerk of works.

24
Institute of Clerks of Works
BUILDING ON QUALITY : A REVIEW OF THE ROLE OF CLERK OF WORKS
1987. pp43
The main conclusions resulting from the study are that the industry is not satisfied with the present level of qualification of clerks of works and that there are varying views on the responsibility that should be given under the contract. A number of recommendations are made in relation to the clerk of works as a profession, and the role of the clerk of works and that of the Institute.

25
Greater London Council
HANDBOOK FOR CLERKS OF WORKS
1983. 3rd edition. Architectural Press. pp130

26
Institute of Clerks of Works
CLERKS OF WORKS APPOINTMENT (INDEPENDENT PRACTICES)
1986. pp21

27
H N E Ward
BUILDING TECHNICIANS AND CLERKS OF WORKS HANDBOOK
1986. Merlin Books. pp153

28
H Tyrvainen et al
INFORMATION MANAGEMENT ON BUILDING SITES
Managing Construction Worldwide. Volume 1 Systems for managing construction.
1987. Spon/CIOB/CIB. pp470-476
Guidelines are presented for the development of computerised information systems for building sites.

29
M P Kim and M J O'Connor
EXPERT SYSTEM FOR DIFFERING SITE CONDITION CLAIMS OF CONSTRUCTION CONTRACTS
Proceedings 4th International Symposium on Robotics & Artificial Intelligence in Building Construction, Haifa, Israel. 22-25 June 1987. Volume 2. pp826-848

30
H R Thomas and G R Smith
CONCEPTUAL DESIGN OF AN AUTOMATED REAL-LIVE DATA COLLECTION SYSTEM FOR LABOUR INTENSIVE CONSTRUCTION ACTIVITIES
Proceedings 4th International Symposium on Robotics & Artificial Intelligence in Building Construction, Haifa, Israel. 22-25 June 1987. Volume 2. pp849-862

31
A Crane
NEXT PHASE IN SITE COMPUTERISATION
Building Technology & Management 1986 *24* October/ November, pp30-33
The application is discussed of computers to planning, information control, materials control, sub-contracting, cost control and valuations.

32
F Anderton
CONFESSIONS FROM A BUILDING SITE
Architects Journal 1986 *184* September 10, pp32-33
Experience of an architectural student on site, illustrate the divide between the sexes - and between design and construction.

33
N Fisher and B L Atkin
COMPUTER WORK STATIONS FOR CONSTRUCTION SITES
International Journal of Construction Management & Technology 1986 *1* (2), pp17-29
Two detailed case studies are presented which highlight the practical problems facing the development of computer work stations on site.

34
A Touran
COMPUTER PACKAGES FOR CONSTRUCTION DATA ACQUISITION AND PROCESSING
Construction Management and Economics 1986 *4* (3), Winter, pp233-243
Problems of construction site data collection are reviewed and their impact on the planning of construction projects and job productivity discussed.

35
W Laing
ENSURING TEAMWORK ON SITE
Building Technology & Management 1985 *23* May, pp3-6
The management aspects of a project comprising an office block, hostel and sports complex are outlined.

36
R Cecil
SITE INSPECTION
Architects Journal 1985 *181* April 3, pp75-76
Guidelines are given for effective site inspection by the architect.

37
L F Heineck and J Rawcliffe
GRAPHICAL PICTURE OF WHAT REALLY HAPPENS ON SITE
Paper to CIB W-65 4th International Symposium on Organisation and Management of Construction, Waterloo, Canada. July 1984. Vol.4, pp1389-1398
A number of house building sites in the United Kingdom were observed through activity sampling; the data bank amassed was explored through a suite of specially developed graphical programs, using either hard copy plotters or colour video monitors. The graphical software is described. The field observations went against the classical assumptions on site programming and control; durations are much greater than expected, the precedence relationship was mainly of an overlapping type, the allocation of resources was erratic and the sequence of work from building unit to building unit was not constant.

38
J G Gunning
SITE MANAGEMENT INFORMATION AND CONTROL SYSTEM
Proc. 4th International Symposium on Organisation and Management of Construction, Waterloo, Canada. July 1984. pp169-178
Those areas are highlighted where site based management can make significant improvements without involving a mass of theoretical managerial concepts or using sophisticated programming or other techniques.

39
ROLE OF THE SITE AGENT
Proc. Supervision of Construction Symposium, ICE, London, 7-8 June 1984 1985. Telford. pp119-125; Discussion pp127-140

40
R H Clarke
SITE SUPERVISION
1984. Thomas Telford. pp250
Related to civil engineering chapters are included covering the contract, running the job, programming, safety, communication, variations, measurement, record keeping, and claims and disputes.

41
P Fitzpatrick
ROLE OF THE CONTRACTOR
Proc. Supervision of Construction Supervision, ICE, London, 7-8 June 1984
1985. Telford. pp111-117; Discussion pp127-140

42
E M O'Leary
SUPERVISION BRIEF
Proc. Supervision of Construction Supervision, ICE, London, 7-8 June 1984
1985. Telford. pp169-175; Discussion pp177-184

43
M J Bresnen et al
EFFECTIVE CONSTRUCTION SITE MANAGEMENT : A REVIEW
ASCE Journal of Construction Engineering and Management 1984 190 (4), December, pp420-436
The contribution of site management to the achievement of construction performance objectives is examined. Drawing upon existing empirical data and linking them with research findings in the social sciences, it is argued that there is often insufficient attention paid towards the range of complex and interdependent variables that can influence construction site activity. As a consequence, the predictive capacity of such research is limited, and the recommendations often derived, particularly concerning appropriate managerial actions, often fail to account for significant variability in circumstances. The applicability of the concept of 'leadership' to the construction site situation, and its relationship with key component variables are singled out for attention.

TRAINING

44
H Try
MANAGEMENT TRAINING. AN AGENDA FOR ACTION
National Builder 1986 67 November, pp324-325
The proposals contained in the Lighthill Report on graduate supply and in the Morley Report on management and supervisory training are reviewed.

45
Joint Committee on Management Development
CONSTRUCTION SUPERVISION AND MANAGEMENT : A NEW TRAINING INITIATIVE (MORLEY REPORT)
1986. pp13
It is concluded that the industry's utilisation of supervisory training courses is inadequate and that a new supervisory framework is needed. The report also recommends that CITB should play a major role in assisting employer organisations to promote supervisory and site management training.

46
G Elliott
SUPERVISORY TRAINING IN BUILDING
National Builder 1985 66 May, pp108-110
A series of short articles reviews the site management training scene centrally and from the viewpoint of Wimpey, Fairclough and Shepherd.

47

A Laufer

ON SITE PERFORMANCE IMPROVEMENT PROGRAMS

ASCE Journal of Construction Engineering and Management 1985 111 (1) March, pp82-97

The various phases are presented of a comprehensive On Site Performance Improvement Program (OSPIP) of a medium size construction project which includes: problem identification; data collection; data analysis and planning the change content; planning the change process; and measuring and evaluating the results. Methods, techniques, and means to put the various program phases into practice are examined, and the problems likely to be encountered are discussed.

48

J Prosper

RIGHT TYPE OF TRAINING

Building Technology & Management 1984 22 May, p24

Criticism is made of training courses which provide idealised approaches to management. The author identifies two key points in managing site effectively; optimisation of working time; and concentration on the most important issues. How these may be achieved is indicated.

SITE ORGANISATION

49

E Ishai

ARCHITECTURAL AND ECONOMIC CONSIDERATIONS IN THE DESIGN OF PREFABRICATED FACADE COMPONENTS (EXTERIOR WALLS)

Construction Management & Economics 1989 7 Autumn, pp189-202

50

K Kahkonen

KNOWLEDGE BASED SYSTEM APPROACH TO DETERMINE DURATIONS OF SITE ACTIVITIES

Proceedings Artificial Intelligence Techniques and Applications for Civil and Construction Engineers Conference, London, September 1989, pp71-76

The system described, Ratu-aj, is currently being used in estimating the durations of large shuttering panel tasks.

51

A Hamiam

KEY APPROACH TO SITE LAYOUT PROBLEMS

Proceedings Artificial Intelligence Techniques and Applications for Civil and Construction Engineers Conference, London, September 1989, pp65-69

The system (CONSITE) described allows the location of temporary support facilities on a site, while satisfying applicable constraints.

52

D T Shaw

KNOWLEDGE-BASED REGULATION PROCESSING FOR SITE DEVELOPMENT

Proceedings Artificial Intelligence Techniques and Applications for Civil and Construction Engineers Conference, London, September 1989, pp57-63

The knowledge-based model (SITE CODE) described was developed to study how regulations can be incorporated into systems for computer-aided site design.

53

W H Arch

STRUCTURAL STEELWORK - ERECTION

1989. BCSA. pp91

Initially attention is given to the roles of the various aims of management, the processes involved in the preparation of a tender, the determination of an erection method and the setting up of a site. Safety is dealt with in some detail and there is a section concerned with plant and equipment and their possible hazards in use.

54

M Evans

HOW TO GET THE MOST OUT OF YOUR PLUMBING SUB-CONTRACTOR

House Builder 1989 48 November, pp107, 109

55

T Whitehead

FRAME DESIGN, CONCRETE v STEEL - THE NEW DEBATE

Contract Journal 1989 November 23, p32

56

D Cole

FLOOR SCREEDS - AN AREA OF CHANGE

National Builder 1989 October, pp37-38

A resume is provided of changes in the construction of floor screeds introduced by BS 8204: Parts 1 & 2 and BS 8203.

57

FAST TRACK CONCRETE CONSTRUCTION

ASI Journal 1989 1 September, pp25-27

58

ARCHEOLOGY AND DEVELOPMENT

Property Journal 1989 14 June, pp16-21

Report of a conference which considers the problem areas, planning and co-operation in the field.

59

J Long

SOLVING MORTAR PROBLEMS

National Builder 1989 September, pp42-43

Some of the problems experienced with mortar on site are examined.

60

R O'Hara

SIMPLE JOBS MAY FOUL THE LAW ON HIGHWAYS

Building Today 1989 197 June 15, pp79, 23-24

Relevant case law is discussed to illustrate where the builder might come into conflict with the law in relation to the site and its access to the highway.

61

S L Goodchild and M P Kaminski

RETENTION OF MAJOR FACADES

Structural Engineer 1989 67 April, pp131-138

The various stages are described of the design development, detailed information being provided on the key factors to be considered, from preliminary investigations to on site installation of temporary works and the construction of the new buiding.

62

S B Tietz

CLADDING: CURRENT PRACTICE IN SPECIFICATIONS, DESIGN AND CONSTRUCTION

Structural Engineer 1989 67 November, p389

Developments in the technology of cladding are summarised. Potential areas of weakness suggest that there is a greater need for analysing structural non-loadbearing elements and regular surveillance.

63

P N Cheremisinoff (Ed)

CIVIL ENGINEERING PRACTICE. VOLUME 4

1988. Technonic. pp685

This volume covers surveying; construction (includes a paper on CPM); transportation; energy; and economics/data acquisition (includes a paper on BASIC programming for civil engineers).

64

R Chudley

BUILDING CONSTRUCTION HANDBOOK

1988. Heinemann. pp539

Provides a summary of basic building processes and techniques, and the regulations which govern them. Chapters are included on site works, plant, sub-structure, superstructure, internal construction and finishes, and domestic services.

65

R K Moore

CHEMICAL ANCHORS IN CONSTRUCTION

CIOB Technical Information Service Paper No.91. 1988. pp4

The development is outlined of materials and application systems, key properties needed in anchor grouts and factors influencing choice of materials.

66

J Burchell and F W Sunter

DESIGN AND BUILD IN TIMBER FRAME

1987. Longman. pp78

A guide for designers, builders and supervisors of timber framed houses and small con mercial buildings. It covers UK practices and reflects the fail safe approach to timber frame construction which has been adopted. Following an introduction there are chapters on elements of construction; installation of services; exterior finishes; special design considerations; obtaining tenders; site erection; and occupation.

67

P Roper

SWEDEN HAS A WAY WITH TIMBER FRAME

Building Today 1987 194 November 12, pp62-63

The approach of Swedish manufacturers to the construction of timber frame is outlined.

68

STANDARD PRECAST CONCRETE BOX CULVERTS - A GUIDE TO SITE PRACTICE

1987. Box Culvert Association. pp3

69

RESTRICTED ACCESS

Contract Journal 1987 September 10, pp16-17,21,22,24,26-35

A series of articles dealing with various aspects of coping with sites of restricted access.

70

A Manesero and K P Chapman

TOWARDS A DECISION SUPPORT SYSTEM FOR CONSTRUCTION METHODS ANALYSIS

Managing Construction Worldwide, Volume 1. Systems for managing construction.
1987. Spon/CIOB/CIB. pp510-516

A computer based decision support system is described for the analysis and selection of construction methods on site.

71

T C Cornick and S P Bull

EXPERT SYSTEMS INTEGRATED WITH CAD FOR COMPONENT ASSEMBLY

Proceedings 4th International Symposium on Robotics and Artificial Intelligence in Building Construction, Haifa, Israel. 22-25 June 1987. Volume 2, pp657-671

72

A Manesero et al

DEVELOPMENT OF A KNOWLEDGE BASED COMPUTER SYSTEM FOR DECISION MAKING ON CONSTRUCTION METHODS

Proc. Application of Artificial Intelligence Techniques to Civil and Structural Engineering Conference, 1987, pp61-65

A system is described which is directed at aiding construction planners with the evaluation of potentially suitable methods of construction and the eventual selection of the optimum method for the current project.

73
M J Mauderley et al
APPROACH TO AUTOMATIC PROJECT PLANNING
Proceedings 4th International Symposium on Robotics and Artificial Intelligence in Building Construction, Haifa, Israel. 22-25 June 1987. Volume 2, pp897-918
A method is described of producing a realistic plan of work for a project based on the project drawings.

74
J E James (Ed)
NO - DIG 87. TRENCHLESS CONSTRUCTION FOR UTILITIES
Proceedings of 2nd International Conference 1987. International Society for Trenchless Technology. pp82

75
Pipe Jacking Association & Concrete Pipe Association
JACKING CONCRETE PIPES
1986. 2nd edition. 11pp
Includes a recommended method of measurement for use with CESMM.

76
CITY SITES : READY, STEADY, SUPPLY
Contract Journal 1988 January 28, pp22-24
The value of using ready mixed concrete on urban sites is outlined.

77
B A Hiscox
CONTRACTOR'S RISK - THE COST OF GETTING IT WRONG
Civil Engineering Surveyor 1987 12 March, pp16,18-20
The danger of damaging underground services during construction works is discussed. The limited options available to the contractor under the ICE Conditions to obtain redress for finance and delay are considered.

78
A Fricker
HOW SITE SUPERVISORS CAN HELP ACHIEVE GOOD TEXTURED FINISHES
Clerk of Works 1987 105 February, pp6-8

79
J G Richardson
SUPERVISION OF CONCRETE CONSTRUCTION
1986. Viewpoint Publications. pp211
An introduction to a number of aspects of concrete construction likely to be experienced on site, setting down general principles of construction methods.
This volume covers planning; health & safety; supervisory skills; mix design; accuracy; joints; falsework & temporary works; formwork; and surface finish.

80
I H Seeley
BUILDING TECHNOLOGY
1986. 3rd edition.
Macmillan. pp250
Covers site work, foundations, various structural elements and building in warm climates.

81
J T Moon
FIXING AND FASTENING METHODS IN BUILDING CONSTRUCTION
CIOB Technical Information Service Paper No.70. 1986. pp6

82
British Archaeologists and Developers Liaison Group
CODE OF PRACTICE
1986. pp4
A means of reconciling in practical terms the needs of the archaeologists and developers.

83
E van der Straeten
WHAT'S NEW IN WOOD-GLUE TECHNOLOGY
CIOB Technical Information Service Paper No.63. 1986. pp7
Recent developments in wood adhesives are highlighted and explained. They are approached from a general background of adhesive types and their selection and specification. Details of specific applications are included.

84
J R Coad and D Rosaman
SITE APPLIED ADHESIVES - FAILURES AND HOW TO AVOID THEM
BRE Information Paper 12/86. 1986. pp4

85
D Clow and R Ramsey
NEW PERSPECTIVE ON OPTICAL FIBRE CABLES
National Builder 1986 67 June, pp168-169
Good practice by contractors in relation to cable damage and the legal consequences of such damage are discussed.

86
G H Curtis
USE OF PERSONAL COMPUTERS TO SCHEDULE REMOTE SITE CONSTRUCTION
Proc. CIB 10th Triennial Congress, Washington, September 1986. Volume 8, pp3471-34 78

87
STEELWORK OF CONSTRUCTION
Construction News Supplement 1986 March 6, pp25-40

88
P Roper
PRACTICAL GUIDE TO BUILDERS' QUESTIONS AND ANSWERS
1985. International Thomson Publishing. pp341
A series of practical questions relating mainly to technical matters are answered.

89
D R Helliwell
TREES ON DEVELOPMENT SITES
1985. Arboriculture Association. pp20
Guidance is given on the retention and planting of trees on development sites.

90
P H Allen
STEELWORK TODAY
CIOB Technical Information Service Paper No.51. 1985. pp7
The development is described of steel and steel products in construction. Current qualities, products and applications are set out and the advantages of its use are identified. Modern developments in corrosion prevention, fire protection, fire engineering and construction method are reviewed.

91
T Mynott
MODERN TIMBER ADHESIVES
Paper to Symposium on Building appraisal, maintenance and preservation.
Bath University, 10-12 July 1985, pp13
Principal properties of the various types of adhesives are summarised and a scheme of adhesive selection is proposed based on a stopwise approach allowing for appropriate interaction at the relevant stages.

92
CONCRETE TECHNOLOGY
Construction News Supplement 1985 March 7, pp21-36
A series of articles is presented covering durability, ready mix, use of pfa, specification, alkali aggregate reaction, hac, formwork and cement compositions.

93
A Warsazawski, M Avraham and D Carmel
UTILISATION OF PRECAST CONCRETE ELEMENTS IN BUILDING
ASCE Journal of Construction Engineering and Management 1984 110 (4), December, pp476-485
Precast concrete components can be used in building construction within a comprehensive 'closed' system, or as separate elements in conjunction with any building method. The feasibility of this second possibility is examined within the framework of a conventional building system and the following alternatives of elements ultilisation: prestressed modular floor slabs, exterior walls, and a combination of slabs and exterior walls. Each of these alternatives is compared to the conventional system without precast

94
Timber Research and Development Association
WOOD PRESERVATION - PROCESSING AND SITE CONTROL
1984. pp3

95
A T Baxendale
USE OF OPERATIONS RESEARCH ON SITE
CIOB Technical Information Service Paper No.36. 1984. pp7
The possibility of applying OR methods on site is investigated and the need discussed for supporting data on construction operation times. Activity sampling is examined as a means of collecting distribution time data and an example given of its use. The use of Monte Carlo simulations in the planning of site operations is also demonstrated.

96
Aggregate Concrete Block Association
CAVITY INSULATED WALLS. PRACTICE NOTE 1. LAYING BRICKS AND BLOCKS
1984. pp4

97
Electrolocation Ltd
PIPE AND CABLE LOCATION THEORY
1984. pp15
The basic theory is explained of the electromagnetic detection of buried pipes and cables and of the twin coil search technique.

98
R D Duffy
ROLE OF PDMS IN IMPROVING CONSTRUCTION EFFICIENCY
Proc. 4th Inst. Symposium on Organisation and Management of Construction, Waterloo, Canada, July 1984. Volume 2. pp379-387
Plant Design Management Systems is designed to minimise construction rework due to interfaces, mismatches, and missing hardware resulting from design errors.

99
A C Sidwell and H Hadhdadi
AN INVESTIGATION OF THE APPLICATION OF COMPUTERS TO CONSTRUCTION SITES
Proc. 4th Inst. Symposium on Organisation and Management of Construction, Waterloo, Canada, July 1984. Volume 2. pp489-497
A flowchart model is presented of a management information system appropriate to construction sites in the UK. The model has been developed following investigation of five case studies of construction firms and the way in which they manage and organise their projects. This work is part of an ongoing research project into the application of computers on construction sites.

100

C P Verschuren

EFFECT OF REPETITION ON THE PROGRAMMING AND DESIGN OF BUILDINGS

Proc. 4th Inst. Symposium on Organisation and Management of Construction, Waterloo, Canada, July 1984. Volume 2. pp651-661

It is shown how a standard method for calculating the 'learning rate' in house construction can be used to influence changes in an urban development plan to give cost savings of 5 to 10%.

101

National Joint Utilities Group

SERVICES ENTRIES FOR NEW DWELLINGS ON RESIDENTIAL ESTATES

1984. pp26

102

G Taylor

CONCRETE SITE WORK. ICE WORKS CONSTRUCTION GUIDE

1984. Thomas Telford. pp59

103

House Builders Federation

CHECK LIST - TIMBER FRAME. SITE 1984. pp1

104

House Builders Federation

CHECK LIST - TIMBER FRAME DWELLINGS. 2. DWELLING

1984. pp2

105

D Blyth and E R Skoyles

CRITIQUE OF RESOURCE CONTROL

Building Technology and Management 1984 22 February, pp29-30

Those factors are identified which militate against good resource control by site management.

106

BUILDING SERVICES INTEGRATION

Building Services & Environmental Engineer 1983 October, pp2

Some of the basic problem areas are identified.

107

Timber Research and Development Association

TIMBER FRAME CONSTRUCTION SITE CHECK LIST

Wood Information Sheet No.8. 1983. pp2

108

Health and Safety Executive

UNDERGROUND CABLE DAMAGE SURVEY. NOVEMBER 12980 - JULY 1981

1983. HMSO. pp33

The survey was carried out in the HSE's London South

Area and indicated that the greatest number of accidents involved pneumatic tools and hand tools; cable locators were not used in 75% of the cases before excavation; LEB plans were unreliable.

109

J W Watts

SUPERVISION OF INSTALLATION: A GUIDE TO THE INSTALLATION OF MECHANICAL AND ELECTRICAL PLANT AND SERVICES

1982. Batsford. pp162

110

National Joint Utilities Group

IDENTIFICATION OF SMALL BURIED MAINS AND SERVICES

1981. pp24

111

National Joint Utilities Group

CABLE LOCATING DEVICES

1980. pp8

112

National Joint Utilities Group

PROVISION OF MAINS AND SERVICES BY PUBLIC UTILITIES ON RESIDENTIAL ESTATES

1979. pp11

COMMUNICATION

113

D J Carter

USE OF STRUCTURED INFORMATION SYSTEMS IN BUILDING CONTRACT ADMINISTRATION

Managing Construction Worldwide. Volume 1. Systems for managing construction.

1987. Spon/CIOB CIB. pp437-447

A study is reported of the information flow system from site to Supervising Officer on a number of hospital building contracts.

114

D J Carter el al

INFORMATION NEEDS OF BUILDING CONTRACT ADMINISTRATION

Proc. CIB 10th Triennial Congress, Washington, September 1986. Volume 8.

pp3452-3457

A study is made of the information flow systems from site to architect/SO on a number of sites. Measures to overcome existing limitations are identified.

115

S Bone

COMMUNICATIONS ON BUILDING SITES

Building Technology & Management 1986 24 April, pp20-22

Existing means of communication both on and to and

from the site are outlined. Future trends considered include the widespread adoption of cordless telephones, FAX machines and mobile offices.

116
R J Biggs
COMMUNICATION
CIOB Technical Information Service Paper No.58. 1985. pp4
The essential elements and practice of communication as it relates to building sites are covered.

117
Co-ordinating Committee for Project Information
PRODUCTION DRAWINGS - DRAFT CODE OF PROCEDURE FOR THE BUILDING INDUSTRY
1984. pp76

118
W A Porteous
BUILDING FAILURE OR COMMUNICATION FAILURE
Paper to AIB/NZIOB Conference 'Communication in our industry', Christchurch, New Zealand, March 1984. pp215-224
Based on a literature search the conclusion is drawn that while there is an established link between building failure and communication failure and much improvement could be made, there is evidence to show that improvement in communication will produce tangible results.

119
D Woolven
EFFECT OF COMMUNICATIONS AMONG MEMBERS OF THE BUILDING TEAM UPON QUALITY IN BUILDING
Paper to AIB/NZIOB Conference 'Communication in our industry', Christchurch, New Zealand, March 1984. pp95-110
The results of a survey indicated that quality standards are set by site management without reference to standards given in the specification. It is suggested that there is a need for contract managers to give specific instructions on the standards of workmanship required.

SITE SURVEYING

120
S G Brighty
SETTING OUT: A GUIDE FOR SITE ENGINEER
1989. 2nd edition. BSP Books. pp206

121
B M Sadgrove
SELLING OUT PROCEDURES
1988. CIRIA/Butterworth. pp123

122
BSI GLOSSARY OF TERMS FOR PROCEDURES FOR SETTING OUT, MEASUREMENT AND SURVEYING IN BUILDING CONSTRUCTION
1988. pp24

123
BSI TERMS FOR PROCEDURES FOR SETTING OUT, MEASUREMENT AND SURVEYING IN BUILDING CONSTRUCTION (INCLUDING GUIDANCE NOTES) BS 6953:
1988. pp24

124
P H Milne
SITE CONTROL AND SETTING OUT BY MICROCOMPUTER
Proc. 3rd International Conference on Civil and Structural Engineering Computing. 1987. Volume 2. pp225-227

FOUNDATIONS AND GROUND CONDITIONS

(a) Site investigation

125
J Bickerdike
SITE INVESTIGATION
ASI Journal 1989 1 September, pp23-24
The main elements of a site investigation are described.

126
M Carter and M V Symons
SITE INVESTIGATIONS AND FOUNDATIONS EXPLAINED
1989. Pentech Press. pp332

127
N F Pidgeon et al
SITE INVESTIGATIONS: LESSONS FROM A LATE DISCOVERY OF HAZARDOUS WASTE
Structural Engineer 1988 66 October, pp311-315
Lessons are drawn regarding the need to consider explicitly contamination, the open ended nature of site investigation, the trade-off between the cost of an investigation and reduction in risk and the need for engineers to be aware of the importance of efficient exchange of information.

128
Building Research Establishment
SITE INVESTIGATIONS FOR LOW RISE BUILDING: DESK STUDIES
Digest No.318. 1987. pp12
A desk study is an important part of the site investigation process to assess the suitability of a site for development.

129
Building Research Establishment
SITE INVESTIGATION FOR LOW RISE BUILDING: PROCUREMENT
Digest No.322. 1987. pp8
The value of site investigation and the steps that should be involved are discussed. Guidance on contractual methods is given.

130
J F Uff and C R I Clayton
RECOMMENDATIONS FOR THE PROCUREMENT OF GROUND INVESTIGATION
CIRIA Special Publication 45. 1986. pp43
The underlying reasons are identified for the shortcomings which exist in the ground investigation industry in the UK, and means of improvement are proposed. The condition of the industry is examined in relation to the need for better quality investigation. The main conclusion is that the fundamental cause of these shortcomings lies in the method of procurement used, because they inhibit the proper use of expertise and allow those involved to take on duties which they are unable to perform. From an analysis of the work required during ground investigation, two suitable systems of procurement are identified. The recommendation that one of these systems should be used to procure ground investigations is coupled with guidance on the choice of system, on the means to achieve good results from both, and on the responsibilities of those involved.

131
R F Fellows and L J Rys
MORE EFFICIENT PROCUREMENT OF SITE INVESTIGATIONS
Building Technology & Management 1986 24 March, pp17-18

132
J Cottington and R Akenhead
SITE INVESTIGATION AND THE LAW
1984. Thomas Telford. pp184
A guide to the ICE Conditions of Contract for Ground Investigations.

133
J B Green
SITE INVESTIGATION FOR STRUCTURAL FOUNDATIONS
Civil Engineering Technology. 1984 8 December, pp2-5

134
R Afia
ELEMENT DESIGN GUIDE. SUBSTRUCTURE 1. DOMESTIC: ENGINEERING
Architects Journal 1986 184 November 26, pp57-62, 65-68, 71
The principles of site investigations and simple foundations for domestic structures are described.

(b) Contaminated sites

135
BSI CODE OF PRACTICE FOR THE IDENTIFICATION OF POTENTIALLY CONTAMINATED LAND AND ITS INVESTIGATION
DD:175. 1988. pp28

136
REDEVELOPMENT OF CONTAMINATED SITES
Property Journal 1988 13 June, pp26-27
Some of the problems are identified which might be experienced with sites used for industrial purposes.

137
D W Lord et al
RECLAMATION OF CONTAMINATED LAND FOR BUILDING DEVELOPMENT
Paper to Institute of Building Control 22nd Weekend School, UMIST. 1988. pp4-8 Attention is given to potential hazards, site investigation, materials analysis, assessment of contamination, methods of redevelopment and legislation.

138
T Nash
CONTAMINATED SITES. THE PROBLEM OF NEGLECT
Building Technology & Management 1986 24 December, pp14-15, 19
The potential risks to personnel from contaminated sites are indicated.

139
T Cairney
DEALING WITH CONTAMINATED LAND
Civil Engineering Technology 1985 9 October, pp2-6
Soil cover reclamation of contaminated land is discussed.

140
P Overill
CONTAMINATION OF DEVELOPMENT SITES: A GUIDE TO POTENTIAL PROBLEMS
1985. Surveyors Publications. pp8

(c) Ground engineering

141
GROUNDWORKS AND DRAINAGE. PART 2
House Builder 1989 48 June, pp43,45-46,50-51,54-55,58,61-62,65-66,69-70, 72,74,96
A special report covering ground preparation of unsuitable sites, vibratory ground compaction, pipes and drains, and hard paving.

142
T Whitley
STAYING AFLOAT ON SITE
Contract Journal 1989 347 January 28, pp30-33
The significance is discussed of ground water control on site.

143
SITE WORK
Building 1986 261 December 19/26, pp49-55
A series of articles is presented on the reclamation of contaminated land, the vibro replacement system of ground consolidation, foundation failures and soil surveys.

144
S H Somerville
CONTROL OF GROUNDWATER FOR TEMPORARY WORKS
CIRIA Report 113. 1986. pp87
Covers the various methods of dewatering control, identification of ground, methods of assessing permeability, design of well systems and approximate costs.

145
J A Charles et al
PRELOADING UNCOMPACTED FILLS
BRE Information Paper 16/86. 1986. pp4
It is shown how the load carrying properties of uncompacted fills can be improved by a temporary preloading with a surcharge of fill.

146
COST EFFECTIVE GROUNDWORK - A SPECIAL REPORT
House Builder 1986 45 March, pp25-64
A series of articles is presented dealing with geotextiles in earthworks; the NWC guide on sewers for adoption; soft landscaping; service supplies to site; hard paving; manholes; plastic pipes in economic drainage and sewerage; hard landscaping; micro diggers and other plant; clay pipes for drainage and sewerage.

147
P C Horner
EARTHWORKS
1988. 2nd edition. Thomas Telford. pp62

(d) Foundations

148
FOUNDATIONS
New Civil Engineer 1989 November 30, pp23-24, 26-27, 29, 31-32, 34-35, 37-38, 40, 42, 45, 47-48, 50-51, 53-55, 57-58, 61, 63, 65-66, 68, 70, 73-74, 77, 79-80
A series of articles is presented covering techniques and plant. Those firms offering foundation services are tabulated.

149
P Gray
DEEP BASEMENT CONSTRUCTION AND FACADE RETENTION
Architect Surveyor 1989 64 November, pp22-25
Some of the practical problems associated with deep basement construction are identified.

150
FOUNDATIONS AND PILING
Building 1989 254 October 20, pp83-85, 89-90
A series of articles is presented covering precision monitoring equipment and dealing with difficult ground conditions.

151
L Domaschuk
INADEQUATE SHALLOW FOUNDATIONS: A DESIGN-COST ANALYSIS OF ALTERNATIVES
Building Research & Practice 1987 15 July/August, pp224-230
Damage to foundations as a result of clay heave in Winnipeg, Canada, is surveyed. An evaluation of five foundation methods suggested that the best solution was an optimised pile system with basement walls designed as grade beams supporting a timber main floor.

152
G H Roscoe and R Driscoll
REVIEW OF ROUTINE FOUNDATION DESIGN PRACTICE
1987. BRE. pp9
Factors governing the choice of foundations and their subsequent performance are discussed. They include cost, ease of construction and ground conditions. Attention is also given to project organisation, site supervision and Building Regulations.

153
PILING AND GROUND ENGINEERING
Construction News Supplement 1987 December 1, pp27-46

154
PILING AND FOUNDATIONS
Contract Journal 1987 February 19, pp16-23
A series of articles is presented covering building on marginal land, the market for piling contractors, and piling technology.

155
Building Research Establishment
FOUNDATIONS ON SHRINKABLE CLAY: AVOIDING DAMAGE DUE TO TREES
Defect Action Sheet 96. 1987. pp2

156
Building Research Establishment
MINI PILING FOR LOW RISE BUILDINGS
Digest No.313. 1986. pp7

157

Building Research Establishment
CHOOSING PILES FOR NEW CONSTRUCTION
Digest No.315. 1986. pp8

158

PILING AND GROUND ENGINEERING
Construction News Supplement 1986 December 11, pp17-32

159

PILING, FOUNDATIONS AND SUSPENDED FLOORING
House Builder 1986 45 February, pp30
The following articles are included: Foundations: a review of techniques by P D Bolton; Methane and redevelopment by D L Barry; Some typical problems in Docklands housing by R Marks; Docklands on firm foundations - the house builders view by J Gates; Growth of precast suspended flooring by T Harding

160

I Ellis
PILING FOR UNDERPINNING
Paper to Symposium on Building appraisal, maintenance and preservation, Bath University, 10-12 July 1985. pp9
Details are given of some case histories to show the wide variety of structures that can be strengthened by pile underpinning.

161

S Thorburn and J F Hutchison
UNDERPINNING
1985. Surrey University Press. pp296
Covers the philosophy of underpinning, simple methods and excavation; conventional piles; 'pali radice' and 'reticulated pali radice'; Pynford method; ground freezing; chemical grouting; and soil improvement by jet grouting.

162

G Ridout
BUILDING A FIRM FOUNDATION WITH MINI PILES AND GROUND TREATMENT
Contract Journal 1985 324 March 28, pp22-24, 27

163

D Greenwood
UNDERPINNING BY GROUTING
Paper to Symposium on Building appraisal, maintenance and preservation, Bath University, 10-12 July 1985. pp16

164

J Pryke
UNDERPINNING - IS IT WORTH IT?
Paper to Symposium on Building appraisal, maintenance and preservation, Bath University, 10-12 July 1985. pp5

165

W G Curtin
TAKE A NEW LOOK AT FOUNDATIONS AND FIND MORE BUILDABLE SITES
Contract Journal 1984 320 August 23, pp30-31
Various foundation techniques are outlined which will allow the use of derelict and abandoned sites.

MATERIALS MANAGEMENT

166

H R Thomas et al
IMPACT OF MATERIAL MANAGEMENT ON PRODUCTIVITY - A CASE STUDY
ASCE Journal of Construction Engineering and Management 1989 115 September, pp370-384
It is demonstrated that formal material management programmes can yield significant cost savings, even to small/medium sized contractors, in terms of reduced work hour losses.

167

D J Munday
STORING TIMBER ON SITE
Building Today 1989 197 March 16, pp32-33

168

J Illingworth and K Thain
MATERIALS MANAGEMENT - IS IT WORTH IT?
CIOB Technical Information Service Paper No.93. 1988. pp5
Arguments are presented to demonstrate the value of materials management and the minimisation of waste.

169

L C Bell
ATTRIBUTES AND COST EFFECTIVENESS OF MATERIALS MANAGEMENT COMPUTER SERVICES
Managing Construction Worldwide. Volume 1. Systems for managing construction.
1987. Spon/CIOB CIB. pp288-295
The systems described are used in an interactive mode to communicate materials information to and from designers, project managers, purchasing managers, stores personnel and craftsmen. The structure of the systems is discussed, together with their cost and benefits.

170

D A Wijesundera and F C Harris
SELECTION OF MATERIALS HANDLING METHODS IN CONSTRUCTION BY SIMULATION
Construction Management & Economics 1989 7 Summer, pp95-102
The development is described of a dynamic interactive simulation model for selecting construction plant.

171

L C Bell and G Stukhart

**COSTS AND BENEFITS OF MATERIALS
MANAGEMENT SYSTEMS**

ASCE Journal of Contruction Engineering & Management 1987 *113* (2) June, pp222-234

In recent years construction contractors have developed integrated, or 'total concept', materials management systems (MMS) that combine and integrate the takeoff, vendor evaluation, purchasing, expediting, warehousing, and distribution functions. These materials management systems produce tangible benefits in the areas of improved labour productivity, reduced bulk materials surplus, reduced materials management manpower, and cash flow savings. The cost of developing and implementing these systems can be substantial, but the benefits outweigh the costs, particularly when the system is used by the crafts for scheduling their work around bulk material availability.

172

J Illingworth and K Thain

CONTROL OF MATERIALS AND WASTE

CIOB Technical Information Service Paper No.87. 1987. pp8

A practical approach is taken to the key issues of materials control. In regard to the organisation and control of materials attention is given to buying schedule, materials scheduling, ordering, progressing, deliveries, quality assurance, materials reconcilation, site measurement and valuations, and introduction of materials management into the firm. The control of waste is considered with regard to the materials adviser, and waste disposal. Finally, the value of good communication and co-ordination is stressed.

173

G Stukhart and D Y M Chang

**AUTOMATED SITE MATERIALS MANAGEMENT
- A PROTOTYPE**

Proceedings 4th International Symposium on Robotics & Artificial Intelligence in Building Construction, Haifa, Israel, 22-25 June 1987. Volume 2. pp721-730

174

E R Skoyles and J R Skoyles

WASTE PREVENTION ON SITE

1987. Mitchell. pp208

An in-depth analysis is given of the principal causes of waste, how it occurs on both small and large sites, its effect on contractual arrangements and its extent for each of the principal building materials. The roles are examined of all personnel in relation to waste before guidance is given on how to attack the problem.

175

J E Gibbons and C Miller

**REDUCTION OF WASTE AND IMPROVEMENT OF
PRODUCTIVITY IN TRADITIONAL MASONRY
HOUSE BUILDING**

Proceedings of the CIB W-65 mini-symposium, November 1985. Istanbul. pp10-1, 10-25

The problems or organisation and mangement of construction in developing countries and international contracting. Design factors leading to enhanced labour productivity on two sites in Scotland are discussed.

176

J Benes and W J Diepeveen

**MATERIAL MANAGEMENT IN CONSTRUCTION
FIRMS**

Proc. CIB 10th Triennial Congress, Sept.22-26 1986, Washington. Volume 3. pp922-929

An interactive program has been devised for material management in building firms, including a warning system for decisions on the building site. The system contains basic questions concerning delivery and the use of materials, as well as an interactive matrix with questions about the progress of work on the project.

177

L C Bell and G Stukhart

**ATTRIBUTES OF MATERIALS MANAGEMENT
SYSTEMS**

ASCE Journal of Construction Engineering and Management 1986 *112* (1) March, pp14-21

The attributes of materials management systems are discussed and the essential elements of a successful system identified. Owner-contractor, engineer-contractor, and home office-project site communications appear to be critical to the success of the materials management effort. Preconstruction materials planning and personnel orientation and training are also important. The complex on-line computer programs that are used to co-ordinate the materials management effort are costly, but essential, if the desired degree of control is to be exerted to prevent potential shortages, surpluses, and cash flow problems.

178

J R Skoyles

**AN APPROACH TO REDUCING MATERIALS
WASTE ON SITE**

CIOB Technical Information Service Paper No.34. 1984. pp7

Data on the wastage of selected materials are presented to illustrate the lack of response to campaigns to minimise the problem. A number of suggestions are made to limit wastage including the use of a 'waste observer'. The 'waste observer' position would be filled by a regular visitor to the site who could provide an on-going evaluation of site performance.

179

E Skoyles

**TRAINING FOR SITE MANAGERS ON WASTE
PREVENTION**

Building Technology & Management 1986 24 February, pp25-26

180
Suspended Ceilings Association
SITE GUIDE TO CEILINGS
1986. pp9
Coverage is given to a summary of ceiling types, data on when and how to programme fixing, handling and storage and the sequence of installation. Special needs of ceiling sub-contractors are also highlighted.

181
K R Good
HANDLING MATERIALS ON SITE
CIOB Technical Information Service Paper No.68. 1986. pp5
Broad principles of material handling are outlined. Guidance notes for handling plaster, plasterboard and bricks are presented.

182
Building Research Establishment
CLAYWARE DRAINAGE PIPES : STORAGE AND HANDLING
Defect Action Sheet No.49. 1984. pp2

183
P Baily and D Farmer
MATERIALS AND MANAGEMENT HANDBOOK
1982. Gower. pp300
A detailed coverage of organisational aspects, strategic planning, control systems, management techniques and 'people' problems. 12 case studies are included.

MATERIALS HANDLING

(a) Concrete

184
H Wylde
MIX YOUR OWN CONCRETE
Building Today 1988 *196* July 14, pp19-20,22
A guide is given to existing concrete mixing plant.

185
H Wylde
BRITISH MIXERS STIR IT UP
Building Trades Journal 1985 *190* November 21, pp21-22
A review is made of the performance and specification of concrete mixers.

186
H Wylde
MIXING CONCRETE ON SITE
Building Trades Journal 1984 *187* May 17, pp20-23
Basic principles of operating a mixer plant on site are discussed.

(b) Concrete pumping

187
M Anson et al
CONCRETE PUMPING IN THE UK AND WEST GERMANY : A COMPARISON
ICE Proceedings 1989 Part 1. *86* February, pp41-57

188
T Cooke
POURING CONCRETE BY PUMP IS OUSTING THE WHEELBARROW
Building Today 1988 *196* September 29, pp20-21
The advantages of concrete pumping are summarised.

189
M Anson et al
PUMPING OF CONCRETE - A COMPARISON BETWEEN THE UK AND WEST GERMANY
1986. University of Lancaster/Polytechnic of Wales. pp34

190
H Wylde
PUMPING INTO THE CONCRETE MARKET
Building Trades Journal 1986 *191* February 27, pp18-19
A survey is made of concrete pumps available from four suppliers. Technical data for 13 different types of pumps are tabulated.

191
D E Egan
HIGH PRESSURE CONCRETE PUMPING FOR MULTI-STOREY BUILDINGS
Constructional Review 1985 *58* May, pp50-57

QUALITY CONTROL

192
J Simpson
WORKMANSHIP ON BUILDING SITES
Building Today 1989 *198* November 30, pp29-31
A review of the major points of BS 8000.

193
J Weller
DEATH OF A PROFESSION:
1. THE ARCHITECTS DIMINISHING ROLE
Architects Journal 1988 *188* November 2, pp81-85
The role of the architect in site quality control is debated.

194
L J Carvalho
QUALITY ASSURANCE IN PRACTICE ON SITE
Structural Engineer 1987 *65A* April, pp137-139

195

D Wells

SETTING SITES ON QUALITY

Contract Journal 1987 January 29, pp16-17

Lovell's approach to quality control is outlined.

196

National Economic Development Office

ACHIEVING QUALITY ON BUILDING SITES

1987. pp62

Problems in achieving quality were found to arise from poor design, poor contractor organisation, inadequate project information and carelessness. Difficulties in putting the problems right were due to unclear responsibilities and lack of team working, motivation and commitment. Client involvement with the project was seen as a key issue in achieving quality.

197

I Ferguson and E Mitchell

QUALITY ON SITE

1986. Batsford. pp176

Those areas of potential failure are discussed in some detail and it is shown how quality can be improved by proper attention to such factors as constructional system, dimensional co-ordination, tolerance and fit, buildability and communication of design intention. Special regard is paid to project documentation and to the problems of site supervision.

198

R Stevens

NO HIGHWAY CODE

Building 1986 250 January 31, p29

A critique of the draft BSI Code 'Workmanship on bulding sites'.

199

R Stevens

BETTER WAY TO SPECIFY

Chartered Quantity Surveyor 1986 8 April, p22

A critique of the draft BSI Code 'Workmanship on building sites'.

200

Cement and Concrete Association

FUNDMANETALS OF QUALITY ASSURANCE ON SITE - GUIDANCE NOTES

1986. pp5

201

R Cecil

SUPERVISION AND QUALITY CONTROL

RIBA Journal 1985 92 January, p71

The issue of quality in building and the respective responsibilities of architect and builder are discussed.

202

P D Allars

ACHIEVING QUALITY BRICKWORK ON SITE

CIOB Technical Information Service Paper No.38. 1984. pp8

A practical approach is taken to the question of obtaining good quality brickwork with particular attention being given to engaging craftsmen; motivation and productivity; brick handling and distribution; working practices; mortars; structural considerations; joints; and damp-proofing and fixings. A checklist for use in interviewing bricklaying sub-contractors is included.

203

BETTER BUILDING. A SPECIAL REPORT

House Builder 1984 43 March, pp23-80

A series of articles is presented which look at means to obtain better quality with particular attention being given to workmanship, external walls, timber frame construction, foundations and floors, and dpcs.

204

Bituminous Roofing Council

FLAT ROOFING - INSPECTION AND QUALITY CONTROL ON SITE

1985. pp4

CONDITIONS OF WORKING

205

R Swan

PRODUCTIVITY NEEDN'T MELT THE ICE

Contract Journal 1989 347 January 26, pp26-28

The need for appropriate facilities for the workforce is discussed.

206

K Watson

OCCUPATIONAL HEALTH. MARKING THE CARDS OF THOSE MACHO MEN

Contract Journal 1987 February 26, pp16-17

The implications are evaluated of the long term health hazards of the construction site.

207

T Niskairen et al

MUSCULAR STRAIN IN 'HEAVY' AND LIGHT BUILDING WORK

Building Research & Practice 1986 14 January/February, pp28-32

The causes and frequency of musculoskeletal strain among concrete reinforcement workers and painters on house maintenance are analysed from observations, from accident report forms and from interviews about unreported minor accidents.

208

T Burch

HEALTH FACTORS AND ABSENTEEISM OF CONSTRUCTION WORKERS

CIOB Technical Information Service Paper No.72. 1986. pp6

209

S Goth

WORKING CONDITIONS AND WORK ORGANISATION IN DANISH SEMI-SKILLED CONSTRUCTION WORK

Proc. 4th Int. Symposium on Organisation and Management of Construction, Waterloo, Canada, July 1984. Volume 2. pp569-578

210

Construction Industry Research and Information Association

EXPOSURE OF CONSTRUCTION WORKERS TO NOISE

Technical Note 115. 1984. pp25

Existing Regulations and Codes of Practice are considered as well as the possible results of moves to extend them to, and enforce them on, sites. A recent EEC directive on noise is summarised in an appendix. Available data are reviewed and a report given on a measurement survey of noise exposure on a small sample of workers.

211

G K Harrison

RELATIONSHIP BETWEEN MANAGEMENT AND MANPOWER IN CONSTRUCTION. REPORT OF A COMPARISON BETWEEN THE CANADIAN AND BRITISH SYSTEMS

1979. pp38

Report of a Cartwright Travelling Scholar which is concerned primarily with the motivation of the workforce and the impact of bad weather conditions.

LABOUR RELATIONS

212

C J Nicholls and D A Langford

MOTIVATION OF SITE ENGINEERS

CIOB Technical Information Service Paper No.78. 1987. pp7

The results are presented of a survey designed to establish the factors motivating site engineers. From this a series of pointers are given which might make a contribution towards obtaining better performance.

213

W F Maloney and J M McFillen

INFLUENCE OF FOREMEN ON PERFORMANCE

ASCE Journal of Construction Engineering & Management 1987 113 (3) September, pp339-415

A sample of unionised construction workers from a major mid-western city is surveyed to gather their perceptions of the behaviour of their foreman. Five dimensions of foreman behaviour are identified: (1) the degree of participation allowed by the foreman in deci-sion-making; (2) the level of support provided by the foreman; (3) the degree of achievement orientation of the foreman; (4) the degree of bias of the foreman; and (5) the level of work facilitation provided by the foreman. The five dimensions have varying relationships with worker motivation, performance, and satisfaction.

214

W F Maloney and J M McFillen

MOTIVATIONAL IMPACT OF WORK CREWS

ASCE Journal of Construction Engineering & Management 1987 113 (2) June, pp208-221

Unionised construction workers in a major mid-western city are surveyed to collect data on the workers' perceptions of their workcrews. Data are analysed using factor analysis and five elements of work crew functioning are identified; general effectiveness, openness, cohesion, goal clarity, and goal difficulty. The relationships between these variables and the workers' motivation, reported performance, and job satisfaction are analysed. The conclusion reached is that contractors must manage their workcrews in terms of planning, organising, staffing, directing, and controlling. By doing this, worker performance and satisfaction can be increased.

215

Luh-Maan Chang and J D Borcherding

CRAFTSMAN QUESTIONNAIRE SAMPLING

ASCE Journal of Construction Engineering & Management 1986 112 (4) December, pp543-556

Craftsman Questionnaire Sampling (CQS) was recently developed for performance measurement and productivity improvement at construction sites. CQS uses a questionnaire to collect data. It has the virtues of the Craftsman Questionnaire Survey, which provided information regarding the sources of delays, the amount of rework performed, as well as creating a participating atmosphere on site. Meanwhile, CQS imitates the sampling procedure from the work sampling method by randomly selecting craftsmen in the field and asking them to determine the activities involved in the immediate past. This procedure empowers the CQS to apply the theory of binominal distribution to estimate ratio delays and to maintain the advantages of relative simplicity and statistical reliability as the work sampling method does.

216

Luh-Maan Chang

INFERENTIAL STATISTICS FOR CRAFTSMAN QUESTIONNAIRE

ASCE Journal of Construction Engineering & Management 1986 112 (4) December, pp492-499

The feasibility of applying inferential statistics to the parameter estimate of manhours lost in the construction field in the Craftsman Questionnaire Survey is discussed.

217

G J Lemna et al

PRODUCTIVE FOREMEN IN INDUSTRIAL CONSTRUCTION

ASCE Journal of Construction Engineering & Management 1986 *112* (2) June, pp192-210

An attempt is made to identify characteristics which differentiate productive industrial construction foremen from less productive industrial construction foremen.

218

W F Maloney and J M McFillen

MOTIVATION IN UNIONISED CONSTRUCTION

ASCE Journal of Construction Engineering & Management 1986 *112* (1) March, pp122-136

A survey was conducted of unionised construction workers in a major mid-western city to collect data on their perceptions of the motivational climate in their jobs. The framework for the collection and analysis of the data was the expectancy model of worker motivation and performance. The findings indicated that the motivational climate is very poor. Contractors rely more on punishment and discipline than they do positive rewards. Little is done to encourage good performance. Discipline is used to discourage poor performance. Contractors provide little in the way of rewards, even when they are not prohibited from providing a variety of rewards by their labour agreements. The workers surveyed reported very little incentive to be highly productive.

219

W F Maloney and J M McFillen

MOTIVATIONAL IMPLICATIONS OF CONSTRUCTION WORK

ASCE Journal of Construction Engineering & Management 1986 *112* (1) March, pp137-151

A survey was conducted on unionised construction workers in a major mid-western city to collect data on their perceptions of the jobs and environment within which they perform their jobs. Construction workers have growth needs that are similar in strength to other blue collar workers. The individuals with stronger needs should respond to jobs that are high in motivating potential. Contractors need to improve worker satisfaction with job context. The qualifications of the workers appears to be more than adequate for the great majority of construction tasks. Contractors need to structure jobs to improve their motivating potential. As currently structured, construction jobs are low in motivating potential.

220

A J Charlett

MORALE ON SITE

Building Technology & Management 1986 24 January, p28

The basic principles of establishing morale within the workforce are considered and their application to construction sites outlined.

221

M J Bresnen et al

LABOUR RECRUITMENT STRATEGIES AND SELECTION PRACTICES ON CONSTRUCTION SITES

Construction Management & Economics 1986 4 Spring, pp37-55

Discussions of the recruitment and employment of labour in the construction industry tend to be aggregate analyses, at the level of the firm or industry. The question as to what firms do when faced with a particular set of labour requirements at the operational level so far remains largely unanswered. This paper focuses upon the patterns of recruitment and selection adopted by main contractors on 43 medium to large construction sites. The balance of directly employed (newly recruited and transferred) and sub-contracted labour is examined and variation noted by size and nature of work, firm size and location of work. Although variable, the resricted degree of direct employment is documented. The recruitment processes adopted on site are identified as relatively informal, adaptive and based upon short-term production needs. Selection processes emphasise criteria such as work history, experience, reliability and conformity, as opposed to formal qualifications. The paper suggests that while the recruitment and selection strategies adopted on sites are both instrumental and rational from the contractor's viewpoint, they may have wider deleterious consequences. In particular the impact of such strategies on training provision, and on the development and maintenance of an adequately skilled workforce is raised.

222

Luh-Maan Chang and J D Borcherding

EVALUATION OF CRAFTSMAN QUESTIONNAIRE

ASCE Journal of Construction Engineering & Management 1985 *111* (4) December, pp426-437

The validity is evaluated of the Craftsman Questionnaire for determining lost manhour estimates. The validity of the Craftsman Questionnaire has been questioned since it was introduced in the construction industry. To examine its validity, the background of the Craftsman Questionnaire is reviewed. Secondly, an analysis of the data collected from past studies is conducted. Herein, sources of inaccurate estimates are revealed. Thirdly, a research method called criteria validity is employed to invetigate the validity of the Craftsman Questionnaire by comparing its measures with those of work sampling.

223

W F Maloney and J M McFillen

VALENCE OF AND SATISFACTION WITH JOB OUTCOMES

ASCE Journal of Construction Engineering & Management 1985 *111* (1) March, pp53-73

Approximately 2,800 workers were asked about the importance they attach to various job related factors and their satisfaction with each factor. The most important set of factors were those relating to the intrinsic nature of the work; working life of a craftsman, performing challenging work, etc. The set of factors with which the workers was most satisfied was that of performance level; high productivity; quantity; and doing your work in a craftsmanlike manner. Individual factors that require attention on the part of contractors to improve worker motivation and satisfaction were identified.

224
T J Gallagher
INDUSTRIAL RELATIONS ON SITE
1984. Construction Press. pp142
A practical guide for site management covering recruitment; employment procedures; joint negotiating machinery; employment legislation; incentive schemes; industrial tribunals; sub-contracting; and management contracting. Specimen letters are given at an appendix.

225
C C Baker
HEALTH SCREENING IN THE CONSTRUCTION INDUSTRY
Safety Practitioner 1984 September, pp4-7
The need is discussed for the provision of occupational health in the construction industry and the associated difficulties. A simple, basic and inexpensive screening scheme is detailed which could be applied to all sites.

226
B C Paulson and J W Fondahl
CRAFT JURISDICTION IMPACT ON CONSTRUCTION
ASCE Journal of Construction Engineering and Management 1983 *109* (4) December, pp369-386
Construction craft jurisdiction, normally exercised by unions, includes both territorial and technological claims to certain categories of work. A research project is reported that had three main objectives; (1) collect and analyse statistics on the incidence of jurisdictional disputes; (2) evaluate existing jurisdictional dispute settlement mechanisms; and (3) survey contractors and owners to determine the impact of craft jurisdictional practices (not only disputes) on costs and schedules. It is concluded that there is inadequate statistical information to support informed decision-making on jurisdictional problems. Most dispute settlement mechanisms are antiquated and ineffective, but there are good plans in some local areas. There are serious cost and schedule impacts on construction projects from observing craft jurisdictional practices.

227
J Benes and W J Diepeveen
MOTIVATION THROUGH SMALL-SCALED COMPOSITE TASK GROUPS
Proc. 9th CIB Congress, Stockholm 1983. Volume 1b. Renewal, rehabilitation and maintenance. pp47-54

Building firms at present tend to reorganise their structure by introducing smaller units of independent profit centres that previously were integrated into centralised structures or big concerns. This can be applied to the organisation of the work on building sites by making use of small-scaled multi-disciplinary task groups or working teams. This may improve the motivation of the labour force considerably.

PRODUCTIVITY

228
A J Stevens
COMPUTER PACKAGE FOR PROCESSING AND ANALYSING DATA OBTAINED FROM ON-SITE PRODUCTIVITY STUDIES
Managing Construction Worldwide. Volume 2. Productivity and Human Factors in Construction 1987. Spon/CIOB/CIB. pp723-734

229
A T Baxendale
MEASURING SITE PRODUCTIVITY BY WORK SAMPLING
Managing Construction Worldwide. Volume 2. Productivity and Human Factors in Construction 1987. Spon/CIOB/CIB. pp812-822

230
E Koehn and R Cook
CONSTRUCTION SITE OPERATIONS - PERCEPTIONS INFLUENCING PRODUCTIVITY
Managing Construction Worldwide. Volume 2. Productivity and Human Factors in Construction 1987. Spon/CIOB/CIB. pp989-1000
The perception of middle managers to their work is seen as contributing to a reduction in the productivity of construction operations.

231
P English
MANAGEMENT RESEARCH NEEDS - SITE PRODUCTIVITY
International Journal of Construction Management & Technology 1987 *2* (3) pp37-40
The major problem with work on site is that of completing the job on time, and the main reason for the difficulties encountered is that design is frequently continued after site work has been started. Research into the delays and extra costs which result from this is required. Research is also needed into the adequacy of technical detail taught on degree courses in order to compensate for reducing levels of trade skills.

232
J G Lowe
LABOUR PRODUCTIVITY AND INVESTMENT IN HAND POWER TOOLS - A REPLY
International Journal of Construction Management & Technology 1987 2 (1) pp5-7 (correspondence)
The methodology used to calculate the enhancement in productivity by the use of power hand tools is questioned.

233
F R Barakat and V K Handa
FIELD MANAGEMENT CONTROL, EFFICIENCY AND PRODUCTIVITY
Managing Construction Worldwide. Volume 2. Productivity and Human Factors in Construction 1987. Spon/CIOB/CIB. pp685-668
Efficiency and productivity as a function of site management and control are considered, particular attention being given to the significance of reporting, cost report frequency, plan management, scheduling, post project analysis, job diaries, use of computers, and organisation set up on site.

234
A R Duff et al
FACTORS AFFECTING PRODUCTIVITY IMPROVEMENT THROUGH REPETITION
Managing Construction Worldwide. Volume 2. Productivity and Human Factors in Construction 1987. Spon/CIOB/CIB. pp634-645
The results are presented to productivity improvement on a number of repetitive construction activities, particular reference being made to factors found to interfere with improvement due to learning. Various formulations of the learning curve are tested on data collected from two sites and it is concluded that the cumulative curve is the most applicable for construction planning and control.

235
R M W Horner
MEASUREMENT OF FACTORS AFFECTING LABOUR PRODUCTIVITY ON CONSTRUCTION SITES
Managing Construction Worldwide. Volume 2. Productivity and Human Factors in Construction 1987. Spon/CIOB/CIB. pp669-680

236
A Baxendale
SITE PRODUCTIVITY. MAKING BEST USE OF RESOURCES
Building Trades Journal 1986 192 July 24, pp19-20
The value is illustrated of work sampling in improving site productivity.

237
R C Whitehead et al
RECORDS DATA
Building Technology & Management 1986 24 August/September, pp6-7
The various techniques available for studying productivity in construction are reviewed.

238
A D Hall and D W Cheetham
LABOUR PRODUCTIVITY AND INVESTMENT IN HAND POWER TOOLS
International Journal of Construction Management & Technology 1986 1 (3) pp52-58

239
A Jaafari and V K Mateffy
WHAT EVERY CONSTRUCTION MANAGER SHOULD KNOW ABOUT CONSTRUCTION COST CONTROL
International Journal of Construction Management & Technology 1986 1 (1) pp21-35
It is shown how the maintenance of a reasonable level of productivity on a job site can be achieved by supplementing cost control with work study techniques, notably random activity sampling.

240
A T Baxendale
MEASURING SITE PRODUCTIVITY BY WORK SAMPLING
CIOB Technical Information Service Paper No.55. 1985. pp6
It is shown how easily productivity studies can be made and how they might be used to point out to site management where work may be carried out more effectively.

241
A D Price and F C Harris
METHODS OF MEASURING PRODUCTION TIMES FOR CONSTRUCTION WORK
CIOB Technical Information Service Paper No.49. 1985. pp11
It is shown that work study can be adopted to meet the requirements of most construction operations, sites and companies. The key lies in the application of site efficiency factors which isolate the basic times for operations and enables computers to be used for recording and handling data. Using the example of a concrete gang of four men pouring two columns and roof slabs it is demonstrated how the data is collected; translated into standard times; which are then used in the synthetic build up of planning times for other operations.

242
H R Thomas
PRESCRIPTION FOR CONSTRUCTION PRODUCTIVITY IMPROVEMENT
Proc. 4th Int. Symposium and Management of Construction, Waterloo, Canada, July 1984. Volume 2. pp629-640
Major theories concerning worker motivation are reviewed. A methods improvement programme is formulated based on the application of work measurement techniques and incentives.

243
H R Thomas et al
IMPROVING PRODUCTIVITY ESTIMATES BY WORK SAMPLING
ASCE Jounral of Construction Engineering and Management 1984 *110* (2) June, pp178-188
Theoretical aspects are presented to evaluate the adequacy of work sampling as a surrogate productivity measure.

244
W D Woodhead
USE OF QUANTITATIVE DESIGN PARAMETERS TO ECONOMISE IN HOUSEBUILDING
Architectural Science Review 1983 *26* September/December, pp99-102
Reduction in costs by reducing on-site labour is discussed. A procedure is described whereby manpower inputs, measured at the sub-element and element levels of a building, may be used to compare different designs.

245
W D Woodhead and G D Salomonsson
COMPUTER BASED SYSTEM FOR RECORDING AND ANALYSING BUILDING SITE LABOUR INPUT
Construction Papers 1983 *2* (2) pp47-52

246
D C A McLeish
IMPORTANCE OF PRODUCTIVITY TO THE ARCHITECT AND CONSTRUCTION MANAGER
Proc. 9th CIB Congress, Stockholm 1983. Volume 1b. Renewal, rehabilitation and maintenance. pp89-101
In a world of limited resources, interest in productivity is important to the architect and construction manager. Productivity may be defined as a measure of the effective use of resources. Research into building productivity based on activity sampling surveys of two traditional house building sites is described. The results support the need for productivity improvement in the construction industry, particularly in house building. Two results were emphasised by the research: (a) in many repeated operations the variation in manhour requirements was commonly 3 to 1 or more, and (b) in the same operations the number of separate visits by operatives to the workplace showed a similar level in variation. A high degree of correlation was found between these results for most operations. The large bricklaying operations analysed in the research tended to differ from the usually much smaller services and finishes operations in three respect: (a) in having less variation in manhours, (b) longer operative visits to the workplace, and (c) proportionally less non-productive manhours. These findings suggest that productivity is likely to be improved by arranging that operations are as large and independent as possible.

247
J Gibbons and L Miller
STUDY IN IMPROVING PRODUCTIVITY AND REDUCING WASTE IN TRADITIONAL MASONRY HOUSEBUILDING
1983. Scottish Development Department. pp12
It is shown that a positive design approach can lead to a reduction in site labour requirements, whilst improving constructional quality and physical standards.

ENVIRONMENTAL FACTORS AFFECTING SITE OPERATIONS

248
J Simpson
BRICKLAYING IN WINTER
Building Today 1989 *198* November 2, pp28-29

249
M Evamy
WINTER WORKING. CONSTRUCTION KEEPS OUT THE WINTER CHILL
Contract Journal 1989 *347* January 26, pp22-25
Means for keeping sites operational during adverse weather conditions are indicated.

250
J R Harding and R A Smith
BRICKLAYING IN WINTER CONDITIONS
BDA Building Note 3. 1986. pp3

251
Meteorological Office
METBUILD - MONTHLY DOWNTIME SUMMARY
1986. pp6
An introduction to the analyses of weather elements having a direct bearing on downtime, namely temperature and humidity, rainfall, snow and wind.

252
Meteorological Office
SERVICES TO THE CONSTRUCTION INDUSTRY
1985. pp10

SITE ACCOMMODATION

253
L R Wernich
SITE ACCOMMODATION - A VIEW FROM THE INDUSTRY
Building Contractor 1988 February, pp15-17

254
H Wylde
BETTER CLASS OF SITE HUT
Building Trades Journal 1987 *193* February 12, pp26-27
Site accommodation is reviewed.

255
INSTANT BUILDING INDUSTRY OFFERS CONTRACTORS WIDE CHOICE
Contract Journal 1985 *323* January 24, pp22-24,27-29
A review is made of available site accommodation.

256
H Wylde
BUILDERS PLANT. IS HIRING THE RIGHT ANSWER?
Building Trades Journal 1985 *190* September 5, pp38-40
The question of hire or purchase is considered in relation to site accommodation.

257
H Wylde
PROVIDING ACCOMMODATION FOR THE SITE WORKFORCE
Building Trades Journal 1984 *188* October 25, pp26-27

258
H Wylde
PROVIDING SPECIALISED SITE ACCOMMODATION
Building Trades Journal 1984 *188* November 1, p20

PLANT

(a) Management

259
R Clayton
PLAN THE LAYOUT OF YOUR SITE BEFORE WORKING ON IT
Building Today 1989 *197* May 4, pp32-33
Some guidance is given on plant and materials location.

260
C M R Chan and F C Harris
DATABASE/SPREADSHEET APPLICATION FOR EQUIPMENT SELECTION
Construction Management & Economics 1989 *7* Autumn, pp235-247
A computerised system of selecting construction equipment is described. A database of technical criteria for the backhoe/loader is assembled and analysed with a spreadsheet technique.

261
A Tavokoli et al
EQUIPMENT POLICY OF TOP 400 CONTRACTORS : A SURVEY
ASCE Journal of Construction Engineering and Management 1989 *115* June, pp317-329
Special attention is given to equipment financing, replacement policy, standardisation, safety and maintenance management.

262
F N Rowley
CONCEPTION, DESIGN AND PROJECT MANAGEMENT OF SPECIALISED CONSTRUCTION SYSTEMS
ICE Proc. 1987 *82* (Part 1) December, pp1073-1088
The project management is described of major items of specialised plant used for civil engineering works.

263
C W Ibbs et al
OWNER-FURNISHED EQUIPMENT (OFE) CONTRACT PRACTICES
ASCE Journal of Construction Engineering and Management 1987 *113* (2) June, pp249-263
Owner-furnished equipment (OFE) procurement is a contract administration technique utilised on many construction projects to save costs and time. The results are presented of a comprehensive research study of OFE practices, problems, and benefits. Fifty-five OFE construction projects are analysed, allowing a series of conclusions and recommendations to be developed. The major points examined include: characteristics of typical OFE projects and products; the quantified extent of project benefits and costs realised; the likelihood and consequences of related contract disputes; effective contract monitoring strategies; and whether owners or third-party contract administrators should manage the supply contract.

264
J Whittaker
EQUIPMENT RATES FROM REVENUE REQUIREMENTS
ASCE Journal of Construction Engineering and Management 1987 *113* (2) June, pp173-178
The revenue requirements method which is used by utility companies for determining rates is applied to the problem of determining rates for construction equipment. The method follows engineering economy principles and explicitly considers operating costs, recovery and return of capital and income taxes. By expressing the result as the required before-tax revenue that a piece of equipment must generate, the method is intuitively appealing and easily comprehended by management. The example used both Canadian and American income tax legislation.

265
I D Tommelein et al
USING EXPERT SYSTEMS FOR THE LAYOUT OF TEMPORARY FACILITIES ON CONSTRUCTION SITES
Managing Construction Worldwide. Volume 1. Systems for managing construction.
1987. Spon/CIOB/CIB. pp566-597

266

H Wylde

PROS & CONS OF PLANT NEEDS

Building Trades Journal 1987 *192* April 2, pp26,28

Some guidelines are given on providing an effective plant service either by hiring or operating an in-house plant department.

267

M C Vorster and G A Sears

MODEL FOR RETIRING, REPLACING, OR REASSIGNING CONSTRUCTION EQUIPMENT

ASCE Journal of Construction Engineering and Management 1987 *113* (1) March, pp125-137

268

J Y Campbell

ECONOMICALLY PLANNED SELECTION OF CONSTRUCTION PLANT

Proc. 4th Int. Symposium on Organisation and Management of Construction, Waterloo, Canada, July 1984. Volume 2. pp353-360

It is shown how computer based simulation of the site performance of plant can assist in plant selection.

269

C A Collier and D E Jaques

OPTIMUM EQUIPMENT LIFE BY MINIMUM LIFE-CYCLE COSTS

ASCE Journal of Construction Engineering and Management *110* (2) June, pp248-265

A minimum cost equipment replacement model based on the present worth of discounted-cash-flow is presented.

270

D Baldry

IS MECHANISATION THE KEY TO INCREASED PRODUCTIVITY?

Building Technology & Management 1984 *22* June, pp10-11

The value is stressed of the careful selection of plant; the dangers of sub-letting and management contracting are that plant divisions will lose the ability to provide a viable service.

271

S Selinger

ECONOMIC SERVICE LIFE OF BUILDING CONSTRUCTION EQUIPMENT

ASCE Journal of Construction Engineering and Management 1983 *109* (4) December, pp398-405

A model is developed for establishing the economic service life of hoists, tower cranes and concrete mixers.

272

H Wylde

LARGE BUILDER BENEFITS FROM OPERATING OWN PLANT FLEET

Building Trades Journal 1985 *190* November 28 Supplement, pp4-7

Jelson's approach to plant management is described.

(b) General

273

W Miller

NEW TREND IN ATTACHMENTS

Constructor 1989 *71* April, pp22-24

A review is made of developments in the attachments available for small (mini) plant.

274

PLANT FOR TIGHT SITES. MINI PLANT MOVES INTO THE SPACE AGE

Contract Journal 1989 July 27, pp21-22,24,27-31

275

Institution of Civil Engineering Surveyors

SURVEYORS GUIDE TO CIVIL ENGINEERING PLANT

1988. pp197

A guide to the current and former models of construction plant classified according to the FCEC Daywork Schedules.

276

A Warszawski and R Navon

DEVELOPMENT OF A ROBOT FOR INTERIOR FINISHING WORK

Proc. 4th Int. Symposium on Robotice & Artificial Intelligence in Building Construction, Haifa, Israel, 22-25 June 1987. Volume 1. pp245-258

277

R B Blackman

CONTROL OF INTELLIGENT MATERIALS HANDLING EQUIPMENT IN TEMPORARY FACILITIES

Proc. 4th Int. Symposium on Robotics & Artificial Intelligence in Building Construction, Haifa, Israel, 22-25 June 1987. Volume 2. pp878-896

278

D W Halpin

ADAPTATION OF FLEXIBLE MANUFACTURING CONCEPTS TO CONSTRUCTION

Proc. 4th Int. Symposium on Robotice & Artificial Intelligence in Building Construction, Haifa, Israel, 22-25 June 1987. Volume 2. pp704-717

An exploratory investigation is described of the transferability of concepts developed for flexible manufacturing systems to the control and automation of fixed plant types of construction processes.

279

C Abel

DITCHING THE DINOSAUR SACTUARY : SEVENTEEN YEARS ON

CAD & Robotics in architecture & construction. Proc.

Joint International Conference, Marseilles, 25-27 June 1986. pp123-132
The author reviews his original prognosis regarding the introduction of computerised, flexible product machinery to revolutionise the industry. Reference is made to recent developments, particularly the use of 'smart' tools in the construction of the Hong Kong & Shanghai Bank. It is concluded that the potential of new building technology may not be fulfilled unless parallel changes are undertaken in architectural practice and education.

280
J Osborne and C Macgowan
PLANT AND TOOLS
Building 1986 *261* October 31, pp48-53
A review is made of the tools that have transformed productivity on site.

281
International Council for Building Research Studies and Documentation (CIB)
RECOMMENDED INTERNATIONAL CLASSIFICATION OF CONSTRUCTION MACHINES AND EQUIPMENT
1985. pp21

282
POWER ON SITE
Construction News Magazine 1984 *10* January, pp32-34
Hydraulic powered plant is reviewed.

283
S Booth (Ed)
SURVEYORS GUIDE TO CIVIL ENGINEERING PLANT
1984 revised edition. ICES. pp198
Lists of plant and their operating capabilities are provided in accordance with the FCEC daywork schedules.

284
G Ive
CONSTRUCTION PROCESS AND THE MARKET FOR, AND CONDITIONS OF PRODUCTION OF, CONTRACTOR'S PLANT AND EQUIPMENT
Proceedings 1982 Bartlett International Summer School, London, 1983. pp3.13-3.20
The application is discussed of mechanisation to assembly, site preparation, materials handling and on-site transformation of materials, before their respective equipment supplying industries are examined.

(c) Hiring and purchasing

285
PLANT HIRE : EQUIPMENT AND SERVICES
Construction News Supplement 1989 October, pp68
Articles are included on powered access, hire rates, computer systems, insurance, security, portable accommodation, safety and concrete pumping.

287
P Marsh
WIDENING PLANT OPTIONS
House Builder 1988 *47* September, pp105,107,110
A review of current developments in plant for hire for house builders is provided.

288
PLANT HIRE : EQUIPMENT AND SERVICES
Construction News Supplement 1987 October 15, pp60
A series of articles is presented covering plant hire conditions, training, crane hire, a listing of the top 50 plant hire firms, security, and skid steers.

289
PLANT HIRE : EQUIPMENT AND SERVICES
Construction News Supplement 1986 October 16, pp64
A series of articles is presented covering the CPA Model Conditions of Contract; dumpers; thefts; site accommodation; tippers; and conveyors.

290
PLANT HIRE : EQUIPMENT AND SERVICES
Construction News Supplement 1985 October *17*, pp64
Included among the contributions are articles dealing with contracts, insurances, investment, access plant, conveyors, drilling, wheeled loaders, platform safety and cranes.

291
H Wylde
POWER TOOL HIRE AS AN ECONOMIC OPTION
Building Trades Journal 1986 *191* June 10, pp24-25

292
PLANT HIRE : EQUIPMENT AND SERVICES
Construction News Supplement 1984 October 18, pp64
The following articles are included: The industry and its assocation; CPA industrial studies: past, present and future; Training and safety aspects of powered work platforms; Telescopic handling equipment; Fetch and carry role and bulk handling keep site dumpers to the fore; Low loader users seek versatility; Portable site accommodation.

293
H Wylde
GROWING RELIANCE ON PLANT HIRE
Building Trades Journal 1984 *188* September 6, pp28,30
Arrangements for hiring plant and the terms and conditions governing those hires are discussed.

294
P Boynton
COMPRESSOR RATES GROUP
Construction Plant and Equipment 1984 *12* January, pp24-25
A general look at plant hire rates is accompanied by a table of hire rates for the Midlands which also contains the changes since June 1983.

295
L Cant
PLANT SUPPLY : SALES AND HIRE
Building Technology & Management 1984 22 June, pp15-16
The relative merits of purchasing and hiring are examined.

(d) Access plant

296
H Wylde
MAST TYPE WORK PLATFORMS RISE IN INDUSTRY'S ESTEEM
Building Today 1989 197 June 29, pp30-32

297
H Wylde
GET A HOIST UP FOR YOUR MATERIALS
Building Trades Journal 1987 194 July 16, pp22,24
A review is made of the various types of hoist available.

298
POWERED ACCESS PLATFORMS
Contract Journal 1986 333 October 2, pp22-23,25-27,30-32,37-38
A series of articles describing the market and range of equipment available.

299
H Wylde
AERIAL VIEW OF WORK PLATFORMS
Building Trades Journal 1986 192 September 11, pp36,39
A review is made of the operating characteristics of available work platforms.

300
SKY'S THE LIMIT FOR GROWTH IN THE WORK PLATFORM MARKET
Contract Journal 1985 325 May 16, pp22-25,27-28
A series of articles is presented covering the powered work platforms.

301
Health and Safety Commission
SAFETY AT POWER OPERATED MAST WORK PLATFORMS
1985. pp15

302
M Darwin
GAINING ACCESS
Construction News Magazine 1984 10 November/December, pp29-31
A review is made of current trends in access platforms.

(e) Breakers and cutters

303
H Wylde
TRIM YOUR PAVING COSTS WITH A BLOCK CUTTER
Building Today 1989 197 March 12, pp26-28
A review is made of commercially available equipment.

304
HYDRAULIC TOOLS GIVE MORE POUNDING FOR THE MONEY
Building Today 1988 196 October 27, pp24-25
A review of hydraulic breakers includes tabulated operating performance data.

305
H Wylde
CUTTING MATERIALS FAST AND ACCURATELY ON SITE
Building Trades Journal 1985 189 February 7, pp27-29
A review is made of available power saws.

(f) Compaction plant

306
H Wylde
COMPACT IN NAME, IMPACT IN DEED
Building Today 1988 195 March 17, pp32-34
A review of light vibrating plates and rammers.

307
H Wylde
TREND IS MOVING AWAY FROM ROLLING YOUR OWN
Building Today 1988 195 February 4, pp21,23-24
A review is made of vibratory rollers.

308
COMPACTION EQUIPMENT
Contract Journal 1985 328 November 7, pp25-28,30,34-35,37
A series of articles covering range and ability of compaction equipment.

309
H Wylde
VIBRATING COMPACTION ON A PLATE
Building Trades Journal 1984 187 February 9, pp36-37
Comparison is made of available vibrating plate compactors.

(g) Conveyors

310
H Wylde
CONVEYING THE USES OF CONVEYORS
Building Trades Journal 1986 191 May 29, pp26-27
A short review is made of commercially available conveyors for site use.

(h) Cranes

311
D Baldry
SWISS SUCCESS OF THE MINI-CRANE
Building Technology & Management 1985 *23* July/August, pp6,9
The extensive use for domestic construction is reported of a lightweight 20m tower crane; a jack up machine, self-erecting with slewing mast and demountable counterweights.

312
H Wylde
CRANING INTO A WIDER MARKET
Building Trades Journal 1985 *189* April 25, pp44-45
A review is made of self erecting tower cranes.

313
W E Rodriquez-Ramos and R L Francis
SINGLE CRANE LOCATION OPTIMISATION
ASCE Journal of Construction Engineering and Management 1984 *109* (4) December, pp387-397
The development is described of a mathematical prescriptive model to establish the optimal location of a crane within a construction site.

314
S Furusaka and C Gray
MODEL FOR THE SELECTION OF THE OPTIMUM CRANE FOR CONSTRUCTION SITES
Construction Management and Economics 1984 *2* (2) Autumn, pp157-176
This paper presents a method of locating a crane on a construction site using mathematical techniques. The optimum location and choice of the crane on a site is seen as one of the most important parts of construction planning. It proposes the algorithm which can define the least expensive cranage cost (the total cost of the hire, assembly and dismantling) by calculating the combined use of different cranes, such as truck crane, crawler crane, travelling based tower crane or fixed base tower crane. Conclusions are drawn as to the relevance of the application of the model to construction projects.

315
K S Rajagopalan
SUPPORTS FOR TRAVELLING CRANES : CASE HISTORY
ASCE Journal of Construction Engineering & Management 1988 114 March, pp114-120
Tight site conditions and a short construction schedule necessitated the use of two travelling cranes during the construction of a hotel. Steep slopes at the south and basement walls at the north posed interesting structural design challenges for these crane supports. Problems encountered in the selection of the crane system and the design of its supports are described; the solutions to these problems are illustrated.

316
A Warszawski and N Peled
EXPERT SYSTEM FOR CRANE SELECTION AND LOCATION
Proc. 4th Int. Symposium on Robotics & Artificial Intelligence in Building Construction, Haifa, Israel, 22-25 June 1987. Volume 1. pp64-75

317
C Gray
CRANE LOCATION AND SELECTION BY COMPUTER
Proc. 4th Int. Symposium on Robotics & Artificial Intelligence in Building Construction, Haifa, Israel, 22-25 June 1987. Volume 1. pp163-167

318
C N Cooper
CRANES - A RULE-BASED ASSISTANT WITH GRAPHICS FOR CONSTRUCTION PLANNING ENGINEERS
Proc. Application of Artificial Intelligence Techniques to Civil and Structural Engineering Conference, 1987. pp47-54
The CRANES rule-based system for selection of tower cranes for multi-storey construction sites is described. The system incorporates a procedural graphics module which allows the user to suggest locations for tower cranes on a graphic display.

319
J Marusic and V Ziljak
SIMULATION AND DETERMINATION OF RESOURCE REQUIREMENTS IN CRANE OPERATION
Proc. CIB 10th Triennial Congress, Washington, September 1986. Volume 9. pp3957-3964

320
Health and Safety Executive
TRAINING OF CRANE DRIVERS AND SLINGERS
1986. HMSO. pp6

321
ERECTING A CASE FOR HOMING IN ON TOWER CRANES
Contract Journal 1986 331 May 29, pp23-24
The reasons are discussed for UK builders ignoring the virtues of self-erecting tower cranes.

322
C Gray and J Little
SYSTEMATIC APPROACH TO THE SELECTION OF AN APPROPRIATE CRANE FOR A CONSTRUCTION SITE
Construction Management and Economics 1985 3 (2) Autumn, pp121-144
One way is described by which expert knowledge of the

criteria used to select appropriate cranage can be made available during the early design process to enable the designer to incorporate the construction implications in his thinking. However the subject is complex since there are many thousands of cranes available in the world market place. Nevertheless the process and criteria for selection are definable and they are described. The description has been incorporated in a computer-based expert system.

(i) Dryers

323
H Wylde
CHOOSING PLANT TO HEAT AND DRY OUT BUILDINGS
Building Trades Journal 1984 188 November 15, pp16-17

324
H Schlick
TEMPORARY HEATING IN CONSTRUCTION CONTRACTS
ASCE Journal of Construction Engineering and Management 1983 **109** (4) December, pp447-459
Three case studies with varying structures and heating systems, show how heating costs are affected by changes beyond the general contractor's control. These costs can be substantial; however, they can be minimised if owner, designer, contractor and mechanical sub-contractor can co-operate.

325
H Wylde
DRY WARM BUILDINGS HELP THE FINISHING TRADES
Building Today 1987 194 December 10, pp18-19
A review is made of the operational performance of space heaters and dehumidifiers.

(j) Excavators and earth moving plant

326
C B Tatum and A T Funke
PARTIALLY AUTOMATED GRADING : CONSTRUCTION PROCESS INNOVATION
ASCE Journal of Construction Engineering & Management 1988 114 March, pp19-35
The partial automation of earthmoving equipment using laser-guided controls is described.

327
J Christian and H Caldera
DEVELOPMENT OF A KNOWLEDGE BASED EXPERT SYSTEM FOR THE SELECTION OF EARTHMOVING EQUIPMENT
Proc. Application of Artificial Intelligence Techniques to Civil and Structural Engineering Conference, 1987. pp55-59

328
B C Paulson Jnr and H Sotoodeh-Khoo
EXPERT SYSTEMS IN REAL-TIME CONSTRUCTION OPERATIONS
Managing Construction Worldwide. Volume 1. Systems for managing construction.
1987. Spon/CIOB/CIB. pp554-565
Attention is focused on the instrumentation and monitoring of earthmoving scraper operations as a way of implementing the non-linear optimisation method 'load-growth curve'.

329
S Burdett
BUILDERS' PLANT, MINI-EXCAVATORS: UK FIGHTS BACK
Building Trades Journal 1987 194 September 24, pp23-24

330
H Wylde
MIGHTY MINIS DIG DEEPER INTO THE MARKET
Building Trades Journal 1987 194 August 13, pp14-16

331
EXCAVATORS
Contract Journal 1986 333 September 18, pp34-37, 39-42, 44-45, 48-51, 54, 56
A series of articles review the market range of excavators. A user's guide to models and makers is included.

332
D G Carmichael
SHOVEL-TRUCK QUEUES : A RECONCILIATION OF THEORY AND PRACTICE
Construction Management & Economics 1986 4 Autumn, pp161-177
Application is discussed of queueing theory to shovel-truck type operations.

333
S Kirupananther and F C Harris
PLANT SELECTION FOR TRENCHING WORK BY COMPUTER GRAPHICS
International Journal of Construction Management & Technology 1986 1 (1), pp68-72

334
H Wylde
DIG INTO THE MICRO EXCAVATOR MARKET
Building Trades Journal 1985 190 September 12, pp24-25
A review is made of micro excavators on the market and their performance.

335
H Wylde
TRENCHING INTO SERVICE UTILITY WORK
Building Trades Journal 1985 190 August 22, pp18-19
The use is outlined of trenching machines as a builder's tool.

336

BIGGER ROLE FOR SMALL EARTH MOVERS

Contract Journal 1985 326 July 11, pp19, 21-22, 24-26, 28-29, 32, 34-35, 37-38, 40, 42

A series of articles is presented describing the range of plant available as excavators, loaders and dumpers.

337

Health and Safety Executive

EXCAVATORS USED AS CRANES

Guidance Note PM42. 1984. pp4

The Note provides an interpretation of the Certificate of Exemption No.CON(LO)/1981/2 (General) 'Excavators, loaders and combined excavator/loaders' issued under the Construction (Lifting Operations) Regulations 1961.

338

L Bernold and D W Halpin

MICRO COMPUTER COST OPTIMISATION OF EARTHMOVING OPERATIONS

Proc. 4th Int. Symposium Organisation and Management of Construction, Waterloo, Canada, July 1984. Volume 2. pp333-341

339

MINI EXCAVATORS

Contract Journal 1984 317 February 23, ppS1-S7, S9, S12, S14-S16, ii-iv, viii-xiii, xv

The survey evaluates the performance and capabilities of 11 leading mini excavators in the 2t-3.5t range.

340

WHERE TO BUY GUIDE PART 1. EXCAVATORS

Construction Plant and Equipment 1984 12 January, pp49-51, 53-54, 56, 59, 61-63, 65, 67-68, 71, 73-74, 76, 79-80

(k) Fork lift trucks

341

H Wylde

FORKLIFTS FOR BUILDERS CAN TAKE A LOAD OFF YOUR MIND - THE CASE FOR ROUGH TERRAIN FORKLIFTS AND TELESCOPIC HANDLERS

Building Today 1988 196 December 1, pp40-41

342

L Tootell

HOW THE 'BACK TO FRONT TRACTOR' REVOLUTIONISED MATERIALS HANDLING

Contract Journal 1985 326 August 29, pp16-18

The development is traced of the rough terrain forklift truck.

343

L Tootell

MATCHING MACHINE CAPABILITIES TO SITE CONDITIONS AND TASKS

Contract Journal 1985 326 August 29, pp19-23,26

A market survey is made of rough terrain fork lift trucks. Tabulated data are presented for vertical mast and telescopic/mono boom machines.

(l) Dumpers

344

H Wylde

BRITISH WORKHORSE PLOUGHS EVER ONWARDS

Building Today 1988 195 April 21, pp30-32

A survey of dumpers and their relative specification.

345

DUMPERS

Contract Journal 1986 November 20, pp31-37

A series of articles is presented covering the development of dumpers over the past 40 yerars. A list of available plant and its specification is tabulated.

346

H Wylde

REVIEW OF THE UK DUMPER MARKET

Building Trades Journal 1986 191 March 20, pp18,20,24

(m) Hand tools

348

D W Cheetham

OVERCOMING THE BARRIERS TO THE USE OF HAND HELD TOOLS ON UK HOUSING CONSTRUCTION SITES

Managing Construction Worldwide. Volume 2. Productivity and Human Factors in Construction. 1987. Spon/CIOB/CIB. pp823-834

349

D W Cheetham and A D Hall

POWER TOOLS : AN INTEGRAL PART OF THE CONSTRUCTION PROCESS

Building Trades Journal 1987 193 June 11, pp23-24

The use is described of hand-held power tools for the housing market.

350

D W Cheetham and A D Hall

KEY TO PRODUCTIVITY IMPROVEMENTS THROUGH THE USE OF HAND-HELD POWER TOOLS

Proc. 4th Int. Symposium on Organisation and Management of Construction, Waterloo, Canada, July 1984. Volume 2. pp547-555

The usage of small hand-held power tools on UK construction sites has been found to be variable and often less than it might be for the achievement of higher productivity. The reasons for this lie partly with management and site attitudes and practices and partly with equipment deficiencies. Though the tools have become highly sophisticated, the power distribution systems have not. The problems are made worse by unsatisfactory communications between the manufacturers and the end-users, mainly due to the marketing system adopted. The influence of the fragmented nature of the construction industry is considered, and some possible remedies to the problems proposed.

(n) Lighting

351
H Wylde
MOBILE FLOODLIGHTING SETS FOR SITE WORK
Building Trades Journal 1985 *189* February 4, pp31-33

(o) Lorries

352
G Anderson
TROUGH-BODY TRUCKS FOR READY MIXED. EXPERIENCES, ADVANTAGES AND PROSPECTS
Building Research & Practice 1987 *15* January/February, pp22-25

353
L Richardson
BUILDER'S TRANSPORT : BENEFITS OF CONTRACT HIRE FOR SMALLER FIRMS
Building Trades Journal 1985 *190* August 1, pp27-28

354
BUILDER'S TRANSPORT : CHOOSING YOUR IDEAL VAN
Building Trades Journal 1985 *190* August 1, pp14,16-17,20
Details of the major makes available are tabulated.

355
BUILDER'S TRANSPORT : LEASING CARRIES THE BENEFITS OF UNRESTRICTED USE
Building Trades Journal 1984 *188* August 2, p17

(p) Loaders

356
H Wylde
SKID STEERS ON TOP OF THE WORLD
Building Today 1987 *194* November 5, pp25-27
Survey of skid steer loaders.

357
R C Ringwald
BUNCHING THEORY APPLIED TO MINIMISE COST
ASCE Journal of Construction Engineering & Management 1987 *113* (2) June, pp321-326
Bunching (Queue's) theory is developed into a quickly applicable set of curves.
When these curves are realistically applied, they assure the most economical match-up of hauler fleet size per loader.

358
BUILDERS' PLANT. REVIVAL OF INTEREST IN TOUGH AND VERSATILE SKID STEER
Building Trades Journal 1987 *194* September 24, pp18-20

359
H Wylde
MORE POWER TO YOUR WHEELBARROW
Building Trades Journal 1986 *192* October 9, pp28,30
Powered barrows are reviewed.

360
COMPACT WORKHORSES LEAD MARKET TREND
Contract Journal 1986 *332* August 7, pp20-21,23,25-27,29-35
A series of articles cover the range and performance of skid steer loaders available.

361
1.5m3 - 2m3 WHEELED LOADERS
Contract Journal 1986 *329* February 27, ppS1-S16,i-xvi
A comparative evaluation is made of one British and six European wheeled loaders.

362
BUILDERS' TRANSPORT
Building Trades Journal 1984 *188* August 2, pp11-13
Summary in tabular form types of transport available and its specification.

363
BUILDERS' TRANSPORT. CONTRACT HIRE IS A 'TOTAL' SOLUTION
Building Trades Journal 1984 *188* August 2, p20

(q) Power supplies

364
H Wylde
MORE POWER FOR DARK DAYS
Building Today 1989 *198* November 2, pp30-31,33
A review is presented of available portable generators.

365
H Wylde
AIR SUPPLY WHERE AND WHEN YOU NEED IT
Building Trades Journal 1987 *193* June 25, pp27-28
A review is made of available air compressors.

366
H Wylde
GENERATING A TEMPORARY POWER SUPPLY
Building Trades Journal 1986 *192* December 11, pp21-23

367
MOBILE SITE POWER
Contract Journal 1985 *323* February 28, pp34-35,37-38,40-41
A review is made of commercially available electricity generators.

368

H Wylde

PRACTICAL BENEFITS OF PORTABLE GENERATORS

Building Trades Journal 1984 *188* July 12, pp20-21
Details are provided of the range of portable generators available.

369

H Wylde

PLANT COMPARISON : A PORTABLE COMPRESSOR IS AN ESSENTIAL TOOL

Building Trades Journal 1984 *187* March 29, pp45-46

370

Health and Safety Executive

ELECTRICITY ON CONSTRUCTION SITES - GUIDANCE NOTE GS24

1983. HMSO. pp4
The precautions are considered which should be taken to avoid electric shock hazards on site.

(r) Pumps

371

H Wylde

PUMPS FOR HIRE

Building Today 1989 *198* October 19, pp32-33
The range of pumps offered on hire by Selwood are described.

372

H Wylde

PREPARING FOR THE WET WEATHER

Building Today 1988 *196* September 29, pp28-29
A review is made of available pumps and their performance parameters.

373

H Wylde

CHOOSING A SUITABLE WATER PUMP TO MEET REQUIREMENTS

Building Trades Journal 1984 *188* September 6, pp34-35

374

R Swan

KEEPING THE SITE DRY

Contract Journal 1987 January 22, pp19-22
A review is made of pumping equipment.

(s) Sweepers

375

H Wylde

MAKING A CLEAN SWEEP

Building Trades Journal 1987 *193* March 12, pp30-31
A short review is made of road sweeping plant.

(t) Washers

376

H Wylde

PRESSURE WASHERS FLOOD MARKET

Building Trades Journal 1986 *191* April 24, pp22-24
A review of commercially available machines.

TEMPORARY WORKS

377

J J Christian and Saif U Mir

USE OF INTEGRATED MICROCOMPUTER PACKAGE FOR FORMWORK DESIGN

ASCE Journal of Construction Engineering & Management 1987 *113* December, pp603-610
A method is described that uses an integrated package in the design and optimisation of concrete formwork. The formwork design method consists of three modules, the input data, the data bank, and the calculations.

378

Construction Industry Research and Information Association

FORMWORK STRIKING TIMES - METHODS OF ASSESSMENT

Report 73. 1987. pp40

379

J Christian and S U Mir

USE OF EXPERT SYSTEMS AND SENSITIVITY ANALYSES IN FORMWORK PRODUCTIVITY AND DESIGN

Proc. Application of Artificial Intelligence Techniques to Civil & Structural Engineering Conference, 1987. pp67-69

380

Health & Safety Executive

CHECKLIST FOR SUPERVISORS AND CHARGEHANDS ERECTING FALSEWORK

1987. pp12

381

R Doughty

SCAFFOLDING

1986. Longman. pp157
A practical approach to every aspect of scaffolding from the materials used to the safety aspects of access scaffolding. There is extensive use of line diagrams.

382

Construction Industry Research & Information Association

PROPRIETARY TRENCH SUPPORT SYSTEMS

Technical Note TN/95. 1986. 3rd edition. pp26
The survey shows that where ground support is essential trench sheeting or piling is still the most flexible and widely used support but it is now often combined with proprietary piling frames and/or a hydraulic strut and

walling system. In open ground, where ground movement may be permitted, proprietary systems are primarily used to protect personnel.

383
G Ridout
TOP OF THE FORMWORK
Building 1986 *261* October 10, pp68-69
The salient points are presented of the new Concrete Society's report Formwork - a guide to good practice.

384
F C Hadipriono and Hana-Kwang Wang
ANALYSIS OF CAUSES OF FALSEWORK FAILURES IN CONCRETE STRUCTURES
ASCE Journal of Construction Engineering and Management 1986 *112* (1) March, pp112-121

385
International Council for Building Research Studies and Documentation (CIB)
MANUAL OF TECHNOLOGY. FORMWORK
1985. Report 85. pp104
The essential criteria governing the quality of formwork are considered. The manual is intended for both designers and contractors. It provides the engineer with guidance on those properties which should be specified and the contractor with information on those factors to be considered in meeting the specification.

386
R C Ringwald
FORMWORK DESIGN
ASCE Journal of Construction Engineering & Management 1985 *111* (4) December, pp391-403
The American Concrete Institute's methodology for designing formwork adequate for its loading is condensed, organised, and simplified into a two-page, step-by-step format of equations and instructions that cover most formwork design situations. Plotted design curves are shown and their limitations discussed.

387
R H Neale
USING COMPUTERS FOR TEMPORARY WORKS DESIGN
Building Technology & Management 1985 *23* July/August, pp29-32

388
Construction Industry Training Board
GUIDE TO PRACTICAL SCAFFOLDING
1985. pp139
The construction and use of basic access scaffold are detailed.

389
C A Clear and T A Harrison
CONCRETE PRESSURE ON FORMWORK
CIRIA Report No.108. 1985. pp31

390
A W Irwin and W I Sibbald
FALSEWORK. A HANDBOOK OF DESIGN AND PRACTICE
1984. Granada. pp178

391
National Association of Scaffolding Contractors
SCAFFOLDERS AND USERS GUIDE TO SAFE ACCESS SCAFFOLDING
1984. pp32

392
Timber Research and Development Association
TIMBER IN EXCAVATIONS
1984. 2nd edition. pp76
Guidance is provided on methods for assessing site conditions, choosing appropriate excavation supports for trenches, shafts and headings, and for checking that the supported excavation is correctly installed and safe to use.

393
D E Egan
CLIMBING FACADE FORMWORK
Constructional Review 1984 *57* August, pp54-61

394
D E Egan
FLYING FORMS
Constructional Review 1984 *57* May, pp52-61
The economic advantages of flying forms are indicated; examples are given of its application on a number of projects.

395
S Trivedi
SCAFFOLDING: IN FROM THE COLD
1982. University of Birmingham. Dept. of Transportation and Highway Engineering. Publication No.60. pp74
Developments in the regularising of scaffolding as a trade activity are traced.
Limitations of the Scaffolders Record Scheme are discussed and the importance stressed of adequate training and inspection, as well as adequate powers of enforcement.

ROBOTICS

396
R Kangari and T Yoshida
PROTOTYPE ROBOTICS IN CONSTRUCTION INDUSTRY
ASCE Journal of Construction Engineering and Management 1989 *115* June, pp284-301

397

M M Cusack

CONSTRUCTION ROBOTS TAKE A STEP FORWARD

Building Technology & Management 1989 27 February/March, pp7-8

A short review is made of developments in robotics for construction.

398

M A Brown

APPLICATION OF ROBOTICS AND ADVANCED AUTOMATION TO THE CONSTRUCTION INDUSTRY

Occasional Paper No.38. CIOB. 1989. pp44

The conditions are examined which have led to Japan becoming the world's most highly mechanised construction industry. This situation is contrasted with that in the UK.

399

R Kangari and D W Halpin

POTENTIAL ROBOTICS UTILISATIONS IN CONSTRUCTION

ASCE Journal of Construction Engineering and Management 1989 115 March, pp126-143

400

A Oakhill

PROGRAMMING THE FRAME OF STEELWORK

Contract Journal 1989 349 June 22, pp14-15

The potential is assessed of automation in constructional steelwork.

401

D Normill

ROBOTIC ROUNDUP - THE UNMANNED CONSTRUCTION SITE

Civil Engineering (US) 1989 59 May, pp76-77

A short review of developments in Japan.

402

G Ridout

YEN FOR ROBOTS

Building 1989 254 October 13, pp98-100

The development by Shimizu of robots and semi-automatic machines for spray fire proofing, finishing floor slabs and to position beams is reviewed.

403

T R Gregg

INTRODUCTION TO ROBOTICS

Chartered Builder 1989 1 September/October, pp30-32

404

D Chevin

ROBOTICS - AUTOMATIC ATTRACTION

Building 1989 254 September 1, pp42-43

A summary of Japanese experience in the field is given.

405

M Skibniewski and C Hendrickson

ANALYSIS OF ROBOTIC SURFACE FINISHING WORK ON CONSTRUCTION SITE

ASCE Journal of Construction Engineering and Management 1988 114 March, pp53-68

The robotisation of on-site surface finishing work, particularly sandblasting, is considered. A possible system design is presented with a cost estimate. Costs and benefits of applying the proposed robot system to on-site sandblasting are outlined and estimated. Net present value of the contractor's investment in the robot equipment is analysed. Conclusions regarding the purchase price of the robot based on its economic feasibility are drawn.

406

R K Venables

ROBOTICS - JAPANESE CONNECTIONS

ICE Proceedings 1988 84 Part 1 December, pp1320-1323

Current developments in robotics for construction and the major influence coming from Japan are discussed.

407

B Johnstone

AUTOMATION ARRIVES ON THE BUILDING SITE

Building Today 1988 195 January 21, pp23-25

The Japanese approach to site robotics is reviewed.

408

A Warszawski

RESEARCH & DEVELOPMENT IN BUILDING ROBOTICS AT THE TECHNION ISRAEL INSTITUTE OF TECHNOLOGY

Managing Construction Worldwide. Volume 3. Construction Management & Organisation Perspective. 1988. Spon/CIOB/CIB. pp274-283

409

C P Verschuren and J van der Ejik

MANIPULATORS IN THE DUTCH BUILDING INDUSTRY

Proceedings 4th International Symposium on Robotics & Artificial Intelligence in Building Construction, Haifa, Israel, 22-25 June 1987. Volume 1. pp103-118

It has been established that processing on the building site and on the shop floor provide the best opportunity for automation, with the operations of carpenter, reinforcement fixer, bricklayer and concrete worker being the most appropriate. Although robotics were considered impractical, sophisticated handling devices were considered to have a chance of success.

410

A Yarnai

SENSOR ARCHITECTURE FOR MOBILE CONSTRUCTION ROBOT

Proceedings 4th International Symposium on Robotics & Artificial Intelligence in Building Construction, Haifa, Israel, 22-25 June 1987. Volume 1. pp119-137

Sensor requirements and form for a semi-autonomous, self-guiding mobile construction robot are discussed.

411
M J Skibniewski
CONSTRUCTION ROBOTICS RESEARCH AT PURDUE
Proceedings 4th International Symposium on Robotics & Artificial Intelligence in Building Construction, Haifa, Israel, 22-25 June 1987. Volume 1. pp138-152
Studies are reported in relation to the design of a prototype automated construction worker for the performance of repetitive tasks; design and evaluation of flexible construction systems; and design of computerised decision aids for robot implementation.

412
I J Oppenheim and L J Petrosky
DYNAMIC STABILITY FOR MANIPULATORS USED IN CONSTRUCTION
Proceedings 4th International Symposium on Robotics & Artificial Intelligence in Building Construction, Haifa, Israel, 22-25 June 1987. Volume 1. pp153-162

413
A H Slocum et al
CONSTRUCTION AUTOMATION RESEARCH AT MIT
Proceedings 4th International Symposium on Robotics & Artificial Intelligence in Building Construction, Haifa, Israel, 22-25 June 1987. Volume 1. pp222-244
Current work is reported viz Wallbots - robots to build interior walls; the Blockbot - a robot to build masonry block walls, and the Shear Studwelder.

414
B Tundu et al
PRESENTATION OF EUREKA/GEO PROJECT: A FACADE WORKING ROBOT
Proceedings 4th International Symposium on Robotics & Artificial Intelligence in Building Construction, Haifa, Israel, 22-25 June 1987. Volume 1. pp94-102
The robot is aimed at carrying out finishing work on facades. The general organisation of the project, the viewed structure of the GEO robot and the way in which it will be programmed are described.

415
A Warszawski and H Argaman
TEACHING ROBOTICS IN BUILDING
Proceedings 4th International Symposium on Robotics & Artificial Intelligence in Building Construction, Haifa, Israel, 22-25 June 1987. Volume 1. pp274-286
Attention is focused on laboratory arrangements adapted to the special features of the building task.

416
J L Crowley
STATE OF THE ART IN MOBILE ROBOTICS
Proceedings 4th International Symposium on Robotics & Artificial Intelligence in Building Construction, Haifa, Israel, 22-25 June 1987. Volume 1. pp364-378

417
To-Choi Lan and Z S Roth
DYNAMICS OF ROBOT MANIPULATION WITH WHEELED BASE
Proceedings 4th International Symposium on Robotics & Artificial Intelligence in Building Construction, Haifa, Israel, 22-25 June 1987. Volume 1. pp379-401

418
B Atkin
POTENTIAL OF ROBOTICS IN DESIGN AND CONSTRUCTION
Chartered Quantity Surveyor 1987 9 January, p9
Developments in the USA and Japan are reviewed.

419
G Atkinson
DESIGNED BY HUMANS - BUILT BY ROBOTS
Building 1987 252 February 20, pp50-51
The potential for robotics on construction sites is discussed.

420
W L Whittaker
CONSTRUCTION ROBOTICS : A PERSPECTIVE
CAD & Robotics in architecture & construction. Proc. Joint International Conference, Marseilles, 25-27 June 1986. pp105-112
Needs, prospects, challenges and goals of construction robotics are presented. They suggest an evolutionary approach to the development of advanced construction machines, building on developments in parallel domains.

421
Y Hasegawa
ROBOTISATION OF REINFORCED CONCRETE BUILDING CONSTRUCTION IN JAPAN
CAD & Robotics in architecture & construction. Proc. Joint International Conference, Marseilles, 25-27 June 1986. pp113-121

422
A H Slocum
DEVELOPMENT OF THE INTEGRATED CONSTRUCTION AUTOMATION METHODOLOGY (ICAM)
CAD & Robotics in architecture & construction. Proc. Joint International Conference, Marseilles, 25-27 June 1986. pp133-149
ICAM is shown to be able to assist on achieving high levels of integration and automation of construction processes using existing technology.

423
R Kangari
MAJOR FACTORS IN ROBOTISATION OF CONSTRUCTION OPERATIONS
CAD & Robotics in architecture & construction. Proc. Joint International Conference, Marseilles, 25-27 June 1986. pp151-158

A systematic approach to the study of the feasibility of robotics in construction is presented. A knowledge-based expert system is introduced which allows contractors to perform a preliminary study for automation.

424
W L Whittaker and E Bandari
FRAMEWORK FOR INTEGRATING MULTIPLE CONSTRUCTION ROBOTS
CAD & Robotics in architecture & construction. Proc. Joint International Conference, Marseilles, 25-27 June 1986. pp159-164

425
M C Wanner
ROBOTICS IN CONSTRUCTION: STATE OF THE ART IN THE FEDERAL REPUBLIC OF GERMANY
CAD & Robotics in architecture & construction. Proc. Joint International Conference, Marseilles, 25-27 June 1986. pp165-168
Three systems in the field of robotics are described; hydraulic excavator; mast for pumping concrete; and a mast with a platform for repair work.

426
L E Bernold
AUTOMATION & ROBOTICS IN CONSTRUCTION: THE SEARCH FOR POTENTIALS
Proc. CIB 10th Triennial Congress, Sept.22-26, 1986, Washington. Volume 3. pp1073-1082

427
R Kangari
SOCIO-ECONOMIC ASPECTS OF ROBOTISATION
Proc. CIB 10th Triennial Congress, Sept.22-26, 1986, Washington. Volume 3. pp1083-1091

428
M J Skibmiewski and C T Hendrickson
ECONOMIC ANALYSIS OF A ROBOTIC CONSTRUCTION SANDBLASTING PROCESS
Proc. CIB 10th Triennial Congress, Sept.22-26, 1986, Washington. Volume 3. pp1098-1106

429
T Ueno et al
CONSTRUCTION ROBOTS FOR SITE AUTOMATION
CAD & Robotics in architecture & construction. Proc. Joint International Conference, Marseilles, 25-27 June 1986. pp259-268
The development is described of robots for fireproofing structural steel, outer balustrade wall finishing, floor finishing and positioning steel beams.

430
M Skibniewski et al
COST & DESIGN IMPACT OF ROBOTIC CONSTRUCTION FINISHING WORK
CAD & Robotics in architecture & construction. Proc.

Joint International Conference, Marseilles, 25-27 June 1986. pp169-186
The application potential is explained of robotics to cleaning, derusting, descaling, coating, painting, sand-blasting and other surface applications.

431
R F Woodbury et al
GEOMETRY & DOMAIN MODELLING FOR CONSTRUCTION ROBOTS
CAD & Robotics in architecture & construction. Proc. Joint International Conference, Marseilles, 25-27 June 1986. pp187-192

432
C F Earl
GRAMMARS, DESIGN & ASSEMBLY IN BUILDING
CAD & Robotics in architecture & construction. Proc. Joint International Conference, Marseilles, 25-27 June 1986. pp193-201
Descriptions of architectural design in terms of building assembly operations are examined and the use of these descriptions for evaluating automatic construction considered.

433
M Kano
SIMULATION METHODOLOGY IN CONSTRUCTION PROCESS
CAD & Robotics in architecture & construction. Proc. Joint International Conference, Marseilles, 25-27 June 1986. pp203-213
A simulation method is described for the construction process on building sites.

434
J L Crowley
NAVIGATION & WORLD MODELLING FOR A MOBILE ROBOT: A PROGRESS REPORT
CAD & Robotics in architecture & construction. Proc. Joint International Conference, Marseilles, 25-27 June 1986. pp215-224

435
W L Whittaker and B Motazed
EVOLUTION OF A ROBOTIC EXCAVATOR
CAD & Robotics in architecture & construction. Proc. Joint International Conference, Marseilles, 25-27 June 1986. pp233-239

436
N Tanaka et al
DEVELOPMENT OF THE MARK II MOBILE ROBOT FOR CONCRETE SLAB FINISHING
CAD & Robotics in architecture & construction. Proc. Joint International Conference, Marseilles, 25-27 June 1986. pp249-257

BUILDABILITY

437
PRACTICAL BUILDABILITY
1989. CIRIA/Butterworth. pp122
The principles and practice of buildability are reviewed. Particular emphasis is placed in the importance of understanding contractors' problems at the design stage and the benefits of early discussion between designers and builders. The value of incorporating buildabilty into the design of projects from the outset is demonstrated through numerous examples and case studies at the level both of the overall design concept and the detailing of elements. From there 16 design principles are identified against which any design or detail can be started.

438
C B Tatum
IMPROVING CONSTRUCTIBILITY DURING CONCEPTUAL PLANNING
ASCE Journal of Construction Engineering & Management 1987 *113* (2) June, pp191-207
The results are given of an investigation of industrial and building projects undertaken to identify the steps progressive owners, designers, and contractors take to improve project constructibility during the early phases of a project. Using information from 15 projects, it describes the major concerns, approaches used, and constructibility improvements related to three key issues: developing the project plan, laying out the site, and selecting major construction methods.

439
J T O'Connor et al
CONSTRUCTIBILITY CONCEPTS FOR ENGINEERING AND PROCUREMENT
ASCE Journal of Construction Engineering & Management 1987 *113* (2) June, pp235-248
Constructibility is the optimum use of construction knowledge and experience in planning, engineering, procurement and field operations to achieve overall objectives. Seven concepts for improving constructibility during the engineering/procurement phase of a project are presented and analysed.

440
A Griffith
INVESTIGATION INTO FACTORS INFLUENCING BUILDABILITY AND LEVELS OF PRODUCTIVITY FOR APPLICATION TO SELECTING ALTERNATIVE DESIGN SOLUTIONS - A PRELIMINARY REPORT
Managing Construction Worldwide. Volume 2. Productivity and Human Factors in Construction. 1987. Spon/CIOB/CIB. pp646-657

441
I Ferguson
BUILDABILITY. THE INFLUENCE OF DESIGN UPON BUILDING METHOD.2. CASE STUDY: A STEEL FRAMED BUILDING
Building Technology & Management 1987 25 February/March, pp32-34

442
A Griffith
BUILDABILITY - THE EFFECT OF DESIGN AND MANAGEMENT ON CONSTRUCTION
Proc. CIB 10th Triennial Congress, Washington, September 1986. Volume 8.
pp3504-3512

443
J T O'Connor et al
COLLECTING CONSTRUCTIBILITY IMPROVEMENT IDEAS
ASCE Journal of Construction Engineering & Management 1987 *112* (4) December, pp463-475
Project constructibility improvement data collection techniques including voluntary survey, questionnaires, interviews, preconstruction meeting notes, and final project reports are discussed and analysed in detail. Constructibility data collected on a large refinery expansion project solicited with various collection techniques from various project personnel are presented.

444
I Ferguson
BUILDABILITY: THE INFLUENCE OF DESIGN ON BUILDING METHOD
Building Technology & Management 1986 24 December, pp42-44

445
C Gray
'INTELLIGENT' CONSTRUCTION TIME AND COST ANALYSIS
Construction Management & Economics 1986 4 Autumn, pp135-150
The concept is researched of taking the contractor's knowledge of buildability and making it available to the design team through an intelligent Knowledge Based System.

446
S Groak
TEACHING PROCESS
Architects Journal 1986 *183* June 4, pp59-61
The BRE teaching package for the construction process, incorporating the essential elements of buildability, is assessed.

447
CAN YOUR DESIGN BE BUILT?
Civil Engineering 1986 *56* January, pp49-51
The concept of 'constructibility reviews' is discussed.

448
J T O'Connor and R L Tucker
INDUSTRIAL PROJECT CONSTRUCTIBILITY IMPROVEMENT
ASCE Journal of Construction Engineering & Management 1986 112 (1) March, pp69-82
Constructibility is defined as the ability of project conditions to enable the optimal utilisation of construction resources. Constructibility improvement ideas collected on a large refinery expansion project are analysed for content, and classification frequencies are observed. Analysis of engineering rework exposes the causes and costs of rework that occurs as a result of constructibility problems.

449
A Griffith
BUILDABILITY - TIME FOR RE-ASSESSMENT
Building Technology & Management 1985 23 October, p30

450
D Bishop
BUILDABILITY: THE CRITERIA FOR ASSESSMENT
CIOB Technical Information Service Paper No.48. 1985. pp7
The paper sets out to identify and to define criteria by which the various ways of achieving buildability might be assessed. It focuses on the operational and economic factors, ie, those factors which most directly affect the productivity of site processess and the efficiency of site management.

451
Building Research Establishment
DESIGNING FOR PRODUCTION: LECTURERS' NOTES: YEAR 1
1985. pp220
One unit of a five part teaching package to demonstrate the essential principles of buildability.

452
J T O'Connor
IMPACTS OF CONSTRUCTIBILITY IMPROVEMENT
ASCE Journal of Construction Engineering & Management 1985 111 (4) December, pp404-410
An analysis of the construction resource utilisation trade-offs, which occur from constructibility improvements, provides insight into the constructibility improvement process. Matrices of construction and engineering impacts likely to result from constructibility improvements are presented. Constructibility improvement collected on a large industrial construction project are analysed for their impact to the job. Frequencies of occurrence of both desirable and undesirable impacts are noted, as are the cost-significances of the various impact types. Constructibility strategies and methods for achieving the most cost-beneficial impacts are presented.

453
W G Curtin
BUILDABILITY - THE NEGLECTED DIMENSION OF CONSTRUCTION DESIGN
Contract Journal 1984 320 July 26, pp38-39

454
DESIGN FOR BETTER ASSEMBLY.5. CASE STUDY: ROGERS' AND ARUPS
Architects Journal 1984 180 September 5, pp87-94
The approach to high tech building as evidenced by the new Lloyds building is discussed.

455
A Griffith
DESIGN RATIONALISATION AND ITS EFFECTS ON BUILDABILITY AND PRODUCTIVITY
Proc. 4th Int. Symposium on Organisation and Management of Construction, Waterloo, Canada, July 1984. Volume 2. pp579-586

456
J R Illingworth
BUILDABILITY - TOMORROW'S NEED
Building Technology & Management 1984 22 February, pp16-19
The need for closer attention to buildability and the reasons why progress in this area has been so slow are discussed.

457
C Gray
BUILDABILITY - THE CONSTRUCTION CONTRIBUTION
CIOB Occasional Paper No.29. 1983. pp33
The results of a study reveal that there is no simple answer to the problem of evaluating the construction implications of a design. A methodology is developed for this purpose based on identifying the relationship and inter-relationship between work packages. The sequences and relationships of these work packages can be built into models of design upon which simulations of the construction process can be made.

SAFETY

General

458
STANDING COMMITTEE ON STRUCTURAL SAFETY. 8TH REPORT OF THE COMMITTEE FOR THE TWO YEARS ENDING JUNE 1989
1989. CIRIA. pp41
Among the recommendations are that: there should be greater involvement of engineers in the design of roofs and walls in domestic and other buildings to reduce the risk of wind damage; engineers should be in involved in the design of claddings of all types; preventative safety precautions are written into contractual arrangements for demolition work.

459

A Galyon III

SITE SAFETY - THE DESIGN PROFESSIONALS DILEMMA

International Construction Law Review 1989 6 April, pp186-195

460

ENGINEERING FOR SAFETY

ICE Proc. 1987 82 (Part 1) February, pp169-221

Following an introduction and a statement of the problem, the work of the Standing Committee for Structural Safety is described. This is followed by a discussion of papers presented at an ICE Symposium on 10 March 1986.

461

Health and Safety Executive

LAW ON HEALTH AND SAFETY AT WORK: ESSENTIAL FACTS FOR SMALL BUSINESSES AND THE SELF-EMPLOYED

1986. pp2

462

Health and Safety Commission

WRITING YOUR HEALTH AND SAFETY POLICY STATEMENT. HOW TO PREPARE A SAFETY POLICY STATEMENT FOR A SMALL BUSINESS

1986. HMSO. pp10

463

Health and Safety Commission

PROTECTIVE CLOTHING AND FOOTWEAR IN THE CONSTRUCTION INDUSTRY

1986. HMSO. pp10

464

D A Langford and P Webster

SITE SAFE 83. NO MAGIC ANSWER

Building Technology & Management 1986 24 August/ September, pp19-22

A review is made of the effectivness of the Site Safe 83 campaign and its success in forging management attitudes.

465

G Socrates

REASONABLY PRACTICABLE DEFENCE

Building Technology & Management 1986 24 January, p26

Reference is made to a recent court case to illustrate the dangers of an employer not complying with the requirements of an enforcement notice issued by an inspector under the Health & Safety at Work etc Act.

466

Health and Safety Executive

WRITING A SAFETY POLICY STATEMENT - ADVICE TO EMPLOYERS

1985. pp16

467

J Hinze and J Roxo

IS INJURY OCCURRENCE RELATED TO LUNAR CYCLES?

ASCE Journal of Construction Engineering & Management 1984 110 (4) December, pp409-419

Data were obtained on all industrial injuries that were reported to the Missouri Division of Worker's Compensation in the years 1980 and 1981. The analysis of these injuries showed that there is no discernable pattern of injury occurrence relative to the lunar cycle.

468

Health and Safety Commission

GUIDANCE ON HEALTH AND SAFETY ADVISORY SERVICES FOR THE CONSTRUCTION INDUSTRY. PART 1. THE NEED FOR ADVICE AND THE SERVICES AVAILABLE

1983. HMSO. pp7

An example is included of a job description for a company safety officer.

469

P Trench

SAFETY CAMPAIGNS AND THOSE LITTLE SILVER BALLS

Building 1983 265 December 23/30, pp24-25

Six major reasons are put forward to explain the industry's poor safety record. They are lax supervision on site; lip service and lamentable support from management; operative stupidity; low significance given to design; liability shuffle; and the limited success of the HSE.

Accident economics

470

A Laufer

CONSTRUCTION SAFETY: ECONOMICS, INFORMATION AND MANAGEMENT INVOLVEMENT

Construction Management and Economics 1987 5 (1) Spring, pp73-90

In order to induce top management of construction companies in Israel to become involved in safety, the present research undertook to prove that in Israel, too, uninsured accident costs are very high. Through in-depth interviews of 50 site managers concerning 210 accidents, complementary interviews with other officials in construction and insurance companies, study of records, and conduct of direct observations of accidents after they occurred, a conclusion contradicting the accepted view in insurance circles is reached. The research establishes that uninsured costs in construction companies with an average safety performance record amount to a mere 0.76% of the payroll. In a parallel research in large, developed companies in the USA, existing safety measurement methods feeding the information system were studied.

471
E Leopold
COST OF ACCIDENTS IN THE BRITISH CONSTRUCTION INDUSTRY
Proceedings 1982 Bartlett International Summer School, London 1983. pp1.28-1.38
The case is argued that employers in construction transfer a proportionally greater share of the 'welfare' costs of employment (including accident costs) on to the public purse.

Accident statistics

472
Health and Safety Executive
BLACK SPOT CONSTRUCTION. A STUDY OF FIVE YEARS FATAL ACCIDENTS IN THE BUILDING AND CIVIL ENGINEERING INDUSTRIES
1988. HMSO. pp44
The survey demonstrates that most fatalities could have been avoided by paying closer attention to the organisation of work.

473
J Lomax
KEEP SAFETY IN YOUR SITES. SAFETY - SOME SUITABLE CASES FOR TREATMENT. PART 1
House Builder 1987 46 July/August, pp66-68
Statistics on the incidence of accidents on house building sites are followed by some illustrative case histories.

474
R Bunn
SAFETY LAST?
Building Services 1987 9 June, pp31-32
Some of the major statistics relating to safety are reviewed.

475
Health and Safety Executive
STATISTICS - HEALTH & SAFETY 1981-82
1985. HMSO. pp74

476
National Association of Scaffolding Contractors
SAFETY REPORT 1983
1985. NASC. pp13
The report presents a series of statistical data covering members of the Association. The incidence of accidents was lower than in 1982 but three fatalities enhanced the fatal incident rate to 44.06.

477
Health and Safety Commission
GUIDANCE ON THE COLLECTION AND USE OF ACCIDENT INFORMATION IN THE CONSTRUCTION INDUSTRY
1983. HMSO. pp6

478
Health and Safety Executive
CONSTRUCTION HEALTH & SAFETY 1981-82
1983. HMSO. pp87
In addition to providing statistical data on construction site accidents, sections are included on co-operating for safety, planning safe systems of work, health problems and the dangers of site transport.

Regulations and legislation

479
B Warner
SAFETY FIRST
Building Today 1989 198 November 9, pp24-25,28
Typical responses by builders to the Control of Substances Hazardous to Health Regulations. A management action plan as detailed in the BEC guide is included.

480
G Ridout
UNDER THE COSHH
Building 1989 254 October 6, pp30-31
A summary of the regulations for the Control of Substances Hazardous to Health and their complications for building sites.

481
R Alban
REGULATIONS. ELECTRICITY AT WORK 2
Building Today 1989 198 October 26, p42

482
R Alban
ELECTRICITY AT WORK 3
Building Today 1989 198 November 2, pp33-34
Safer systems of work, competence and training are discussed in relation to the Electricity at Work Regulations 1989 Act.

483
S Hazell
SITE SAFETY - A NEW APPROACH
1988. Titmuss, Sainer and Webb. pp7
The main areas are highlighted of concern surrounding construction site safety. Several proposals currently being considered or implemented in the UK and EC to improve and harmonise site safety standards are outlined.

484
G Socrates
NEW ACCIDENT NOTIFICATION REGULATIONS
Building Technology & Management 1986 24 March, pp7,10
The regulations coming into force in April relate to the reporting of injuries, diseases and dangerous occurrences.

485
Health and Safety Executive
**SAFE ERECTION OF STRUCTURES. PART 4.
LEGISLATION AND TRAINING**
1986. HMSO. pp16

486
Building Advisory Service
CONSTRUCTION REGULATIONS 1961 and 1966
1985. BEC/BAS. pp62
A guide to the metricated Construction Regulations.

487
Health and Safety Executive
**GUIDE TO THE ASBESTOS (LICENSING)
REGULATIONS 1983**
1984. HMSO. pp27

488
WHEN SAFETY REGULATIONS DO NOT APPLY
Times 1984 June 8, p13
In the case of *Gardiner v Thames Water Authority* it was
held that Regulation 38 of the Construction (Working
Places) Regulations which was expressed to apply only
where compliance with the requirements of certain
other regulations was impracticable, does not apply
where compliance with those other regulations was, on
the facts, excused or excluded by the regulations.

489
B R Norton
**RESPONSIBILITIES OF CONTRACTORS,
SUB-CONTRACTORS AND OTHERS UNDER THE
HEALTH & SAFETY AT WORK ETC ACT 1974**
CIOB Technical Information Service Paper No.40.
1984. pp7
The inter-related responsibilities as identified under the
Act are discussed, reference being made to the imp-
lications of the *Swan Hunter* case.

490
Health and Safety Commission
**ELECTRICITY AT WORK (OTHER THAN AT
MINES AND QUARRIES). DRAFT ELECTRICITY
AT WORK REGULATIONS 198-. DRAFT
APPROVED CODE OF PRACTICE AND
GUIDELINES 1984.**
HMSO. pp26

Building Operations

491
T Niskanen and J Lauttalammi
**ACCIDENT PREVENTION IN MATERIALS
HANDLING AT BUILDING CONSTRUCTION
SITES**
Construction Management & Economics 1989 *7*
Autumn, pp263-279

A practical philosophy towards safe materials handling
on site is presented. Appendices provide a check list for
safety management and recommendations for planning
and technical factors in safe materials handling.

492
R Alban
BURIED CABLES CAN KILL
Building Today 1989 *198* September 21, pp36-37,40-41
The dangers of buried cables and means whereby acc-
idents can be avoided are discussed.

493
H Wylde
SHORE IT, DON'T CHANCE IT
Building Today 1989 *198* August 10, pp25-26
Safe support systems for trench work are reviewed.

494
Health and Safety Executive
**AVOIDING DANGER FROM UNDERGROUND
SERVICES**
1989. HMSO. pp24

495
J Hinze
QUALITIES OF SAFE SUPERINTENDENTS
ASCE Journal of Construction Engineering & Manage-
ment 1987 *113* (1) March, pp169-171
The relationship between managerial characteristics
and safety records is examined.

496
J Lomax
**THERE'S BEEN AN ACCIDENT - WHAT SHOULD
YOU DO?**
House Builder 1987 *46* November, pp84-86
A case history is presented to illustrate the action that
should be taken in the event of an accident on site.

497
E Sawacha and D Langford
**ATTITUDES TO SAFETY REPRESENTATIVES
AND COMMITTEES ON CONSTRUCTION SITES**
International Journal of Construction Management &
Technology 1987 *2* (4), pp3-11
There has been, in the UK, an ever-increasing set of
statutory regulations regarding safety. Research in-
dicates the importance of attitudes to safety, the use of
a safety representative on site and that the attitudes of
operatives to safety are vital.

498
S Ashley
HOW TO MAKE BUILDING SAFE
Building Services 1987 *9* June, pp32-34
Some of the industry's blind spots in its approach to
safety management are highlighted.

499

National Joint Council

SITE SAFE AND YOUR HEALTH: A GUIDE TO HEALTH AND WELFARE PRACTICES FOR THE SITE TEAM

1988. pp54

500

BSI CODE OF PRACTICE FOR SAFETY IN ERECTING STRUCTURAL FRAMES

BS 5531: 1988. pp36

501

Health and Safety Executive

CONSTRUCTION SUMMARY SHEET:

Scaffolds, pp2; Safe use of ladders, pp2; Safety in excavation, pp2; Tower scaffolds, pp2; Safety in roofwork, pp2; Safe use of propane and other LPG cylinders, pp2; Portable electric tools and equipment, pp2; Avoiding danger from buried services, pp2.

502

J Hinze and P Raboud

SAFETY ON LARGE BUILDING CONSTRUCTION PROJECTS

ASCE Journal of Construction Engineering & Management 1988 *114* June, pp286-293

503

HEALTH AND SAFETY EXECUTIVE

Build Safety 1988, pp16

A series of features for smaller firms indicating the benefits of applying a sensible and responsible approach to safety on site.

504

J Stokdyk

PRIVATE INVESTIGATIONS

Building 1988 *253* October 21, pp92-93

Practical safety precautions on site as observed by safety professionals are discussed.

505

J Stokdyk

WAS THE BLITZ WORTH IT?

Building 1988 *253* October 7, pp72-73

The approach by the Health and Safety Executive is evaluated of setting up a campaign of site visits to check on safety precautions.

506

G Ridout

SAFETY IN NUMBERS

Building 1988 *253* October 14, pp60-61

The industry's apparent disregard for safety precautions - with a resultant continuing increase in site accidents - is discussed.

507

British Gas Corporation

PRECAUTIONS TO BE TAKEN WHEN CARRYING OUT WORK IN THE VICINITY OF UNDERGROUND GAS PIPES. ADVICE TO MANAGEMENT AND SITE PERSONNEL

1986. pp5

508

R Ormerod

BUILDING ON SITE. SAFE AS HOUSES

Building 1986 *261* October 17, pp61-64

Report of a visit to a housing site where safety is under the supervision of consultants.

509

A Laufer and W B Ledbetter

ASSESSMENT OF SAFETY PERFORMANCE MEASURES AT CONSTRUCTION SITES

ASCE Journal of Construction Engineering & Management 1986 *112* (4) December, pp530-542

Various methods for the measurement and classification of safety performance at construction sites, eg, timing relative to the moment of accident, data collection method, safety effectiveness criterion, performance measure, and frequency and severity of the measured event, are analysed. The effectiveness of the various methods and the extent of their use at construction sites are examined. Attributes that are investigated include efficiency, reliability, and validity and diagnostic capacity of the measure in order to identify the cause for success or failure, respectively, of the safety programme at a site.

510

M B Hill

HEALTH AND SAFETY AT WORK - DLOs AND THE CONTRACTOR

Building & Maintenance 1986 (4), pp18-20

The responsibilities are outlined of a manager for his subordinates and advice given on setting up effective safety procedures.

511

Y J Lovell (Holdings) plc

SAFETY PROCEDURES FOR SUB-CONTRACTORS

1986. pp53

In addition to giving sound practical advice to sub-contractors on statutory requirements, the company's safety requirements are defined, that need to be complied with as part of the sub-contractor's contractual undertaking.

512

P Guest

BETTER SAFE ...

Building 1986 *261* October 17, p33

The manner in which management contracting is contributing to safety problems is discussed with reference to three sites.

513

Health and Safety Executive
**SAFE ERECTION OF STRUCTURES. PART 3.
WORKING PLACES AND ACCESS**
1986. HMSO. pp19
Advice is given on methods to minimise the need to
work at heights; on choosing means of safe access, egress and working at heights; and on systems, methods
and devices to assist in preventing accidents at heights.

514

Health and Safety Executive
**SAFE ERECTION OF STRUCTURES. PART 2. SITE
MANAGEMENT AND PROCEDURES**
1985. HMSO. pp19

515

Health and Safety Executive
**AVOIDING DANGER FROM BURIED
ELECTRICITY CABLES**
Guidance Note GS33. 1985. HMSO. pp10

516

Building Employers Confederation
**GUIDANCE NOTE ON SAFETY
RESPONSIBILITIES FOR SUB-CONTRACTORS ON
SITE**
1984. pp8

517

Health and Safety Executive
**SAFE ERECTION OF STRUCTURES. PART 1.
INITIAL PLANNING AND DESIGN**
1984. HMSO. pp8

518

R Ord
**PROJECT MANAGEMENT - ORGANISING FOR
SAFETY**
Building Technology & Management 1984 22 July/
August, pp9-10
The approach to be adopted by a management contrac-
tor towards site safety is described.

519

V Powell-Smith
BEWARE OF TRESPASSERS
Contract Journal 1984 320 August 16, p25
The responsibility for an unauthorised visitor's safety
on site is discussed in relation to the Occupiers Liability
Act 1984 which came into force on 13 May 1984. At-
tention is focussed on the cases of *British Railways
Board v Herrington; Wheat v Lacon & Co Ltd*; and *Pan-
nett v McGuiness & Co Ltd*.

520

National Joint Utilities Group
**RECOMMENDATIONS ON THE AVOIDANCE OF
DANGER FROM UNDERGROUND ELECTRICITY
CABLES**
1979. pp14

521

Health and Safety Executive
**HEALTH AND SAFETY IN DEMOLITION WORK.
PART 1: PREPARATION AND PLANNING**
1984. HMSO. pp7

522

Health and Safety Executive
**HEALTH AND SAFETY IN DEMOLITION WORK.
PART 3: TECHNIQUES**
1984. HMSO. pp11

523

Health and Safety Executive
**HEALTH AND SAFETY IN DEMOLITION WORK.
PART 4: HEALTH HAZARDS**
1984. HMSO. pp7

524

A Lindblad
**BETTER WORKING ENVIRONMENT AT
CONVERSION OF HOUSES**
Proc. 9th CIB Congress, Stockholm, 1983. Volume 2.
Building Technology, Design and Production.
pp511-517
The contractor and his site manager normally have the
responsibility for the safety and the working environment. In recent years Bygghalsan in three different
studies, in Gothenburg and Malmo, have penetrated the
possibilities to influence the working environment at
the planning and design stage. The main opinion is that
the biggest working environment problems are hygienic
risks from dust and ergonomical problems in material
handling, heavy transport and bad working postures.

Plant

525
BSI SAFE USE OF CRANES. PART 1. GENERAL
BS 7121; Part 1. 1989. pp34

526
J Rowley
SAFETY. LOOK TO YOUR HOISTS
National Builder 1989 October, pp36

527
M Morris
**LEARNING TO LIVE WITH SAFE SCAFFOLDING.
SAFETY IN ITS ERECTION AND USE**
House Builder 1989 48April, pp52-53, 56, 60

528
Health and Safety Executive
**KEEPING OF LPG IN CYLINDERS AND SIMILAR
CONTAINERS**
1986. HMSO. pp20

529
T Smith
AVOIDING COMPRESSOR HAZARDS
Building Trades Journal 1984 188 September 6, pp30,32

530

S R Curwell and C G March (Eds)
HAZARDOUS BUILDING MATERIALS
1986. Spon. pp139
The use is reviewed in detail of materials known or suspected to involve health hazards. Alternative materials are evaluated for safety, performance and cost, providing architects, specifiers and others with the facts upon which to base an informed selection, thereby minimising risks to occupiers.

531

P W P Anderson
BRIEF REFERENCE GUIDE TO SOME CONSTRUCTION INDUSTRY HARMFUL AGENTS
1984. Institution of Occupational Safety and Health. pp24
The guide lists in tabular form the harmful agents; their properties; likely use and location; control data; and further reading.

532

RISKS AND REMEDIES - INDOOR AIR QUALITY
Architects Journal 1987 *185* June 10, pp55-61, 63-64
A series of articles covering the scope of the problems, viz allergy and infection, sick building syndrome, carcinogens, and toxins and comfort factors. Remedies deal with standards for indoor air pollution, methods of control, asbestos and the sick building.

533

RISKS AND REMEDIES. INDOOR AIR QUALTIY. 2: THE REMEDIES
Architects Journal 1987 *185* June 17. pp57-63
Attention is given to standards for indoor air quality and methods of control. Two case studies are presented on asbestos and the sick building syndrome.

534

M A Brown and H Semper
ASBESTOS IN CONSTRUCTION - A REVIEW
CIOB Technical Information Service Paper R2; 1987 pp6
A review is made of the major publications covering the use of asbestos in construction and the precautions to be taken in regard to its safe use.

535

W Penney
ASBESTOS IN BUILDING
Building Technology & Management 1987/88 *25* December/January, pp16-17
A short guide to the safety precautions to be taken in relation to asbestos used in construction.

536

Asbestos Institute
DOES THE EPA REALLY HAVE A CASE AGAINST CHRYSOTILE ASBESTOS?
1987. pp48

537

Health and Safety Executive
WORKING WITH ASBESTOS: A GUIDE FOR SUPERVISORS AND SAFETY REPRESENTATIVES
1985. pp36

538

R Doll and J Peto
EFFECTS ON HEALTH OF EXPOSURE TO ASBESTOS
1985. HMSO. pp58

539

Institute of Maintenance and Building Management
ASBESTOS II
1985. pp9
Provides in turn: a discussion paper on the sealing or encapsulation of asbestos materials in buildings; guidance notes for public authorities in the training of operatives for direct works departments.

540

Health and Safety Executive
PROBABLE ASBESTOS DUST CONCENTRATIONS AT CONSTRUCTION PROCESSES
Guidance Note EH35. 1984. HMSO. pp3

541

Health and Safety Executive
WORK WITH ASBESTOS CEMENT
Guidance Note EH36. 1984. HMSO. pp4
Information is provided on the risks of exposure to asbestos dust when working with asbestos cement and on the precautions required for personal protection.

542

Health and Safety Executive
WORK WITH ASBESTOS INSULATING BOARD
Guidance Note EH37. 1984. HMSO. pp7
Covers the identification of the material, the nature of the risks arising from its presence and from the most common work activities involving it in maintenance or other construction work; and on the precautions required.

543

International Labour Office
SAFETY IN THE USE OF ASBESTOS
1984. pp116
Guidance is given on monitoring, preventive measures, protection and supervision of the health of workers, and the containment and disposal of asbestos waste. More detailed information is given on the limitation of exposure in specific activities.

544

Institute of Maintenance and Building Management
**ASBESTOS: 1. SETTING UP DIRECTLY
EMPLOYED REMOVAL TEAMS. 2. CODE OF
PRACTICE FOR ASBESTOS REMOVAL**
1984. 2nd edition. pp49

545

L D West
**ASBESTOS IN BUILDINGS - TAKING A BALANCED
VIEW**
Paper to BMCIS Seminar 'Every maintenance man's
nightmare' September 1984. pp7-9

546

M C Osborne and T Brennan
**PRACTICAL PROBLEMS IN REDUCING RADON IN
HOUSES**
Building Research & Practice 1986 (6) November/
December, pp363-366

547

M C O'Riordan et al
**HUMAN EXPOSURE TO RADON DECAY
PRODUCTS INSIDE DWELLINGS IN THE UK. A
MEMORANDUM OF EVIDENCE TO THE ROYAL
COMMISSION ON ENVIRONMENTAL
POLLUTION**
1983. National Radiological Protection Board. pp35

548

Health and Safety Executive - Construction Industry
Advisory Committee
HAZARD INFORMATION SHEET 1. CEMENT
1985. pp2

549

Construction Industry Advisory Committee
**HEALTH HAZARD INFORMATION SHEET 4.
LEAD**
1986. Health and Safety Executive. p1

550

Construction Industry Advisory Committee
**HEALTH HAZARD INFORMATION SHEET 5.
SOLVENTS**
1987. Health and Safety Executive. p1

551

Construction Industry Advisory Committee
**HEALTH HAZARD INFORMATION SHEET 6. AIDS
1987.**
Health and Safety Executive. p1

552

Health and Safety Commission
SKIN HAZARDS
Construction Health Hazard Information Sheet No.7
1987, p1

553

Health and Safety Executive
**HEALTH HAZARD INFORMATION SHEET NO.2
COLD WEATHER**
1985. p1

554

Health and Safety Executive
HEALTH HAZARD INFORMATION SHEET: NOISE
1986. p1

SECURITY

555

M Evanny
KEEPING THE SITE SAFE AS HOUSES
Contract Journal 1988 January 28, pp25-27
Approaches to site security are described.

556

F Morris
**YOU CAN STILL WIN SOME - AND LOSE A LOT
LESS. AN OVERVIEW OF THE SECURITY
PROBLEM IN THE CONSTRUCTION INDUSTRY**
House Builder 1987 46 June, pp49,52,54

557

F Morris
SITE SECURITY. WHAT AND HOW
House Builder 1987 46 November, pp78-80

558

A Charlett
**VANDALISM AND TRESPASS BY CHILDREN ON
BUILDING SITES**
CIOB Technical Information Service Paper No.60.
1986. pp8
The different types of vandalism and trespass and their
possible motives are examined; the ways these affect the
builder on his site are considered.

559

A J Charlett
SECURITY OF BUILDING SITES
CIOB Occasional Paper No.33. 1985. pp19
Some of the factors are examined which influence the
incidence of theft and vandalism on site.

560

F Morris
UPS AND DOWNS OF SCAFFOLDING SECURITY
National Builder 1985 66 May, pp96-97
Some suggestions are made regarding preventing scaf-
folding losses.

561

K Smith
**TIGHT AND TIDY SITES KEEP PROFIT ON THE
COMPANY INVENTORY**
Contract Journal 1984 320 August 23, pp14-15

Cost effective sinmple procedures are described to prevent the theft of equipment on site.

DEFECTS

562
A J Bryan
MOVEMENTS IN BUILDINGS
CIOB Technical Information Service Paper No.R3.
1989. pp8
A framework is provided for the analysis of movements that occur in buildings by considering the causes, evaluation and assessment of movements associated with the well being of the building throughout its life. Current literature is reviewed which provides guidance and information on the use of specific materials or the design, detailing and construction of particular elements.

563
BRE
SIMPLE MEASURING AND MONITORING OF MOVEMENT IN LOW RISE BUILDINGS. PART 2. SETTLEMENT, HEAVE AND OUT OF PLUMB
Digest No.344. 1989. pp4

564
J Hinks and G Cook
WHAT CAUSES FAILURES IN DOMESTIC FLAT ROOFS
Building Today 1989 *197* March 16, pp20-26
Common faults are examined with domestic flat roofs finished with built up bitumen felt.

565
J Hinks and G Cook
DEFECTS IN FLAT ROOFS. PART 2: MASTIC ASPHALT
Building Today 1989 *197* April 13, pp40-43
Causes of defects and their remedy are identified.

566
BRE
SIMPLE MEASURING AND MONITORING OF MOVEMENT IN LOW RISE BUILDINGS. PART 1. CRACKS
Digest No.343. 1989 pp8

567
C Millett
HOT SUMMERS CAN MOVE FOUNDATIONS
Building Today 1989 *198* November 16, pp20-22
The problems likely to be experienced with foundations on clay soil following a drought are discussed. Measures to prevent the problem with new construction are described and illustrated.

568
C Briffett
DIAGNOSTIC APPROACH TO SOLVING BUILDING DEFECTS
Professional Builder (Singapore) 1989 4 September, pp5-8
An investigation procedure is outlined which allows the information obtained regarding defects to be as accurate and comprehensive as possible.

569
A Griffith
BUILDING PERFORMANCE IN TIMBER-FRAMED HOUSING
Chartered Builder 1989 1 November/December, pp5-8
The magnitude, origins and degree of severity of problems affecting performance of timber framed housing are assessed and the factors influencing their occurrence are considered. The study is based on a survey of housing in Scotland and the USA.

570
J Hinks and G Cook
DEFECTS IN PAINTWORK
Building Today 1989 198 September 21,22-23,26
Some common causes of failure are discussed.

571
T Whitehead
HOW LONG BEFORE THE CRACKS SHOW?
Contract Journal 1989 October 5, pp16-17
Concerns are expressed regarding the quality of buildings. It is suggested that existing contractual methods and design systems could lead to an avalanche of defects in the future.

572
J E Cheney
25 YEARS HEAVE OF A BUILDING CONSTRUCTED ON CLAY, AFTER TREE REMOVAL
Ground Engineering 1988 21 July, pp13-14,17-18,21-22,25-27
Long-term observations have demonstrated the longevity of problems associated with clay heave. It is stressed that a record should be made of tree species, height, position and date of removal on sites where it is intended to build.

573
M Denton
RECOGNISING PROBLEMS IN THE GROUND
House Builder 1989 48 April, pp89,92,96
Some commonly occurring problems on the building site related to ground conditions are reviewed.

574
BRE
UPDATE ON ASSESSMENT OF HIGH ALUMINA CEMENT CONCRETE
Information Paper IP 8/88. pp4

Current guidelines are summarised for assessing the conditions of buildings containing HAC concrete.

575
D Wells
LATENT PLANS FOR A FUTURE DEFECTS POLICY
Contract Journal 1988 November 3, pp14-15
NEDO's suggestion that building owners should pay to insure their properties against latent defects (BUILD) is discussed.

576
G Ridout and M Spring
CRISIS, WHAT CRISIS? LATENT DEFECTS
Building 1988 253 September 30, pp34-39
A range of options is canvassed regarding the current boom in construction and the likely incidence of latent defects. Main problem areas are identified and the solutions available are noted.

577
D F Cutler and I B K Richardson
TREE ROOTS AND BUILDINGS
1989. 2nd edition. Longman. pp71
Based on the Kew Tree Root Survey essential guidance is given on the likely damage to buildings from tree roots and on the safe planting of trees. Information is given on life expectancy, pruning tolerances, in addition to more general data on the relevant aspects of tree growth and husbandry.

578
D Bishop
SEARCH FOR A BETTER SOLUTION TO THE OVERALL MANAGEMENT OF THE AFTERMATH OF LATENT DEFECTS IN BUILDINGS
Managing Construction Worldwide. Volume 1. Systems for managing construction.
1987. Spon/CIOB/CIB. pp350-359
It is suggested that project based material damage insurance without recourse to subrogation is a practical possibility that would benefit client and the industry alike.

579
F C Hadipriono and C L Lim
RULE BASED DIAGNOSIS OF FAILURE EVENTS
Proceedings 4th International Symposium on Robotics & Artificial Intelligence in Building Construction, Haifa, Israel, 22-25 June 1987. Volume 2. pp561-574
The development is described of an expert system for failure diagnosis.

580
J R W Robinson
INCIDENCE OF RECTIFICATION OF FAULTS IN RESIDENTIAL BUILDINGS IN MELBOURNE
Australian Institute of Building Papers 1987 2, pp31-48
As assessment is made of defects in houses in Melbourne to demonstrate the geographic ageing and environmental factors which cause deterioration in building materials. The extent of rectification undertaken by home owners is also analysed.

581
R G Kinnear
ASSESSMENT OF DEFECTS IN STRUCTURES
Paper to World Congress of Building Officials, London, October 1987. pp4
The basis for a number of typical defects in structures is discussed and a plea made for better feedback to eductional esablishments.

582
M Ross
LATENT DAMAGE: PROBLEMS UNRESOLVED
Construction Law Journal 1987 3 (2), pp71-82
The principal difficulties likely to be encountered in applying the Latent Defects Act are identified, reference being made to the interaction of tort and contract; criteria to determine the existence of damage; the measurement of the plaintiff's knowledge; burden of proof concerning limitation issues; contribution proceedings; and successive purchasers.

583
D E Jones and K A Jones
TREATING EXPANSIVE SOILS
Civil Engineering (US) 1987 August, pp62-65
It is demonstrated how the damage caused by expansive soils can be prevented or mitigated by preconstruction soil treatment.

584
K Smith
JAPAN MISSION FINDS IN FAVOUR OF A MORE CONSTRUCTIVE APPROACH TO THE ASR PROBLEM
National Builder 1987 68 September, pp298-299
A report of a study visit to Japan to see how the problems of the alkali-silica reaction in concrete is being tackled. The major drive is towards developing a coating to seal structures against water ingress. This is based on the premise that structures can be succussfully protected, and only in extreme cases need strengthening.

585
J Fielding
KIND OF DETECTIVE STORY. DIAGNOSING BULDING DEFECTS
Building Research & Practice 1987 14 March/April, pp93-96

586
J L Smith
FRAMEWORKS FOR DEFECTS DIAGNOSIS. EXPERIENCE IN DEVELOPING AN EXPERT SYSTEM
Building Research & Practice 1987 14 March/April, pp85-87

587

B G Richards et al

SOILS OF ADELAIDE AND THEIR IMPORTANCE IN FOUNDATION BEHAVIOUR

Construction Supervisor 1987 21 February, pp35,37,39

588

P L Berry et al

SETTLEMENT OF TWO HOUSING ESTATES AT ST.ANNES DUE TO CONSOLIDATION OF A NEAR-SURFACE PEAT STRATUM

ICE Proc. 1987 82 (Part 1) February, pp223-241 (Discussion)

589

CONSTRUCTION RISKS AND REMEDIES. MOVEMENT

Architects Journal 1987 185 January 21, pp57-64.67-68

The causes and effects of movement in buildings are discussed.

590

CONSTRUCTION RISKS AND REMEDIES. MOVEMENT. PART 2: THE REMEDIES

Architects Journal 1987 185 January 28, pp49-57

591

L R Davison and C J Davison

BUILDING DEFECTS ASSOCIATED WITH GROUND CONDITONS: A REVIEW

CIOB Technical Information Service Paper R1. 1986. pp11

Following consideration of the public's perception of subsidence and limits of acceptable settlement attention is given to ground movements due to loading, with particular reference to the characteristics of soil types, other causes of movement are then explored. Assessment of defects follows and the paper concludes with a short section on liability.

592

H S Staveley

DESIGN AND CONSTRUCTION DEFECTS

Building Technology & Management 1986 24 October/November, pp22-24

Some practical guidance is given on minimising the risk of defects.

593

R Johnson

FOUNDATION PROBLEMS ASSOCIATED WITH LOW-RISE HOUSING. PART 1

CIOB Technical Information Service Paper No.61. 1986. pp12

Designed to assist those involved in house building identify possible hazards on sites zoned for development. Although a comprehensive list of solutions to each problem is not feasible some are described which will enable the assessment to be made of the viability of a site.

Starting with a site checklist attention is then given to the sloping site; made ground or the filled site; clay sites; the alluvial flood plan; soils with a high sulphate and acid content; and contaminated soils.

594

R Johnson

FOUNDATION PROBLEMS ASSOCIATED WITH LOW-RISE HOUSING. PART 2

CIOB Technical Information Service Paper No.62. 1986. pp8

Covers mine and mineral workings; inner city infill sites; effect of frost on soils and foundation concrete; the hard rock site; and the non-suspended oversite slab.

595

W Allen

FAILURES ARE GOOD OPPORTUNITIES FOR LEARNING

Keynote presentation CIB 86 'Advanced Building Technology'. pp15

Three common types of building failure are examined to demonstrate ineffective transfer of knowledge, missed opportunities to identify practical problems upon which research is needed and a failure to think scientifically about large-scale technological change thrust upon the industry.

596

J Dixon

TREE ROOTS, BUILDINGS AND LOCAL AUTHORITES

Building Control 1986 (16) May/June, pp39-40

A short review is made of the effect of shrinkable clays and tree roots on foundations. The responsibility of local authorities is also considered.

597

F C Hadipriono et al

EVENT TREE ANALYSIS TO PREVENT FAILURES IN TEMPORARY STRUCTURES

ASCE Journal of Construction Engineering & Management 1986 112 (4) December, pp500-513

The enabling and triggering events that caused failures in temporary structures are examined. Using event tree analysis the propagation of component failures is analysed. The concept of fuzzy sets is used to manipulate human-based subjectibve judgments, which are abundant during the construction of temporary structures, and to minimise the complexities of conditional events in the trees. Ranking of critical paths, based on the highest probability of failure, is also presented for quality control purposes.

598

J Taylor

AVOIDING DEFECTS IN BUILDING CAUSED BY INCOMPATIBILITY OF MODERN MATERIALS

Building Technology & Management 1986 24 February, pp11-13

599
BRE
**INFLUENCE OF TREES ON HOUSE
FOUNDATIONS IN CLAY SOILS**
Digest No.298. 1985. HMSO. pp8
Design guidance is given on minimising the effects of
soil shrinkage or swelling on clay soils in the presence
of trees or following tree removal.

600
C Richardson
**STRUCTURAL SURVEYS. 2 DATA SHEETS:
GENERAL PROBLEMS**
Architects Journal 1985 *182* July 3, pp63, 65-67, 70-71
Typical symptoms, probable causes and principles of
repair of a number of commonplace structural problems
found in buildings are examined. The influence of social
and economic history on structural defects are outlined.

601
C Richardson
**STRUCTURAL SURVEYS. 3 DATA SHEETS: THE
INDUSTRIAL REVOLUTION**
Architects Journal 1985 *182* July 10, pp45-47, 49, 51-52
Typical defects to be found in buildings constructed
between 1750 and 1850 are decribed.

602
C Richardson
**STRUCTURAL SURVEYS. 4 DATA SHEETS:
COMMON PROBLEMS 1850-1939**
Architects Journal 1985 *182* July 17, pp63-65,67-69

603
C Richardson
**STRUCTURAL SURVEYS. 5 DATA SHEETS: THE
POST-WAR BUILDING BOOM**
Architects Journal 1985 *182* July 24, pp63-67,70
The causes are illustrated of typical defects in post war
buildings.

604
W A Porteous
**PERCEIVED CHARACTRISTICS OF BUILDING
FAILURE - A SURVEY OF THE RECENT
LITERATURE**
Architectural Science Review 1985 *28* (2) June, pp30-40
The results of an analysis are presented of the differen-
ces and similarieiss between the perceptions of building
failure. A list shows how the perception of such failures
has changed over the years.

605
W G Curtin and M Priauix
**TIME TO REMIND SOCIETY THAT IT WILL ONLY
GET WHAT IT PAYS FOR**
Contract Journal 1985 *327* September 19, pp14-15
It is argued that the failings in buildings are not down
to the materials but their misapplication through fin-
ancial pressures on builders and designers to come up
with cost-effective answers.

606
A Lord
**NEED TO RELATE FABRIC DEFORMATIONS TO
BOTH THE SOIL CONDITONS AND HISTORY OF
THE BUILDING**
Paper to symposium on Building appraisal, maintenance and preservation, Bath University, 10-12 July
1985. pp11
Examples are considered of a wide range of cracked
buildings from which it is concluded that, with a few
notable exceptions, the deformations seldom arise
solely from deficiencies of the foundations or weakness
of the soil.

607
Building Maintenance Cost Information Service
**DESIGN/PERFORMANCE DATA. BUILDING
OWNERS REPORTS: 1. EXTERNAL WALLS**
1985. pp56
Examples are presented of building failures to external
walls, cladding, curtain walls, windows, external doors
and screens.

608
W Allen
OVERCOMING THE PROBLEM
Paper to OYEZ/IBC seminar, 'Building defects. What
can be done and who picks up the bill?' February 1985,
London. pp74-76

609
W Curtin
**CUTTING THE COST OF HIGH-RISE - WHERE
PROFESSIONAL SKILL COUNTS**
Contract Journal 1985 *326* August 29, pp14-15
The skills and techniques associated with the diagnosis
of defects in high-rise buildings are described.

610
Institute of Clerk of Work
**ELEMENTS OF QUALITY CONTROL AND
WORKLOADS OF CLERKS OF WORKS**
1985. pp3
In addition to specifying the volume of work that may
be handled by a clerk of works and the cost, some
thoughts are expressed on how the COW might be best
utilised in limiting building failures.

611
F G Coffin
**BUILDING DEFECTS: THE TYPES OF PROBLEMS
ENCOUNTERED AND THEIR SCALE**
Building and Maintenance 1985 1 April, pp4-6
Attention is given to foundations, reinforced concrete
and external cladding.

612
FEARS FOR THE WORST
Building 1986 *268* June 7, pp28-29
A BRE report on the quality of timber-frame housing is
summarised.

613

N Royce

THE PROBLEM

Paper to the OYEZ/IBC seminar, 'Building defects. What can be done and who picks up the bill?' February 1985, London. pp6-13

615

T J Coombes

TREES IN RELATION TO BUILDINGS AND DRAINS IN CLAY SUB-SOILS

Building Control 1985 (8) January/February, pp25-27
A review is made of the potential danger of tree varieties on foundations and drains.

616

M Bar-Hillel

HOW DIANA WATT'S WORLD CAVED IN

Chartered Surveyor Weekly 1985 *10* January 17, p135
A case study of the building defects encountered in a newly converted flat, built according to the Building Regulations and surveyed professionally.

617

C Davies

HAMMERSMITH'S HOUSES OF HORROR

Architects Journal 1985 *181* February 6, pp56-61
Defects in high rise housing is reported. It is concluded that the main cause of failure was bad workmanship and lack of detailed supervision at the work face.

618

A Kennaway

ERRORS AND FAILURES IN BUILDING - WHY THEY HAPPEN AND WHAT CAN BE DONE TO REDUCE THEM

International Construction Law Review 1984 *2* October, pp57-79

619

R Matthews

WIDENING THE GAP

Building 1984 *267* August 31, pp22-23
Some typical causes of damp penetration across cavity walls both filled and unfilled with insulation are considered. The practical advantages of employing a 100mm cavity are indicated.

620

T Hall and D Cheetham

PROBLEMS IN THE MAKING

Building 1984 *266* February 10, pp30-31
Defects that have occurred in timber framed housing built over 15 years ago are considered. Sole plates and differential movement are significant features.

621

J R W Robinson

INCIDENCE AND RECTIFICATION OF FAULTS IN RESIDENTIAL BUILDINGS IN MELBOURNE

Proceedings CIB W55 3rd International Symposium on Building Economics, Canda, 1984. Volume 1. pp145-154

622

Association of Metropolitan Authorities

DEFECTS IN HOUSING. PART 2: INDUSTRIALISED AND SYSTEM BUILT DWELLINGS OF THE 1960s AND 1970s

1984. pp70
The origins of industrialised building and the programme of its use are discussed; typical defects are identified. Recent Government action to redress the problem and the cost implications of repair are also considered.

623

Association of Metropolitan Authorities

DEFECTS IN HOUSING. PART 1: NON-TRADITIONAL DWELLINGS OF THE 1940s and 1950s

1983. pp28
The background to the introduction of non-traditional housing and the extent of its use are discussed. Typical defects experienced are identified and action by Government and local authorities is outlined.

624

C Sjostrom et al

DEFECTS OF BUILDING MATERIALS AND COMPONENTS IN SWEDISH DWELLINGS BUILT 1955-79

Proc. 9th CIB Congress, Stockholm, 1983. Volume 2. Building Technology, Design and Production. pp221-232

625

G Atkinson

DEFECTS REVEALED

Building 1983 *265* December 23/30, pp32-33
The findings are evaluated of the BRE study into the failure of prefabricated concrete housing.

626

H A Kagan

EVALUATING CONSTRUCTION FAILURES

ASCE Journal of Construction Engineering and Management 1983 *109* (4) December, pp460-472
The failures decribed involved light framed timber trusses, one failure involved precast tees, and one involved overloaded steel floor joists. Each collapse centres around a failure to recognise the weakness of a partly completed structure. Human failure was the heart of the problem. Greater attention to details by those responsible for construction procedures is suggested.

627

British Board of Agrement et al

CAVITY INSULATION OF MASONRY WALLS - DAMPNESS RISKS AND HOW TO MINIMISE THEM

1983. pp21

POLLUTION

628
J Miller and J Charles
RISK & REMEDIES. 2. NOISE: THE REMEDIES
Architects Journal 1988 *187* February 10, pp55-63

629
R A Harris et al
METHOD FOR ANALYSING CONSTRUCTION HAUL NOISE IMPACTS
ASCE Journal of Construction Engineering & Management 1987 *113* (1) March, pp6-16
A technique is described that compares the potential noise impacts of construction hauling for a number of project alternatives. It is based on a modification of the level weighted population method to account for the duration of the hauling activity along various haul routes.

630
CIRIA SOUND CONTROL FOR HOMES. WORKED EXAMPLES
Report 115 1987. pp56
In four parts the manual illustrates techniques used in the planning of a site for new dwellings and in the design of the building envelope to control external noise; illustrates noise control techniques used in the internal planning of various types of new dwelling and the selection of appropriate forms of construction; illustrates design methods used to ensure reasonable sound insulation on converted properties; and contains a series of short examples illustrating methods used to solve noise problems in existing buildings.

631
E Silke
LEGISLATORS SPELL OUT RULES FOR THOSE WHO WILL NOT HEAR
Special Publication Contract Journal 1986 *330* March 20, pp20-22
New regulations relating to noise control due to be initiated on 26 March are discussed.

632
Construction Industry Research and Information Association
USE OF SCREENS TO REDUCE NOISE FROM SITES
Special Publications 38. 1985. pp4

633
Health and Safety Commission
NOISE IN CONSTRUCTION: GUIDANCE ON NOISE CONTROL AND HEARING CONSERVATION MEASURES
1986. HMSO. pp10

634
Health and Safety Executive
NOISE FROM PORTABLE BREAKERS
1986. HMSO. pp2
Dangers of noise from portable breakers and its minimisation are outlined.

635
Construction Industry Research and Information Association
SOUND CONTROL FOR HOMES
Report 114. 1986. pp71
Provides a guide to the principles of acoustics and noise control, including measurements, characteristics of sound sources, sound reduction criteria for dwellings. Considerations for noise control when planning and designing housing schemes are fully dealt with, with particular regard to the provisions of the Building Regulations 1985 for England and Wales and other official requirements. Details of construction for good sound insulation are covered, including all types which appear in the Approved Document. Checklists covering design and site inspection are given for each type. A list of information sources is appended.

636
F Fricke
CONSTRUCTION SITE NOISE
Architectural Science Review 1985 *28* March, pp12-22
The noise from construction and demolition sites is discussed. Starting with a definition of acoustic terminology, simple theory in outlined, together with the simplifying assumptions usually made. Measures used to predict hearing damage and annoyance are indicated. Finally, the main sources of construction site noise are identified, control measures discussed and the cost effectiveness of the controls assessed.

637
British Standards Institution
NOISE CONTROL ON CONSTRUCTION AND OPEN SITES. PART 2. GUIDE TO NOISE CONTROL LEGISLATION FOR CONSTRUCTION AND DEMOLITION, INCLUDING ROAD CONSTRUCTION AND MAINTENANCE
1984. pp5

DEMOLITION

638
DEMOLITION
Construction News Supplement 1988. November 17, pp32
A series of articles is presented which cover recycling, European markets, tippers, plant, listed buildings and facade retention, technology, training and safety.

639

NEDO THINKING ABOUT DEMOLITION AND DISMANTLING: A GUIDE FOR PROPERTY OWNERS

1988. pp6

Guidelines are provided for the owner covering assembling information on the building and its previous users, taking professional advice and choosing a contractor.

640

Construction Industry Research and Information Association

PRESTRESSED CONCRETE BEAMS: CONTROLLED DEMOLITION AND PRESTRESS LOSS ASSESSMENT

Technical Note 29. 1987. pp42

641

J d'Arcy

KNOCK DOWN BOOM

Contract Journal 1987 June 19, pp22-23

The major issues facing the demolition industry are discussed.

642

P Lindsell and S H Buchner

MONITORING DEMOLITION OF PRESTRESSED CONCRETE

Building Research & Practice 1986 14 May/June, pp160-163

643

C de Pauw and E Rousseau

EQUIPMENT PERFORMANCE AND MIXING TECHNIQUES

Building Research & Practice 1986 14 May/June, pp164-169

The performance, comparative cost and practicality on site of a number of methods of demolition are compared.

644

C Molin

CONTROLLED BLASTING OF CONCRETE. FULL SCALE STUDIES OF LOCALISED CUTTING

Building Research & Practice 1986 14 May/June, pp170-174

645

G Ridout

INFERNAL TOWER

Building 1986 261 October 3, pp412-44

The demolition of Ronan Point is described.

646

E K Lauritzen

MINI-BLASTING FOR REPAIR WORK. CONTROLLED USE OF SMALL EXPLOSIVE CHARGES

Building Research & Practice 1986 14 September/October, pp274-280

The use is described of local explosive blasting to remove defective concrete in a multi-storey housing block.

647

M Hague

DEMOLITION/REFURBISHMENT EONOMICS AND RELATED CRITERIA

Paper to 8th National/1st European Building Maintenance Conference, 18-19 September 1985. pp5

Some criteria are discussed designed to assist in the decision making process of whether to demolish or refurbish.

648

G Ridout

BLAST FROM THE PAST

Building 1985 269 December 6, pp42-45

Some details are given of a repair contract let in the traditional manner which presented some demolition problems. A restricted site and the insertion of new floors created their own problems.

649

R Davies

LOCAL AUTHORITY HAS FOUR-YEAR TIME LIMIT TO ENFORCE DEMOLITION

Financial Times 1984 May 11, p15

In the case of *Peacock Homes v Secretary of State for the Environment* it was held that where planning permission is given to erect a building on condition that it is demolished at the end of a specified period, an enforcement notice requiring demolition is invalid unless issued within four years of breach of the condition.

650

DEMOLITION

Construction News Supplement 1984 March 15, pp23-50

A series of articles is presented covering training, safety, prestressed structures and plant.

651

Health and Safety Executive

HEALTH & SAFETY IN DEMOLITION WORK. PART 2: LEGISLATION

1984. HMSO. pp13